ONTOPOIETIC EXPANSION IN HUMAN SELF-INTERPRETATION-IN-EXISTENCE

ANALECTA HUSSERLIANA

THE YEARBOOK OF PHENOMENOLOGICAL RESEARCH

VOLUME LIV

Editor-in-Chief:

ANNA-TERESA TYMIENIECKA

The World Institute for Advanced Phenomenological Research and Learning
Belmont, Massachusetts

For sequel volumes see the end of this volume.

ONTOPOIETIC EXPANSION IN HUMAN SELF-INTERPRETATION-IN-EXISTENCE

The I and the Other in their Creative Spacing of the Societal Circuits of Life

Phenomenology of Life and the Human Creative Condition (Book III)

Edited by

ANNA-TERESA TYMIENIECKA

The World Phenomenology Institute

Published under the auspices of
The World Institute for Advanced Phenomenological Research and Learning
A-T. Tymieniecka, President

KLUWER ACADEMIC PUBLISHERS
DORDRECHT / BOSTON / LONDON

Library of Congress Cataloging-in-Publication Data

Phenomenology of life and the human creative condition / edited by Anna-Teresa Tymieniecka.
p. cm. -- (Analecta Husserliana ; v. 52-54)
Papers presented at the Second World Phenomenology Congress, Sept. 12-18, 1995, Guadalajara, Mexico.
"Published under the auspices of the World Institute for Advances Phenomenological Research and Learning."
Includes index.
Contents: bk. 1. Laying down the cornerstones of the field -- bk. 2. The reincarnating mind, or, The ontopoietic outburst in creative virtualities -- bk. 3. Ontopoietic expansion in human self-interpretation-in-existence.
ISBN 0-7923-4445-6 (hardback : alk. paper : bk. 1). -- ISBN 0-7923-4461-8 (hardback : alk. paper : bk. 2). -- ISBN 0-7923-4462-6 (hardback : alk. paper : bk. 3)
1. Phenomenology--Congresses. 2. Husserl, Edmund, 1859-1938--Congresses. 3. Life--Congresses. 4. Creative ability--Congresses. I. Tymieniecka, Anna-Teresa. II. World Institute for Advanced Phenomenological Research and Learning. III. World Congress of Phenomenology (2nd : 1995 : Guadalajara, Mexico) IV. Series.
B3279.H94A129 vol. 52-54
[B829.57]
142'.7--DC21 97-2276

ISBN 0-7923-4462-6

Published by Kluwer Academic Publishers,
P.O. Box 17, 3300 AA Dordrecht, The Netherlands.

Sold and distributed in the U.S.A. and Canada
by Kluwer Academic Publishers,
101 Philip Drive, Norwell, MA 02061, U.S.A.

In all other countries, sold and distributed
by Kluwer Academic Publishers Group,
P.O. Box 322, 3300 AH Dordrecht, The Netherlands.

Prepared with the editorial assistance of Robert S. Wise

Printed on acid free paper

TABLE OF CONTENTS

PART FOUR: THE SENSE OF CITIZENSHIP, POLIS, RIGHTS IN THE MORAL INTERPLAY BETWEEN THE SELF AND OTHERS

PART FIVE
THE CLASSIC PHENOMENOLOGICAL THINKERS CONVERGING TOWARD THE HUMAN LIFE PERSPECTIVE

PART SIX
ORTEGA Y GASSET'S PHILOSOPHY OF HUMAN SPIRIT AND CULTURE

ACKNOWLEDGEMENTS

We owe thanks to all who made possible our Second World Phenomenology Congress, September 12–18, 1995, Guadalajara, Mexico, the work of which is presented in these volumes. First of all, we thank Dr. Santiago Méndez Bravo, Rector, and Dr. Cesário Hernández Hernández, Vice-Rector, of the Universidad del Valle de Atemajac, as well as the Chairman of the Department of Philosophy there, Dr. Xavier Rodríguez, for having welcomed the Congress with warm hospitality. We also thank Dr. Sergio X. Vázquez and Dr. Miguel Jarquín Marín, who extended the hospitality of the Instituto de Terapia Guestalt Región Occidente (INTEGRO) and led the Organization Committee for preparing ideal conditions for the Congress. A special note of appreciation must also be offered to both of them for their offering INTEGRO's premises for a reception for Ibero-American scholars gathered for the historic purpose of founding a new affiliate society of The World Phenomenology Institute, the *Sociedad Iberoamericana de Fenomenología*. Our thanks go to Dr. Jorge García-Gómez of Long Island University for having accepted the responsibilities of Secretary General of the society.

My thanks go to my assistant, Dr. Tomasz Panz, for his attentive and dedicated work during the organization phase of the Congress. My thanks are also due to our Assistant Editor, Mr. Robert Wise, for his work before and during the Congress in producing the Congress program, registration, and numerous other tasks, reprising the role he played at the First World Phenomenology Congress at Santiago de Compostela, Spain in September, 1988. He was ably assisted in the registration and reception of scholars by our Administrative Assistant, Mr. Louis T. Houthakker, who keeps our membership and event participation records and is responsible for the distribution of our review, *Phenomenological Inquiry*.

We are indebted to Mr. Wise, Miss Isabelle Houthakker, and Mr. Mitchell Seagrave for the careful copy editing, proofreading, and indexing of this volume.

A-T. T.

A-T. Tymieniecka (ed.), Analecta Husserliana, Vol. LIV, ix.

PHENOMENOLOGY WORLDWIDE

The Creative Spacing of the Self and the Other Within the Societal Web of Life

Given the crisis of philosophical thinking, which goes so deep as to denounce the validity of reason in all the prerogatives it assumes in the practice of life as well as in the philosophical tradition of the West, on the one hand, and in the spread of phenomenologically inspired thought in various fields interpreted *sui generis*, which once presented spread further into all fields of knowledge, on the other, phenomenology would seem to be dispersed among hermeneutics, semiology, analytic thought, etc. It was so mixed with existential thought (principally with that of Kierkegaard and Nietzsche to begin with) that one has the overall impression that phenomenology as a cogent core of intuitions, insights, concepts, precepts and convictions, as found in this entire classical stream from Husserl to Merleau-Ponty, has dissolved, and this innermost core of identifying elements disintegrated.

The Second World Congress of Phenomenology which The World Phenomenology Institute celebrated in Guadalajara, Mexico, September 11–18, 1995, has manifested the facileness of any such assumption. Phenomenology did not lose its identity; it merely transformed itself.

Indeed, crowning the twenty-five years of the work of the Institute, this gathering of scholars from all around the world to consider the philosophical framework of the intuitions, insights, and discoveries that the Institute by following by and large the *ontopoietic outline* of my phenomenology of life and of the Human Creative Condition has been further elaborating, manifested the phenomenological spirit at its most vigorous. Of course the classical ideas of Husserl, Scheler, and others retain their legitimate claims albeit in novel formulations and within a novel framework. But in this novel era, the era in which phenomenology has at last arrived or is arriving at its destination, the classical ideas with their claims are left rather in the shadow while new incipient ideas are being laid down as cornerstones of a gigantic edifice to be slowly built. It is indeed, from an overturn of priorities, with cognition and its specific mental intentionalities now assuming second place, while the creative act

A-T. Tymieniecka (ed.), Analecta Husserliana, Vol. LIV, xi–xiii.

of the human being assumes first place, that this novel edifice receives the orientation for its blueprint. Furthermore, although a proper place remains allotted to transcendental consciousness, the subject with its kinesthetic extensions is not just displaced in terms of its classical priority, but is turned out of its directing position to a limited realm for the newly recognized *ontological principles of Life*, which is that of *self-individualization-in-existence*; this latter serves as the key to a new phenomenological territory of the inner workings of life, nature, bios . . . and to the specifically human transcendental sphere of mind itself.

With this access to the depth of life itself the so-projected philosophical inquiry opens a vast field in which the rationalities of nature and of the bios are integrated with those of the specifically human, *creative rationalities* of the full-fledged mind within the ontopoietic self-individualizing progress of human *self-interpretation-in-existence*.

Thus at the very center of the newly opened field of investigation we find clues for the genuinely phenomenological *mathesis universalis*, that was the initial dream of Husserl and which, having gotten lost in his meanders, he failed to devise.

By thus going back to the creative virtualities of the human being and the principle of self-individualization of life, an entire new field opens in which all rationalities *emerge in their own right*. Here is situated the work of the present volume. Indeed, the wealth of ideas, points of view, and insights which come from otherwise diversely influenced conceptions and reflections offer us in this new RAINBOW OF PERSPECTIVES a cogent investigation of some major issues which reappear *now* in new formulations: principally the issue of a human individual's self-interpretation-in-existence, insofar *as he or she becomes and evolves with respect to the Other*. What has its classical flavor of "intersubjectivity" is clad now in modalities of the nuanced and modulated genetic interplay of forces which allow the human being *to create himself while co-creating the other*, and this in the societal, vitally conditioned, significant networks of *human trans-action*.

Human rights with respect to other individuals – human and animal – the polis, etc. go back to the fundamental virtuality of the Human Creative Condition of the moral sense, whence the morality, ethics, and regulations which the human beings establish for their co-existence take off.

However, the above-indicated set of principles which has overturned classical phenomenology and philosophy as such, while at the same time standing counter to the deflated tendencies of contemporary philosophy, has been operative to various degrees throughout more than two decades during which the attention of numerous scholars has been drawn to its novel vistas and principles. The Second World Congress has attracted attention anew to these approaches. During at least these two decades we have been witnessing in our conferences and publishing programs the inception of a reinterpretation of the classical thinkers, finding in their thought ideas congenial to ours which have hitherto been overlooked, both in their dominant trends of interpretation as well as in excavating reflections and insights pertinent to our problems.

It is in this spirit that our work has been advancing. Thus in this volume too – in Part V – some of the classical phenomenologists are interpreted anew in the perspective of how their thought converges on our phenomenology of life.

The volume closes with a section devoted to the Spanish precursor of the phenomenology of life, Ortega y Gasset, and in particular to his contribution to the theme of the "space between the self and the other" in this collection.

I conclude this presentation hoping to have shown that phenomenology worldwide and discipline-wide attests to the philosophical faith by its fulgurating spirit: there is no reason to bemoan the fallibility of reason and to succumb to irrationalism; this is a shallow bend of mind manifesting the "fin de siècle" mood of a time when long-cherished convictions and expectations brought stepwise to face the tests of progress have not been able to maintain their validity or well-foundedness. But in the period of thought which has emerged with this progress the old dreams and presumptions – disavowed in their validity – do not vanish; there is a "grain of salt" in them lively enough to lay a novel foundation for their metamorphosed fulfillment. Phenomenology worldwide is the promise and part-realisation of this fulfillment of this century's long-lasting philosophical dream, its fulfillment in the phenomenology of life and of the Human Creative Condition.

ANNA-TERESA TYMIENIECKA

PART ONE

THE SPACE BETWEEN THE SELF AND THE OTHER

UNIVA, the site of the congress.

ANTONIO DOMÍNGUEZ REY

THE SPACE OF THE OTHER

Mens agitat molem et magno se corpore miscet.
– Virgil: *Aeneid*, Vl. 727

Daily speech "die alltägliche Rede" is the field of the manifestation of the other. When we speak, the other is introduced to us and we introduce ourselves as other. This experience has a double face in Husserlian philosophy: an immediate one, as it is presented – the aspect thing of language – and a mediate one, in reality hidden, inaccessible. The immediate part of language is its body, already invested of a certain degree of spirituality – *Leib* – but this spiritual hint is precisely what is hidden behind the body, and we can only perceive it by intuition, mediately. In the verbal experience of the other we obtain in the first instance, his material word, not the internal one, that of his spiritual experience. Its "Innerlichkeit" remains hidden. We reach a "Mit da", not a "Selbst-da". The intimate experience of the other, its vital current, never offers itself as originary presence, but as a collateral experience, already present.

Although there is a similarity between the perception of the other and of a thing – in both cases this is related to the common mediacy of intentionality – in the experience of things we can reach the originary, a completeness through successive presentations, whereas, in the experience of otherness, the inner part of the other is always, a priori, contained, impossible – "apriori ausgeschlossen". It is the other who incessantly speaks about that which is here, in its sensitive materiality. The language matter repeats itself in successive presentifications. Sensitivity reproduces sensitivity and it gets exhausted in it. Now again, the "Körper" speech is flesh pierced with sense – "Leib" – a body invested with spirit – "begeistete Objekte" –. This is what makes it basically different from the matter perception, as it relates to an absence from a "Mit da". It is about the perception of the "alter ego" of Husserl.[1] Language belongs, in fact, to that perception.

The transfer from my "Körper" to "Leib" or "Leib-Körper" is evidenced in the synthetic unity of my inner acts and potentials, in which I discover myself as the organizer – organised centre of my own organic

A-T. Tymieniecka (ed.), Analecta Husserliana, Vol. LIV, 3–26.

constitution. It is the perceptive transformation of the organ in object and of the latter, in its turn, in the former,[2] providing in this way an inter-corporeal "reflection" that is constituting, in reality, the inner space of the body as something objective in opposition to the living current.[3]

This organized spaciousness or "Leib-Körper" is the mediation that Husserl uses to take the analogic jump – a jump without time – towards the animist corporeity of the other. My Leib is an original fundament of the pairing – "Paarung" – and of the intentional "transgression" which is produced in the passive synthesis that originates it.[4] When the presence of another "Körper" appears in this living sphere of mine – shoots into my experience – in that shot my "Körper" also becomes a "Leib" –[5] another present, already past of my body and, in its own virtue, also accompanied by other former presents, both presences join, one alive, the other revived, and the sense mutually transferred. As my "Körper" was already "Leib-Körper", in the mediation is also implied, as an inference of associative transit, without logic, the transfer of a "Leib" to the "Körper" of the other now present in my vital sphere. A different matter is, as we have said, the inner part of the other, its spirit, only perceived by collateral intuition through the manifestations of its "Leib-Körper".

If this scheme is applied to language, what is deduced is that the communication of speech is always mediate, a continuous interpretation of signs and symbols of these organic bodies. Nevertheless, its "Leib" is also somewhat different from any "Körper" or thing – "Ding". When M. Theunissen analyses the initial function of language inside the constitutive system of the other, he underlines the impression produced, as we introduce ourselves in the soul of the other from his organic body.[6] Speech "Rede" serves in this way as a guide, even a base, to the "originary instauration" of the other. Language also belongs to the "constitutive systems, mutually correspondent and concordant" of the intersubjective harmony.[7] Husserl does not develop this "Leib-Körper" of language in his *Cartesian Meditations*. He influences us with the transfer of the "Mit da" to the "Selbst-da", never reached, of the living current of the other, but the dichotomies expression/indication, signification/manifestation of the *Logical Investigations*, still prevail in him.[8] The "Leib" of language remains reduced in the worldly sphere of the "Körper", even though it continuously acts in the analogic transfer of the "Übertragung".

If speech, like other concordant activities of the intersubjectivity, for example, work, offers us the synthesis of the expression and of the expressed, as Husserl says in *Ideas II*,[9] language will also have an internal function. Its enclave belongs, at least, to the reduced structure of the objective world, to the scope of the odd. However, the concept world is a homonym in Husserl. On the one hand, it belongs to the sphere of the self, it is constituted in it, but on the other, it transcends it. It is a substratum of the experiences, something that belongs to me and that, in the same instant, surpasses me. In it, the living experience discovers that which is not my own essence.[10]

As transcendent world, language can only be established as experience, like the alter ego. That is why it offers in its structure a scheme of transit.[11] However, as inherent world of the sphere of the self, it could only belong to the organic corporeity, objective and spacial. In the reduction of the body to the sphere of the self, as facing the strange, Husserl resorts to the objective function of the performing organ and to the organic function of the respective object: "das wird dadurch möglich, daß ich jeweils, mittels der einen Hand die andre, mittels einer Hand ein Auge usw, wahrnehmen, kann, wobei fungierendes Organ zum Objekt und Objekt zum fungierenden Organ werden muß".[12] Confident in the method's mechanism, in this reduction, all that is not real property is excluded – "*eigenheitliche Reinigung*" – of oneself. Therefore, says Husserl, I lose even my natural sense of being I, while any meaningful relation of a possible us – "Uns" – or we – "Wir" – from which we started has been eliminated.[13] It is the sphere of the spiritual property. Nevertheless, in this reduction enters likewise "the space-time form" discovered in the relation organ-object-organ, from which we deduce the spaciality of each one of the objects reduced to it.[14] The "predicates of value and virtue" also belong to it,[15] where language should be placed, therefore, in the sphere of the self.

As the "we" is excluded and the sphere of the self is confined to a counterview I-world inside my psychophysical being, in which the I is also a part of this reduced world, what remains of language is only an halo of itself, because communication disappears and, with it, its indicative value. The existing relation between expression and expressed is not, as Husserl says in *Logical Investigations*, an indicative mark – "anzeigen" – of existence, such as that of daily speech, but rather an exhibition – "hinzeiger". Here, "Das Dasein des Zeichens motiviert nicht das Dasein, oder genauer, unsere Überzeugung vom Dasein der Bedeutung".[16] It is the monologue dimension which Husserl opposes to speech. In the sphere of talking to oneself there is not a communicative interchange. We represent ourselves as speakers and communicants, "man stellt sich nur als Sprechenden und Mitteilenden vor".[17] We attend alone the scene of speech. That language is pure imaginative representation of sound, graphemes or interlocutors.

The surprising thing is that Husserl keeps talking about expression in the sphere of the self, when in reality he rejects any of its differentiated signs. Why then, the necessity of an unreal word, but not the imaginary, only its imaginative representation. That necessity refers to the act of the imagination itself as pure imaginative form, to the point that "noesis" and "noema" coincide. It is not the case of a meaning with well-defined features, and opposed to that of the other even less of some materially formalized substance. The expression synchronizes in Husserl both sense and meanings, or he constitutes them by piercing them. It is the self fluency of the sense in its own constitution and excedence, or the intentional direction that constitutes sense. When this one is formalized, what appears is the *Bedeutung*, the meaning, and its support is precisely the dynamic of that intention of the I going towards itself and discovering itself as sense in the sphere of the self, that is, as part of the world, organic body and personal I. In fact when Husserl asks himself what changes the expressive lived contents experience when the expression interferes, he remits them, to its intentionality, to its immanent sense, to the difference that separates, on the one hand, this sense and the eidetic moments of the pre-expressive element, and on the other, to the significance of the expressive phenomenon in itself, with the moments that belong to it.[18] He also defines expression as the form that adapts itself to each "sense" as noematic nucleus, and introduces it in the realm logos, of the conceptual and general.[19]

J. Derrida saw in this function "une double sortie hors de soi du sens (*Sinn*) en soi, dans la conscience, dans l'avec-soi ou l'auprès-de-soi".[20] This double vector would be represented by the expression and the effective intention, not always simultaneous. J. Derrida also underlines, when he explains the ideality of the *Bedeutung*, the non coincidence between *expression*, *Bedeutung* and *object*.[21]

This is established here: a semiotic space clearly delimited by the three vertexes. It only remains to add the sense as nucleus of the triangle whose constitution it presides over. It is the processual germ of the objective apprehension, so either it culminates in intuition, with a really real, well represented object, or it only signals the objective intention, its "visée", not complimented by a full intuition. The expressive level refers thus, under the dynamic presence of the sense, to a meaning – notes it – or to a real or represented object. Such a semiotic space is that of the objective constitution of consciousness. The pointing of expression towards meaning, through the voice, corresponds to the "*acts of giving sense*" or "*significative intentions*", which are essential to the expression while the completion of those significative intentions compete with other kinds of acts which are not essential, but which fulfil, "verify", "strengthen" or "illustrate" the expression. They are the acts that fulfill the significative intention or the implied sense.[22] He calls them *acts of fulfilment of the sense or significative fulfilment*; they anticipate, in our understanding, the pragmatic. All of them get together when the reference of the expression to the objectivity is completely realized.[23]

With these divisions, Husserl establishes the bases of semiotic and of systematic linguistics. According to the relations among vertexes, he distinguishes three kinds of expression. First, the absolute, which corresponds to what we know now as linguistic sign or "sound animated with sense";[24] second, the "mere expression", the pure verbal sense, the sensitive or simple "voice", where the word "ceases to be a word";[25] and thirdly *the plain expression*, in which case the expressed is the *act of completing*,[26] including in it the reference, as we would say today.

Behind the triangle there is, however, a pre-expressive layer whose form introduces us to the scope of the imaginary representation. Its formal character is imprecise. If it is limited to a neutral reproduction, non-positional, it will be pure psychic fluency. The word is not important then, or it gets confused with that current in the same way as the poetic idealism of the romantic creators, a hypothesis that Husserl does not develop but whose ferment we perceive by intuition in the most recondite part of the transcendental reduction. Husserl's ultimate Logos contemplates words as a scene that reproduces itself in front of his eyes, without him mediating in it as actor or author, without he himself being the audience or presentative part of it. Does not language function, at least, as the psychic correlate of mental fluency, as meaning or boundary, not yet specific, between sense and that flowing act? Does not the dialogue, its ex-structure, slide under the monologue? Am I not my "you" in this case? How could I be I without him?

J. Derrida interprets this bordering form of the inner expression as a transparent signifier that erases its body on behalf of the pure form of the consciousness: "Cet effacement du corps sensible et de son extériorité est *pour la conscience* la forme même de la présence immédiate du signifié".[27] This invisible signifier became somewhat methodological. Husserl reveals it until he identifies it with the form of the concept. When making its productive character explicit, he points out that its noematic action "erschöpft sich im *Ausdrücken* und der mit diesem neu hereinkommenden *Form des Begrifflichen*". It even maintains a perfect identity of thematic essence with the pre-expressive layer, and for that same reason, such is the expressive representation,[28] that is, a diaphanous unproductive transparence of the presence, as explained by J. Derrida.

L. S. Vygotsky analyzed these implications between thought and word. "The rule that governs the inner language", he says, "is the predominance of the sense over the meaning, of the sentence over the words and of the context over the sentence".[29] The inner speech is like "an autonomous function of the language". Words live jointly with thought and die with it.[30] Even if they do not coincide, they mutually motivate. "The relation between thought and word", he goes on to say, "is a living process; thought is born through words. A word without thought is a dead thing, and a thought destituted of words remains in the shadow".[31] He also notices its processual and evolutive connexion.

If we relate this observation to Husserl, we could say in advance that, although we can perceive by intuition thoughts behind the reduced words, as a consequence of the method, it never happens totally without them, even if it only counts on its shadow. Because it is an intentional relation, it refers, at least, to the word which, constituted, also constitutes it in some way. The question imposes itself: Can thought, through a reductionist method, become free of this verbal shadow? It is not enough to say that a sign is evoked in our imagination but that in reality it does not exist. That real inexistence carries out a reproductive function which allows one to think in silence. Would there be a monologue if there has not been a dialogue? In fact, Husserl attributes new intentional functions to the expression, although correlatively tributary to those that operate in the intentional base.[32] M. Dufrenne remembers, to this purpose, the Kantian scheme and the *Bild* by Heidegger. It would be a mute voice, bare, but schematic. He places it in the depths of the body, where the words are outlined and the voice is unable to make them sound.[33]

Husserl's delimitations clash, on the one hand, with the ideality of meaning and sense, and on the other, with the fluency of the pure living content. These polarities have determined in linguistics the recurrence of the value, as differential formality emanating from the system. Such is, for example, the position of Saussure and, after him, of other linguists, such as Hjelmslev. Saussure invents a new concept, the "pensée-son", to illustrate the synthesis, still mysterious, of thought and matter, already sung by Virgil. "Il n'y a donc ni matéralisation des pensées, ni spiritualisation des sons, mais il s'agit de ce fait en quelque sorte mystérieux, que la 'pensée-son' implique des divisions et que la langue élabore ses unités en se constituant entre deux masses amorphes".[34] Once "thought-sound" has been synthesised, only its formal differences are of any interest,[35] and thus skipping a great deal of the problem in favour of the method.

Husserl's intention consists in the retention of some echo of expressive substance formalized in its interior but without resorting to the external, to the foreign. Once the sensitive character of the word is reduced – its natural attitude and the worldly thesis of existence from which participate all the other phenomena of self-affection – such as touching and seeing oneself, there remains, however, in listening to and hearing oneself, a kind of significant substance. Its pertinence consists in reducing to the maximum the distance between signifier and signified, between the experience and its psychic content. When the signifier becomes self-expressive presence, it facilitates the subjectivity, the *pour-soi*. This purely representative sign auto-represents itself. Such is, according to J. Derrida, the attribute of the phenomenologic voice: "l'être auprès de soi dans la forme de l'universalité, comme conscience. La voix est la conscience".[36]

The process is equally and immediately repeated in the intersubjectivity, where the signifier remains also practically reduced – it is a simple medium – to relate two phenomenologic origins of pure self-affection. To talk to someone is to hear oneself talking and to make the other *repeat immediately* in himself the same form of hearing myself talking that I have produced. J. Derrida signals that receiver "reproduces the pure self-affection without the help of any exteriority".[37] This is true, but with two exceptions. Before hearing himself the receiver hears the talk of the other. In his immediate experience he is not himself the object of his reception or as Condillac would say, he does not find himself in what he hears.[38] He only receives a talking, not a "hearing himself talking". And that at a distance. The air transmitter remains in the middle. And it does not perceive the same form. The self-affective inner vibrations are lost, that is what D. Charles calls "bruit de forme" in the voice, the very turbulences of the speaker, for example, the timbre.[39] Children happily repeat these turbulences which vibrate in the body and that we can even verify with the hand. The voice is registered and written in the flesh. The signifier does have in this case an exteriority, even when the listener withdraws from it the same idealized form, the phonemes. Prior to the abstraction of language, is its belonging to the world of gesture or, as J. Ortega y Gasset says, to the organic living experiences with regard to the gesticulation of the flesh. This philosophy recriminates the science of linguistics for the fact of forgetting that gesture is the root of language and of the body speaking. The scientific consideration of language is born and then reduced "as an amputation of the saying".[40]

Concerning the signified, Husserl's observations about the perception of speech in the *alter ego* aspect or about inadequate intuitive representation are still valid. Among interlocutors, there is no equality of living experiences. None of them live language in the same way. One cannot get inside the other. They remain on a level of interpretative assumption.[41]

The individual that hears himself talking also participates in the synthesis organ-object-organ, but in a different manner. When we speak, we make the sound tactile. The phonetic organs produce, by contact and resonance, a sonorous object. We perceive it by means of another organ through the air and the inner resonances of the voice in the body. Although the process is of immediate perception, there is a transition between the articulate and the acoustic components, as well as an auditive selection and an abstraction of sound that S. Vygotsky places as a new phase, that of "thinking in words", after the egocentric language in which they are only articulated.[42] There is even an organic identity common to the mouth and the ear. The branchial organs structure the middle ear and the oral pharynx, "l'oro-pharynx". A. Tomatis sees in this common origin, motor and sensorial, of organic intimity, the base of its functional unity.[43] Phonating, on the other hand, is produced with the respiratory movement. The speaker assimilates and transforms himself inside an external object that almost immediately returns to the world. The phonetic tact is beside space-time articulation. To the punctual space of the affected organ is added the discrete character of the articulate linearity. There is a culminative differentiated succession. In those points are amalgamed time and space. All that is received, in its turn, in varying gradations of timbre, duration and intensity, and so on, that homogenize with a different criterion the articulate concretions. The coincidence of the discreet character in sound and phoneme is not clear.

All we want to say is that the body as objective space-time unity, is already language. The speaker articulates a space and creates time. It's one of the first conformations of the world in the consciousness. In that way a vinculum of a first stage of the symbolic configuration of sound, is created. We can apply to it the conclusion of Merleau-Ponty about the corporal auto-reference: "Il y a un rapport de mon corps à lui-même qui fait de lui le vinculum du moi et des choses".[44] In such a vinculum is fulfilled the "ek-stase" of the heteronymous and antonymous appropriation when the primordial being, is joined with its own body and that of another.[45] In the phonic vinculum I perceive something external which is mine, my own voice, but surrounded and surrounding something else, that something else given previously as noise of the world and its own fundamental sense, from which derive history and knowledge.[46] And also as its own voice that is accompanying mine, as a primordial, *on*,[47] in the articulate sense. In such suppositions the voice prevents a recurrence of the Other to the Same, because it breaks the capsule of the I even when it hears itself. Sound in that way overtakes, as pointed to by E. Levinas, the given. It is a laceration in the world from which this world is extended to another scope which is inconvertible into vision.[48] Perhaps because in it resounds the world's echo, the sounds of nature seem more pleasant and less disappointing. The only dissonant ones are those that are not consonant with that extension of the saying in the voice that says. The constitution of the articulate sound also responds to phenomenological analysis. Its self-perceptive and verbomotor characters unify to differentiate themselves from the concomitant sounds, in such a way that the global sense reached confers a pertinent function to the particular movement that produces it. In it, there is always a globalizing structure, also subject to anticipating implications, that orientates the proprioceptive sensation – "propiozeptive Empfindung" – as G. Lindner points out.[49]

Therefore, when the subject perceives his own sound, he verifies the objective constitution too, identifying the perceived with the produced. He lives himself. Not only do I recognise the self. In it I discover something that transcends me, elaborated by myself. The ear establishes, as M. Dufrenne says, the reference oneself, but it also needs, an inner part, the voice.[50] In that sense, it is important to remember that the phoneme is already "unité complexe". It results from the combination of a heard and a spoken unit.[51] It carries in itself, apart from self-affection, a comparison basis, of course, of the first conceptual stage. The "acoustic image" is associated with the corresponding articulated act, which, in its turn, implies educated muscular and motor images. We can also verify this in the phonic game of children when they repeat their own sounds and in the articulate correction when we make a mistake.

The phenomenologic constitution of the phonic space-time is linked to this recognition. It is not a question of mathematic time, of "equal times" in the verbal chains, but of "temps homogènes",[52] of impressive units conformed according to the synesthesic mode of perception.[53] When we compare them with the space of articulation, at that distance, added to the time itself of the articulated act, plus the intracorporal vibrations, the space-time phenomena are configurated better than anywhere else. That is why language should be also considered an attribute of the self, together with the world, the body and the I. It is its base of constitutive inter-relation and foundation, in its turn, of intersubjectivity in the mode of fundamental perception.[54]

The psycophonetic confirms these liminal operations of the conscience. The regular or irregular vibratory movement of the vocal folds, the more or less forced pass of air through the glottis, are translated into auditive qualities, dark or clear, of the voice. In the phonic games, between the age of two or three, the child discovers with pleasure that he is himself the source of the phenomena. In this same years he discovers with surprise, his own image in the mirror. I. Fónagy associates even those periodic movements of the voice with the recreation of a prenatal atmosphere dominated by the cardiac rhythm of the mother. The relation between larynx and ear is also translated in to the intonation, by a spatial segment of laryngeal mimicry.[55]

In his living experience of the voice the individual becomes conscious of himself. A. Tomatis associates this process with that of humanizing: "Ce repli sur lui-même, c'est au langage qu'il le doit. . . . C'est par l'écoute de sa propre voix que la notion de vie le pénètre. C'est par ce jeu incessant avec le verbe que son corps prend une image".[56] Language is created on a layer of self-affection whose vibrations contribute to form the signs and its symbolic units. The world is configurated in them jointly with its production. Its reflection is, for example, as D. Charles quotes when he comments on Heidegger, the tone, the peculiar accent of an area. Heidegger thought about these implications apropos the *Mundarten* or "patois". In such vocal modes speaks the Earth of origin. Body and mouth, says the German philosopher, belong to the flux and development of the Earth in which we, mortals, grow up and from which we receive genuine roots – "das Gediegene einer Bodenständigkeit".[57] In the voice we articulate the world. With respect to this Ortega y Gasset says: "the gestures with which a language is pronounced symbolize the way of life that a nation prefers". This spontaneous attitude reflects an originally melodic and lyric behaviour of men.[58]

There is another component in these synthetical units of objective conformation. Today we know that the fetus recognizes the voice of the mother. Her phonic vibrations reach the body growing inside her body. Hydrophonic and microphonic techniques detect background noises in the uterine abode, among which stand out the vascular and intestinal ones, the mother's voice transmitted through the thorax and the abdomen. A. Tomatis reproduced an amniotic medium and he verified that children reacted with pleasure when they received the sounds through that medium. They even pull out their lips in a sucking gesture.[59] In these answers, a kind of collateral and implicit perception is likewise configurated. In the fetus mediated vibrations of the world act, the "air" of the voice that's in the form of oxygen, also penetrates in its blood. The body is world and the world becomes corporal. From a phenomenologic point of view, a function corresponds to the voice in this constitutive grammar.

Its reflection would be, in its turn, the organic "reflection" of survival phonemics. With this term B. Ucla names the symbolic relation between signifier and signified, the signic motivation of the phoneme – a recollection of the Socratic kinesthesia of the sounds – and, with an anthropologic base, supported by the lateralization of the brain and language, the projection of ontogeny over phylogeny.[60] He remembers, for example, the maternal semantic field established by B. Malmberg in connection with the sound m, oro-naso-labial sound associated with the semantic chain mother – breast-food.[61] In opposition we find the n, for whose pronunciation, the tongue, liberated from its suction immobility, can percuss the palate or the teeth. The n reflects the first indication of autonomy, of identity, and it is characterized, like the dentals, by its external opening. To this joint field of opposition m/n another external field is opposed, in its turn, that of the dental sounds, which represent the father's world. The n is the linking element between the subjective and the growing objective. Nasal and dental represent the internal world and the deictic world, subjectiva indicative of a corporeal semantic in which the labial sounds correspond to the first feeding activity.[62] In the humanization process, while the hand is liberated of its prehensile character, the mouth develops, at the same time as its feeding function, a gestural field, deictic and signalizing, of which some of these phonemic "survivals" would be the echo until a degree of symbolic organization is reached.

Of all exposed until now, three important facts stand out: the discovery of the self-affection in the phonoauditive unit, the processual organization of the signifier in motive and synesthesic and motivated layers and thirdly, as an implication of the other two, the symbiosic world-body in that unit of auditive articulation. From this point, the semantization of the body remains configurated as expressive function of such phonoacoustic unit. The signifier generates meaning by recurrence and amalgams of sensory and motive functions. The body somatizes and the soma becomes sema. There is no meaning outside the pragmatic performance of the signifier. When meaning is withdrawn from its pragmatic functions and converted, therefore, in new signifiers or when these are objectified and they remit to others, conceptualized, there emerges the authentic internal dynamic of the sign, which is a stage of superior organization compared to the index and sensoeromotive signal. Such an epoche is processual and coincides, in its first conjecture, with initial pryings of sensorial and conscious autonomy.

All that happens in the phonoauditive unit in relation to the constitutive discovery of the world-body unit, happens, at the same time, with the revelation of the I inside it. What is articulated is the soma. The subject becomes thematic when he perceives himself as distanced from himself, distance in which he understands and organizes himself as sense. Being myself means becoming or seeing myself as somebody else in that nature that happens to me, or hearing the other who is invading and modeling me, revealing me as his interlocutor. A mode that generates a model with Husserl. It is to feel and interpret the "mute concretion" of *The Crisis*.[63] However, that concrete constantly co-requires of other presentations that leave it behind an athematic, anonymous rest,[64] and this one configures our selves from the base like slime. To that separation-encounter with oneself belongs the phenomenologic origin of language. It is an experienced phenomenon, the surging of feeling in what is sensed. It talks to me. Therefore, the "mute concretion" is revealed as a rather methodologic remnant, that which I perceive by intuition from a silent speech as its own culmination.

For Husserl, these units respond to the *mode of fundamental perception*, already quoted, or to the *originary restoration*.[65] In the first case, "the psychic fluid" of the signifier, of that already given, contributes, in its own fluency, to constituting another kind of objectivity, that of spiritual objects, also already mentioned. Its sensitive, sonorous flesh, for example that of a poem, is not existing flesh.[66] It suffers a metastasis. When the body is animated and the spirit somatized, these units also acquire an ideal character. Nevertheless, the word remains the anchor and basis of the sense. The sonorous signifier is dragged by it and insufflates it with pre-indication, with discursive orientation. Inside the word is the rhythm, the dynamism.[67] It converts it into a special matter, into *Leib*, as we saw. In the signifier a formalized substance exists which is not pure substance. The expression is divided into two forms, one purely expressive, insufflated in the phonic matter, and another conceptual, which delimits the fluidity of the sense in semantic concretions. The substance of the content reserves for itself a previous level of expression, whose border, transparent and schematic results are imprecise. It does not leave traces of its productivity inside the substance. Upon this basis and on the consequent reduction of Saussure, Hjelmsley organizes the structure of the linguistic sign and the semiotic function.[68]

Husserl transfers the implicit conception of the sign to the field of subjectivity. There is in it a structural scheme of discourse or *articulation of sense*: an animated unit is integrated into another of a superior order. It starts from it to explain the perceptive constitution of men which is second with respect to the I and the other. I understand the other when I perceive myself in the same way as I perceive them, by presentation. My direct I, revealed in front of my body as its own inner object, coincides with the other apperceived I. That coincidence is called "being men". Husserl converts this unit into a sign. The signifier is the body and the signified, the soul. They are integrated into a superior unit and they function likewise as signifier in front of another sign, which would elucidate its corresponding signified. In that way is formed the society or supreme discourse of signs: the field of the we. Only in it am I an authentic I, and the other, an authentic other. I and other have sense only as we.[69]

Such an identity of I's, that perceived directly and that perceived by the other, coincides with the identity of the elucidation in the "Origin of Geometry". When we perceive, "a form in formation" is produced in the other and, by its virtue, both processes, mine and his, are the same.[70] They are derived processes of a unit of globalizing sense, of "intentional transgression" of pairing: men integrate in body and soul: each of these units synthesizes corresponding perceptions: society includes men, and so on. Man is for society the correlate of sense for the word, in the same way that speech is for society. There, as well as here, little corresponds to the somatic aspect, even less to the corporeal, because in reality, it means a continuous interpretation of that which coexists, of the co-represented, of that presented.

Something similar, analogical, happens in *The Crisis*. To perceive "the strange" is a mode of self-perception, of self-variation, such as my being somebody else in the lasting chain of the continuous de-presentation. The ways of offering themselves, the other, the mine and the other inside myself, are analogic. And such is also the perception of my being, as object, in the other.[71]

Husserl's sign reduces to ideality the affective reference of its first substance. It eradicates from the self that which fundamentalizes it. In the constitutive conscience there is a unit differentiated at the same time from world, body, I and functional dependencies that the intimate structure of time imposes as perceptive substance. All contents, already synthesized, enter a preceding flux of motivations which irradiates, in its turn, its own elements. They incarnate in the flux. The static of the perception-noematic becomes dynamic-noetic in that way.[72] The linguistic has its equivalent in the syntagmatic and paradigmatic relations. The noesis go beyond its noemes, transcending them.

In the *intimate form of time* occurs the fold of consciousness itself without any kind of opaque intervention. Language disappears from it with its obstacles. It shines through in something else. The unit of sense absorbs the signifier that made it possible. However, in its temporal dynamic we perceive by intuition its reflection. It projects a transparent shadow. From where does the proto-I take its declinable character, the being I of the I or in the I, the constituting "from oneself and in oneself the transcendental inter-subjectivity", the "I of the transcendental others?"[73] From the auto-donation, from the auto-constitution, from the auto-temporalization. But this always came from the strange, from the foreign that invades me, as much in its negative dimension – that other part of myself as I discover myself – as in its positive: in the deployment of my acts I cannot make decisions without horizons, without somebody else inside myself. Even my own constitution is something strange in the immanency of my I. The alienation of the strange epoche.

Could language then be the duplicity of that fold, the product of auto-explicative dynamism? The mark of the auto-distancing, the space of the other, his energy?

The passing from the *Vorstellung* to the object, which in Husserl corresponds to the expression, compels one to consider an imaginary language. We do not know of what consists that word of the imagination outside a verbal image, of the remnant that language determines in consciousness affecting its own discursive forms. It is difficult to conceive an unpolluted consciousness which is not affected, in its distancing, by the plasticity of the spiritualized object that it constitutes. How could an absolute substance, always identical, give an account of the phenomenic complex, varying and growing, that we discovered in the correlation world-I-language? G. Santayana rejects this possibility and substitutes it for an "actual plastic form" in the body and the mind. With it he explains the permanence of the thought and the synthetical units of the experience.[74]

The continuous interrogation that the apogee realizes presupposes a kind of dialogue. Husserl insinuates that the question is linguistic, but not the answer obtained. The transcendental I escapes in that way the scheme of language that enables it to be a question. It is the deep solitude of the apogee as methodologic exigence.[75] The eye that sees is seen. Could it not be possible that the language that speaks, the saying, is being spoken, said? That the solitude of the epoché is the constitutive silence of the word, the slime where it germinates? That is why we could still ask: does Husserl's expression reflect that plasticity? Does the transcendental I remodel itself as it is constituting, clarifying its perceptive units? Is it pure explanatory principle? The auto-temporalization lets us glimpse a plastic mode of engendering without explaining the objective correlative of the expression. Pure form? Winged substance? An *extra* or plus of consciousness? The identity needs forms of retention: the consciousness, inscriptions, marks. It is not enough to think about the present as punctual form elapsing. If it elapses and has to be known, it crystallises in the form of some conscious mode. To feel oneself is to take notice of oneself, to have evidence of oneself, a perceptive unit, even if it is liquified or ethereal. The phonoauditive is processual, emergent. It would not exist without a body. It leaves marks that we call meanings. In the same way that it is articulated in a first level invested of sense, so it is also articulated from this initial sign, in other more complex meanings. Its interiority is a signifier.

Humboldt pays attention to this processual determination of the thought objectifying itself in words. Language is "das bildende Organ des Gedanken",[76] whose place of inscription is in the articulated movement. Articulation predominates over reception on the onomasiologic level because it is on it that, like a primordial source, the objective constitution is produced. The sound of language is a created object. It did not exist in the world outside of human nature. It appears because consciousness clarifies it or because the arising of consciousness is already sounding. We would know nothing of its interiority without the recurrent forms, without its mutual resonance. Its sound is aerated concept. Humboldt underlines in this way the originary organization, in indissoluble agreement, of thought, phonation and ear as something characteristic of the human nature.[77] Speech is not merely the condition of conceptual existence. It acts on the mode of the concept, stamping a mark on it.[78] The concept is born spontaneously with language, although not always adjusting entirely to it the richness of its content that also overflows the verbal matter. Even the objective representation is indebted to the incipient language.[79] That is why the content is not pure subsistent noeme, susceptible of apprehension in an intuitive stroke or by successive approximations. The fixed laws that rule it are shown in its process, but not its field of application or the nature of its product, which remains indeterminate.[80] In the noematic application, the noemata arise as their own marks of inscription.

Such a productive function is the *Darstellung*, or step from the representation to the object. There is in this term a kind of connaturality with Husserl's expression, but Humboldt centres his activity in the articulation, architectonic key or lever of the intimate knot which establishes the connected work of language and thought.[81] It is an irreducible, originary source, and it identifies itself with the signifying intention rather than with the sound properly considered. The intention and its significative capacity constitute by themselves the articulate sound.[82] Humboldt underlines the signifying capacity of the sound over its sonorous materiality. This articular intentionality is very important. Without it the sonorous material will never reach its character or sign. It is "foné semantiké", as defined by Aristotle in the *Peri Hermeneias* (16 a.2). Humboldt describes its process through the voice emerging from the chest as "live music" or "breath of life", to which it insufflates sense, where is it concentrated, in the same way that speech restores in the represented object the emotion that accompanies it, as its contemporaneous. It is for that reason that intentionality surged in a live medium. In that way are merged in it world and men, activity and human receptivity.[83] It even relates the linguistic sound with the hominal verticality, raised in some way by it. Language is a constitutive synthesis of a world. It includes the activity of nature through the energy of the body and in its activity, man assimilates at the same time as he opens the world.

Therefore, the object constitutes itself in the linguistic space, and it reintegrates with it the affectivity of the voice. The objective constitution happens in the emergence of interiority knowing itself as word. The object becomes object when it reintegrates itself in the subject, and the emission, in reception.[84] It is the basis for what O. Hansen-Love calls the Copernican Revolution of language. However, Humboldt emphasizes the pure sonorous donation as emergence of articulate meaning, preceding, in appearance, the constitution of the object to its affective reception by the ear. This does not imply that feeling accompanies the object in the second instance. He says that it restitutes it and they are contemporaneous. It was then hidden in some way. It came first in latency and emerges at the same time that the object is constituted. Such a latency is incrusted in the voice, in life as the voice, which is the one that gives sense. The object constitutes itself in a layer of sense that emerges from the body. And the voice "accompanies, even beyond language: the pain, the happiness, the laziness, the desire, all of them designations of objectivated affection. The individualized voice: the concept, which springs up from its linking, abstracts. Thanks to the sound, both poles merge into one with differentiated values depending on the application and nature of each language.

Now, here Humboldt emphasizes the pertinence of language in the objective constitution without omitting subjectivity. What has been articulated resounds and returns to audition. Without this preamble, concept does not exist, nor, for the same reason, thought. Speech is even an essential condition of monologue.[85] The objectivity is reinforced when it's returned to us as a foreign sound, that is, when somebody else repeats it. What we carry and produce inside ourselves is like a modification of what's common to everybody, because speech precedes us. Language is the union of the human beings. It discovers the same essence because we understand each other.[86] Even among individuals of different communities the linguistic capacity awakens depending on the relationship among men. The constitutive objectivity is the tribute of the individual to a common essence that transcends him. Language, according to

Saussure, is that systematic and social, common to all individuals: the speech, which thus is peculiar to all individuals.[87] But one does not exist without the other.

The Humboldtian constitution shows itself as transcendent. Speech is "a strange object" and its product comes from something different from that which it is applied over.[88] However, it belongs to the sphere of the self, where it is objectivized, and in it transcends it. Humboldt discovers the transcendental value of language in similar terms as those that Husserl uses to reduce the I of the same kind. Language objectivizes in the individual and transcends him, because it precedes him. The speaker is autonomous but he obeys superior imperatives nevertheless. That which rules him in the language comes, in its turn, from an associated nature. It has been inside others as inside myself, but not with the same modification: "und das Fremde in ihr (der Sprache) ist daher dies nur für meine augenblickhlich individuelle, nicht meine ursprünglich wahre Natur".[89] Is this an implication of the I of the epoché, the momentous nature, primordial, and the transcendental and inter-subjective I? Humboldt anticipates phenomenologic questions and converts language in the objectivation of thought.[90]

If we summarize what is exposed so far, we find out that in the speech – "Rede" – the affectivity is awakened and language refers, on the one hand, to the amniotic nucleus, which speaks to us from outside and, on the other, to the air of the world, which returns our own articulation and constitutes a synthesis of a new nature. All these external claims discover the inner transitivity of the I in its pretentions of folding on itself and reaching itself as pure form. In such folds its constitutive otherness is proved. To look for or express oneself implies covering a hole that distances the I from itself in successive modes of presence, none of them fulfilling. F. Pessoa asks himself in *The Book of the Restlessness*, "What is this interval between me and me?"[91] This transitive space alters and makes the self feel strange, originating a diverse heteronym. Each name names the unattainable other. When we speak, we answer to something that has previously invested us – the speech of the other. The presence is initiated as a question or astonishment in the presence of the strange. One of those names is the space of the other or function of the body in the verbal transit of the world. In that fissure – *RIB* – W. Iser bases, for example, the scenic action – "Inszenierung" and "Performanz" – of language and reading, because man in his desire for knowledge, leaves an ontic hole or never reaches himself according to the measure of his wishes. In that way is produced an *Ereignis*,[92] that for us has been previously found in the speaker and author of the text. The other readers always remain, at least, in a second place in relation to this phenomena. The fissure refers us to the abyss – *Kluft* – that Humboldt situates between being and nothingness. It happens at the precise instant of the emergence – "Erscheinung" – or appearance of the phenomena. This emerging, constantly retreating, is also in consonance with the rending and fissure – RIB – of the event – *Ereignis* – of Heidegger in *Denken und Dichten*, as interpreted by J. Derrida, that is, as inscription, a cut – *entame* – or trace – *trace* – that happens in effacing.[93]

The expression identifies itself with the presence of the embodied world in consciousness. It is the differentiated form of the world in that body and not in another, its determinate formal principle, the conception of the concept which is inscribed in the space-time differential of animated matter organized as human and felt substance. Consciousness is conceived in and from the strangeness of itself, because something makes it feel strange, in the same way, as, for example, the *facticity* of J. P. Sartre,[94] because it is never the foundation of its being, although it is the space-time of its sense. That strangeness is a voice that arrives in the world and resounds like speech, claiming it, but comes in and from inside itself, objectivizing itself. That which is concrete and differential in itself presupposes a felt identity formally differentiated in that same instant, for which reason its openings gape, at the same time, in fissures or hole, that can only fill, always distantly and in an inadequate form, the suture of a sign generated in that space-time. It is the fissure of language. Since then, consciousness and sign are mutually engendered. They belong one to the other. In that gap J. Kristeva sees the reconstitution of the originary thing through the sign or negation – "*dénégation*" (*Verneinung*) – of the maternal loss. Such a negative principle constitutes the language as suture or *transposition* of the sign in the originary absence. Language translates the maternal separation as the emergence of sense.[95] Now again, that translation is born for us in the same form as the fissure. The mold of the sign is its emptiness.

M. Henry extends the autodonation of the affection, as essence of life, to the sensitive, imaginative, and conceptual levels.[96] They are three aspects of the formal functioning of affectivity, understood as immanent and radical identity of the content with its form.[97] He identifies the function of *cenesthesis* with the *facticity* of J. P. Sartre, and he describes it as "the inevitable form" from which consciousness "se lève vers le monde et construit ses projets". Its present "*n'est une présence que par la distance où la tient la conscience, elle-même identique avec cette distance comme telle*".[98] In this distancing language is engendered for us as its own function.

Thought was never such beyond the word. If there is a previous layer of expression, as Husserl argues, we conceive it from its own expression. Outside of this, thought is lost in the oblivion whose anonymous character is the base, for example, of the different interpretation of the Husserlian synthesis by E. Levinas. What's previous to the expression is also auto-constitutive. The search for the origin retains and supports it. The form surges as the thinking function of consciousness. It has not been there for ever and ever. It is the budding of the world, with these and not other determinations. It depends as much on the ideality as on its material sensorial being, because it is a function of the abstracted matter that already acts in the constitution of the sensitive. It identifies itself with language. In the monologue, the sensitive layer of the signifier does not disappear. Man would disappear as a perceptive totality. The sensitivity is in suspense, permitting the dynamic of conscious flux, the space-time that this flux determines over an apparent emptiness which is abysmal silence. The distance from oneself to oneself in consciousness also has meaning. It is part of the signifier which engenders its function, where the fundamental apperception of Husserl is precisely constituted. Signifier and signified are a reflection of the expressive function, its own constitutive parts, not in isolation. The remnant of its acts leaves a mark or trace that, being in suspense in the sensitive – an essence which is present and acts in any instant – is produced in the imagination. There is, then, a functional image of the constituted language. This is what allows for the inner language, silent, and inside it the nomadic voice also germinates; errant, it carries away that I by the folds of its functions. In the auto-donation, always "fissured", of consciousness, the form is engendered as its functional expression. This is what constitutes the transcendental character of the language as self-primordial form and resonant dimension of the Other.

UNED, Madrid

Translation: Rosa María Suávez Castro with suggestions by Louis Bourne

NOTES

[1] Husserl, E., *Cartesianische Meditationen* (Hamburg: F. Meiner Verlag, 1987), p. 114.
[2] *Ibid.*, p. 99.
[3] *Ibid.*, p. 107.
[4] *Ibid.*, p. 115.
[5] Theunissen, M., *Der Andere* (Berlin: W. de Gruyter, 1977), p. 65.
[6] *Ibid.*, p. 59.
[7] Husserl. E., *Cartesianische Meditationen, op. cit.*, p. 110.
[8] Husserl. E., *Logische Untersuchungen, zweiter Band.* Husserliana. Bd. XIX. 1, pp. 30–110. The first investigation, "Ausdruck und Bedeutung".
[9] Husserl, E., *Ideen II.* Husserliana. Bd. IV, p. 236.
[10] Husserl, E., *Cartesianische Meditationen, op. cit.*, p. 109.
[11] On this level, only the body of the word, its phonic materiality, would be the object of the originary reproductive perception, and the signified, in its turn, presented, what is hidden liable to inter-subjective consideration and commercial with approximations, more or less certain, but never concluding. The experiential plenitude only corresponds to the signifier.
[12] Husserl, E., *Cartesianische Meditationen, op. cit.*, p. 99.
[13] *Ibid.*, p. 100: "so habe ich meinen natürlichen Sinn eines Ich insofern verloren, als ausgeschieden bleibt jeder Sinnbezug auf ein mögliches Uns oder Wir und alle meine Weltlichkeit im natürlichen Sinne".
[14] *Ibid.*
[15] *Ibid.*, p. 110.
[16] Husserl, E., *Logische Untersuchungen, op. cit.*, p. 42.
[17] *Ibid.*, p. 43.
[18] Husserl, E., *Ideen I.* Husserliana, Bd. III, 1, p. 287.
[19] *Ibid.*, p. 286.
[20] Derrida, J., *La Voix et le Phénomène* (Paris: PUF, 1983), p. 35.
[21] *Ibid.*, p. 101. It is based on the ninth paragraph of the logical investigation, 1. In the fourteenth are found the three components: "Die beziehenden Reden von Kundgabe, Bedeutung und Gegenstand gehören wesentlich zu jedem Ausdruck" (*Logische Untersuchungen*, I, *op. cit.*, p. 56). Those notified are psychic living experiences, those signified, the sense or content and those named, the object of the representation (*ibid.*, p. 38).
[22] Husserl, E., *Logische Untersuchungen*, I. *op. cit.*, p. 56.
[23] *Ibid.*, pp. 44–45.
[24] *Ibid.*, p. 44.
[25] *Ibid.*, pp. 41–42.
[26] "Der erfül lende Akt". *Ibid.*, p. 45.
[27] Derrida, J., *La Voix et le Phénomène, op. cit.*, p. 86.
[28] Husserl, E., *Ideen I, op. cit.*, p. 287.
[29] Vygotsky, L. S., *Pensamiento y Lenguaje* (Buenos Aires: La Pleyade, 1984), p. 189.
[30] *Ibid.*, pp. 191, 192.
[31] *Ibid.*, p. 196.
[32] Husserl, E., *Ideen I, op. cit.*, p. 288.
[33] Dufrenne, M., *L'Oeil et L'Oreille* (Paris: Jean-Michel Place, 1991), p. 51.

[34] Saussure, F. de, *Cours de Linguistique Générale* (Paris: Payothèque, 1983), p. 156.
[35] *Ibid.*, p. 166.
[36] Derrida, J., *La Voix et le Phénomène, op. cit.*, p. 89.
[37] *Ibid.*
[38] Condillac contemplates the corporal auto-reflection, the process organ-object-organ, in the hypothesis of the statue. He only recognizes that there are other bodies when he does not find himself in those he touches.
[39] Charles, D., *Le Temps de la Voix* (París: J.-P. Delargue, 1978), p. 19.
[40] Ortega y Gasset, J., "Comentario al Banquete de Platön," in *Obras Completas*, T. IX. Alianza Editorial-Revista de Occidente. Madrid, 1983, pp. 755, 759–761, 767.
[41] Taking into account how important and actual this phenomenologic statement is for the theory of the reception, we transcribe here the words of Husserl: "Es ist der große Unterschied zwischen dem wirklichen Erfassen eines Seins in adäquater Anschauung und dem vermeintlichen Erfassen eines solchen auf Grund einer anschaulichen, aber inadäquaten Vorstellung. Im ersteren Falle supponiertes Sein, dem Wahreheit überhaupt nicht entspricht. Das wechselseitige Verständnis erfordert eben eine gewisse Korrelation der beiderseitigen in Kundgabe und Kundnahme sich entaltenden psychischen Akte, aber keineswegs ihre volle Gleicheit" (*Logische Untersuchungen*, I, *op. cit.*, p. 41).
[42] Vygotsky, L. S., *Pensamiento y Lenguaje, op. cit.*, p. 177.
[43] Tomatis, A., *L'Oreille et le Langage* (Paris: Seuil, 1978), p. 69.
[44] Merleau-Ponty, M., *Signes* (Paris: Gallimard, 1960), p. 210.
[45] *Ibid.*, p. 215.
[46] Merleau-Ponty, M., *Eloge de la Philosophie* (Paris: Gallimard, 1953–1960), p. 49.
[47] Merleau-Ponty, M., *Signes, op. cit.*, p. 221.
[48] Levinas, E., "La Transcendance des mots. A Propos des biffures," in *Hors Sujet* (Montpellier: Fata Morgana, 1987), p. 219.
[49] Lindner, G., *Hören und Verstehen* (Berlin: Akademie-Verlag, 1977), pp. 67, 69, 73, 166.
[50] Dufrenne, M., *L'Oeil et l'Oreille, op. cit.*, pp. 51–63.
[51] Saussure, F. de, *Cours de Linguistique Générale, op. cit.*, p. 30.
[52] *Ibid.*, p. 64.
[53] *Ibid.*, p. 30.
[54] Husserl, E., *Ideen II, op. cit.*, p. 238.
[55] Fónagy, I., *La Vive Voix*. Essais de Psycho-Phonétique (Paris: Payot, 1983). Conf. pp. 117–121.
[56] Tomatis, A., *L'Oreille et le Langage, op. cit.*, p. 65.
[57] Heidegger, M., *Unterwegs zur Sprache*. Gesamtausgabe, I. B. 12. V. Klostermann GmbH. Frankfurt a. M., 1985, p. 194 (p. 205 in G. Neska, Pfullingen, 1959).
[58] Ortega y Gasset, J., "Commentario al Banquete de Platón", *op. cit.*, pp. 760, 762.
[59] Tomatis, A., *L'Oreille et le Langage, op. cit.*, p. 72.
[60] Ucla, B., *Phonème et Latéralité. Les Origines du Langage* (Marcilhac-sur-Céle: TextIMus, 1990), pp. 125, 88.
[61] Malmberg, B., *Introducción a la Lingüistica* (Madrid: Cátedra, 1982), pp. 191–192.
[62] Ucla, B., *Phonème et Latéralité, op. cit.*, pp. 109–111, 122.
[63] Husserl, E., *Die Krisis der Europäischen Wissenschaften und die Transzendentale Phänomenologia*. Husserliana, B. VI, p. 191.

64 *Ibid.*, p. 111.
65 Husserl, E., *Cartesianische Meditationen, op. cit.*, p. 114.
66 Husserl, E., *Ideen II, op. cit.*, pp. 239, 243.
67 *Ibid.*, p. 241.
68 Conf. Hjelmsley, L., "Expression et contenu," in *Prolégomènes à une Théorie du Langage*, in *Essais Linguistiques* (Paris: Minuit, 1971), pp. 45–77; "Entretien sur la Théorie du Langage," in *Nouveaux Essais* (Paris: PUF, 1985), pp. 69–86.
69 Husserl, E., *Ideen II, op. cit.*, p. 242.
70 Husserl, E., *Die Krisis, op. cit.*, p. 374.
71 *Ibid.*, pp. 188–189.
72 Husserl, E., *Ideen II, op. cit.*, p. 227.
73 Husserl, E., *Die Krisis*. pp. 188–189.
74 Santayana, G., *Reason in Common Sense* (New York: Dover Publications, Inc., 1980), p. 71.
75 Husserl, E., *Die Krisis, op. cit.*, pp. 187–188.
76 Humboldt, W. von, *Schriften zur Sprachphilosophie* (Stuttgart: Cotta, 1988 (1963)), p. 426. The quotations of this edition between the pages 144–367 correspond to "Ueber die Verschieden-heiten des menschlichen Sprachbaues"; furthermore, the pages 368–756 refer to "Ueber die Verschiedenheiten des manschlichen Sprachbaues und ihren Einfluss auf die geistige Entwicklung des Menschengeschlechts".
77 *Ibid.*
78 *Ibid.*, p. 433.
79 *Ibid.*, p. 154.
80 *Ibid.*, p. 431.
81 *Ibid.*, p. 192.
82 *Ibid.*, p. 440.
83 *Ibid.*, p. 428.
84 Conf. Hansen-Løve, O., *La Révolution Copernicienne du Langage dans l'Oeuvre de W. von Humboldt* (Paris: Vrin, 1972), pp. 63–64.
85 Humboldt, W. von, *Schriften zur Sprachphilosophie, op. cit.*, p. 429.
86 *Ibid.*, p. 432.
87 Saussure, F. de, *Cours de Linguistique Générale, op. cit.*, p. 30.
88 Humboldt, W. von, *Schriften zur Sprachphilosophie, op. cit.*, p. 437.
89 *Ibid.*, p. 438.
90 It would be of some interest to revise, for example, part b) of paragraph 54 of *Die Krisis* (*op. cit.*, pp. 187–190) with the section "Natur und Beschaffenheit der Spracha überhaupt", of Humboldt, in the concrete nineteenth point, from which we extract the following paragraph: "Die Thätigkeit der sinne muß sich mit der inneren Handlung des Geistes synthetisch verbinden, und aus dieser Verbindung reisst sich die Vorstellung los, wird, der subjectiven Kraft gegenüber, zum Object und kehrt, als solches auf neue wahrgenommen, in jene zurück. Hierzu aber ist die Sprache unentbehrlich. Denn indem in ihr das geistige Streben sich Bahn durch die Lippen bricht, kehrt das Erzeugniss desselben zum eignen Ohre zurück. Die Vorstellung wird also in wirkliche Objectivität hinüberversetzt, ohne darum der Subjectivität etnzogen zu werden. Dies vermag nur die Sprache . . ." (Schriften zur Sprachpilosophie. *op. cit.*, pp. 428–429).
91 Pessoa, F., *Libro del Desas Siego* (Barcelona: Seix Barral, 1984), p. 45.

[92] Iser, W., *Das Fiktive und das Imaginäre. Perspektiven literarischer Anthropologia* (Frankfurt am Main: Suhrkamp Verlag, 1991), pp. 157, 481, 486, 512, etc.

[93] Derrida, J., "Le Retrait de la métaphore," in *Psyché. Inventions de l'Autre* (Paris: Galilée, 1987), p. 89.

[94] Sartre, J. P., *L'Etre et le Néant. Essai d'Ontologie Phénoménologique* (Paris: Gallimard, 1943), pp. 121, 356.

[95] Kristeva, J., *Soleil Noir. Dépression et Mélancholie* (Paris: Gallimard, 1987), pp. 52–59.

[96] Henry, M., *L'Essence de la Manifestation II* (Paris: PUF, 1963), p. 649.

[97] *Ibid.*, p. 648.

[98] *Ibid.*, p. 637.

KONRAD ROKSTAD

A BRIEF REFLECTION ON LOVE AND THE CONSTITUTION OF THE OTHER

In his last great work, the *Crisis*,[1] Husserl develops the *life-world* as a fundamental concept in his, in some way new[2] reflection on the whole project of transcendental phenomenology. In the years before he had made extensive investigations into the question of *intersubjectivity*, which was established as a fundamental[3] issue at least from the time of his *Cartesian Meditations*. In this rather brief reflection it is my intention to give some indications[4] as to *how* this may constitute an interesting opening for reflecting on *life-worldly experiences* related to the concept of *Love*, given Husserl's analysis of intersubjectivity in the *Cartesian Meditations*. However I will not be proposing any clear-cut thesis.

The *Constitution of the Other* is about the possibility of objectivity and intersubjectivity quite generally, and in the context of Husserl's phenomenology this is dealt with *transcendentally* – in the most profound and genuine manner so that it is meant to have decisive impact on the whole project of transcendental phenomenology. As a universal human phenomenon *Love* is also a quite far-reaching matter – the concept may have many meanings – in literature, religion etc., but above all in life itself. Our understanding, based on feelings and experiences of different kinds, relates to a diversity of senses[5] that invade us – each of us in our communal life. Primarily, as we live and look at life, love is, however, not a concept, but a powerful reality and a phenomenon intimately integrated in our experience of living together as individuals. This may be obvious, but in our context it might, nevertheless, be important from the very beginning to underline the point that life has priority: As a universal human phenomenon, love is first and foremost an *expression of life* – and not of language, even though it, of course, has to be that as well.

Making one subject of two complex problems, let us first consider some plain facts that we all know from our everyday life. Love is related to special types of relations between people; most often, between a woman and a man, children and parents, etc. The connection implied in the title of this paper may have some rather obvious instantiations on that ground. In the sexual sense the act of love between a man and

A-T. Tymieniecka (ed.), Analecta Husserliana, Vol. LIV, 27–36.

a woman sometimes results in the "creation" of another individual – the child; this usually includes taking care of it, giving the child impulses and environmental conditions for its development, becoming a socialised person, which might be seen as a "constitution" of an other as a person. We know, of course, that it is natural processes that "create", and that the conditions for enabling the newborn child to become a fully grown up person are rather complicated, involving society, etc. But many (if not all) of us want to think about all this as really having to do with love.

This, however, is not the line of thought that will be developed in this paper. Nevertheless, associations along these lines are not without relevance and importance. As a matter of fact such associations connecting what we are focusing on as a fundamental *philosophical* problem, to the experiences, feelings and thoughts of everyday life are of rather great importance in the perspective I will try to explicate, looking into some aspects of the problem of the Constitution of the Other in a more *life-worldly* grounded reflection.

* * *

As we from this perspective, now turn to the *Cartesian Meditations*, our focus will be the concept of "pairing" in the Fifth Meditation; this is introduced as the reflection comes to the constitutive "realities" in the analysis of intersubjectivity. Let us very briefly summarise the situation that has developed at that stage. The text is now entering into its decisive phase; the meditations before have laid the phenomenological ground constituting subjectivity transcendentally as exactly that very self-constituting Ego, but this also calls for another subjectivity of essentially the same kind, yet not being the same. The puzzle might perhaps be described in this way: Subjectivity not only constituted transcendentally, but also necessarily being in itself a transcendental Ego, itself giving the most originative functions that constitute, will be in need of taking hold of the achievements that have been reached so far, giving them "objectivity". And being so, it also has to wipe out all the circumstantial steps that might have been taken on the road to its self-constitution – and thereby (yet) hold itself as the self-produced eidetic result of a necessary process. In a way the whole process has to be repeated – it even has to be repeatable with no restrictions – facing not only (the structures of) the phenomenologically reduced and described world which it now has established for itself as its essential correlate, but also the quite concrete (constituted) possibility of being constituted in essential correlation to an Other – being or taking part in a transcendental Ego. As the world however has to be the one common world of them both (and of everyone), and this other Ego also has to be constituting the world, quite decisive efforts will be demanded in this constitutional process of intersubjectivity. If therefore this communal constitution succeeds, really establishing "transcendental" intersubjectivity (in Husserl's project strictly transcendental), it constitutes (the essential preconditions of) the common world – thereby solving the most profound problem of phenomenology, bringing the Ego back from its necessary (methodical) "solitude" to that which is common and objective.

This brief characterisation of the problem of intersubjectivity, which is to some extent already life-worldly-impregnated,[6] points to the very inclusive connection between the Fifth Meditation and those before, and as the reduction of transcendental experience to the sphere of "ownness" takes place, it is in a sense a repetition – explicitly directed at the problem of the alien. In the sphere of ownness there are processes and structures that reflect "immanent transcendence" and a "primordial world", which also gives the field for grounding the phenomenological experience of what is not my own. As this framework has been laid, we first encounter the concept of "appresentation" – and then that of "pairing".[7]

Of course Husserl is now reflecting transcendentally, but his central point, from which all the descriptions and intentional implications develop, is nevertheless the *body* – the animated body of my ownness, wherein it also encounters another animated organism similar to mine. It might not be easy to follow all the steps that Husserl takes in developing the transcendental experience that demonstrates the Other's necessarily being an Ego like myself; and this would certainly also necessitate going back to results gained earlier – which makes it even more complicated. Instead of going into an "immanent" explication and criticism, we call on the concept of love – reflected through life-worldly experiences related not least of all to sexual love.

The rather obvious interpretation of "pairing" in a sexual sense, may imply a much more profound structural similarity between Husserl's problem of transcendental intersubjectivity and the act of love. As persons make love and really feel this is you and this is me – as what we most of all really want together – then the body does indeed demonstrate its own deep-seated, pulsating universe of sense. The act of love is in no way exclusively a set of mechanical movements mixed up with feelings that come from outside the act itself. Primarily the act is bodily mediated feeling engaged in reaching something that I myself am not. In a pretty powerful and self-experienced manner, the body might in this way concretely demonstrate that one is one person by becoming one pair; and one, or two, are a pair by – for a moment – being in the strict sense *one*. In the act of love this overarching of sense might take place and have its moment in realising the *feeling* of fulfilment itself – so to speak universalising the individuality, confirming the originary sense of unity and belonging that means very much in the life of most people.

My body is not primarily an object, but a centre of action and the pulsating meaning that penetrates what I see and feel as it functions in relation to my surroundings. This also holds when it comes to "my ownness" in Husserl's sense. The Other "presented" and first given (maybe) as an object is at the same time "appresented" with senses that are not directly given. I might nevertheless be motivated to look into more of those senses, so to speak creating in fantasy senses that originally are my own – though correlated with that outside "object" with its appresented possibilities. At first those possibilities are however "only" mine, given in my "primordial" world. But even if I never can be quite sure, I might spot similarities as to what I feel motivated to (do); in my ownness there are drives directed at its transcendence into what I might objectify as the attractive, beautiful, the good, . . . and all this may also be powerfully felt to such an extent that a "unity of similarity" is created. It is me creating it – nevertheless realising the distinctness of the "data" going into the unity. As a matter of fact, the whole driving power activating fantasy and possible other actions I might take can only be motivated by that which I myself am not – that is by the appresented "reality" of the possibilities that are presented as they take hold of me. My realising the sense of what is attractive, stimulating, beautiful and good has as its object what is not mine, and therefore also drives me out of myself, headed for a possible "living mutual awakening and an overlaying of each with the objective sense of the other . . . , a total or a partial coincidence . . .",[8] to put it in Husserl's words. The idea of the beautiful and good is constituted "in" me, in my present relation to this attractive, stimulating "object" – and this is so as I actually feel and really want a "unity of sense" to be fulfilled. My power of judgement is not set out of function, it is rather functioning in a free creative manner, reflecting the primal relation between my bodily mediated will and the "composition" of that object or organism over there – now being constituted as beautiful and attractive exactly through this relation.

But of course this would only be half the story; this is still only my reflected feeling having no objectivity outside my ownness. The "reality" to be constituted implies more complicated, concrete processes, that at the same time have to establish the "object" as a subject like myself. This implies me getting into what that reality as an animated organism just might appresent, and if we turn more directly to Husserl himself again – after he has explicated his reflection using the concept of pairing in demonstrating a mutual transfer of sense that founds intentional over-reaching – he right away also asks if the problem is so transparent . . . if this reality will not rather be "a second organism of my own"[9] and not that of another, thereby annulling the whole transfer (or should we rather say "affair"). Husserl proposes[10] a closer look at the intentional situation, and the decisive point he is hinting at seems to be that "appresentation . . . is combined with an original presentation . . ." where ". . . the Other's animated body and his governing Ego are given in the manner that characterises *a unitary transcending experience*". He further points to the general fact that "every experience points to further experiences that would fulfil and verify the appresented horizons . . .", and as it comes to an experience of someone else "it is clear that its fulfillingly verifying continuation can ensue *only by means of new appresentations that proceed in a synthetically harmonious fashion*, and only by virtue of the manner in which *these appresentations owe their existence-value to their motivational connection with the changing presentation proper, within my ownness*, that continually appertain to them". And Husserl then gives us "a suggestive *clue*", saying, "The experienced animate organism of another continues to prove itself as actually an animate organism, solely in its changing but incessantly *harmonious behaviour*".

Let us stop at this point, not going into the text's further analysis demonstrating the constitution of the Other transcendentally. Regarding our context some decisive points may already have been stated. First of all, it seems to be behaviour that has to constitute the actual rather obvious field for confirming the *reality* of my loving that beautiful attractive "organism" my affection is oriented toward – not only my behaviour, but just as much that of the other. She or he has to respond and react in certain ways, and that does not always happen. But it does happen. And when it happens, it might be experienced as a genuine, strongly felt overlaying of sense. Let us look at that situation in light of some of the other points made in the above quotations.

The other body entering the field of possible harmonious behaviour has its physical side indicating something psychic appresentatively, now reacting and responding as my presence and my intentions directed at her or him seem to have been realised. However I really do not know that – but as she (or he) is acting I will know a bit more. How then should the adequate response and reaction to my intention appear? Presumably the other should feel as I feel towards her, having desire, feeling attracted and maybe introducing myself into her picture of the beautiful and desirable – that all of her really wants to be part of. But will there possibly be any action expounding her feeling that way towards me? And how can I act, adequately expounding my feelings and intention towards her? This has to be, it seems, a question of how not *only one* body in combination with its governing Ego can be given in the manner that characterises a "unitary transcending experience", but of how *two bodies* do this, at the same time realising *the union* that constitutes their governing Egos. That union has to be the world – *their world as experienced, felt* for at least a moment, mutually – in a possible *common experience*[11] that is *presented to both and lived through* as "harmonious" action and behaviour. Could there then possibly be a situation grounding that type of experience? In the life of real people such a situation may certainly not be an everlasting situation, fixed permanently as a stable object of theoretical study;[12] but as living life it might occur – experienced as felt in the mutual act of loving someone else, creating feelings and affections with power that may last for a lifetime.

How then is this situation to be characterised? It may be described and experienced in many different ways, but its exact character and just exactly what the decisive move might be are very difficult to decide. It depends on circumstances – historically situated personal and common – and on specific conditions that may impact the way people feel and experience each other. But then, here Husserl also underlines that experience – "its fulfillingly verifying continuation can ensue only by means of new appresentations that proceed in a synthetically harmonious fashion . . . (as) these appresentations owe their existence-value to their motivational connection with the changing presentations proper, within my ownness. . . ." Of course, in our context it will be a problem of what really "my ownness" might designate and mean. But as they are bodily mediated and situated in the historicity of human life,[13] the presented appresentations constitute the fascination that continually appresents – in that way making "the human condition" open and essentially ambiguous. Perhaps this may characterise something universal that constitutes every such situation in the most genuine manner, reflecting the core of those deep-seated historical structures most people experience as felt – in mutually loving each other – in that way verifying the always pulsating sense and the meaning that continues and (re)creates human life.

* * *

This has been just a brief reflection, and as there was no clear-cut thesis, there will be no such conclusion. But hopefully it has presented some interesting problems that might invite and motivate further communication and discussion. In the discourse of phenomenology more generally, this may have consequences along at least these three lines: 1) It may quicken understanding of the relation between the *Crisis* and the *Cartesian Meditations* and perhaps, even further, understanding of their relation to Husserl's works from earlier times, even the seeking of a possible developmental unity that might have been, or might (now) in retrospect crystallise as a genuine core of insight working through (maybe all?) the stages of phenomenology's development.[14] 2) It may prompt joint consideration of the life-world and phenomenology as an unceasing self-reflecting philosophy; is there an interesting and essential inter-constitutive connection between them that might have been there even before the concept was systematically introduced? 3) It may deepen the problem of intersubjectivity as such – what about the life-world *and* the possibility of treating it as a strictly ("pure") transcendental problem?

All these lines of thought really do interconnect, and this constitutes a very extensive field of problems, that have not been dealt with in this paper. Nevertheless it sets up the *horizon of problems* that situates our quite limited undertaking. In analysing some presumably common experiences related to the concept of Love, we may realise life-worldly potentialities that might fit in extremely harmoniously with the decisive problems that Husserl is trying to solve by strict transcendental analysis and description, in order to constitute and legitimise the transcendental intersubjectivity he is headed toward, while also giving it a sort of idealistic (self)interpretation. As the life-world, however, has become the central theme of phenomenology, realising life-worldly potentialities grounding the most advanced transcendental analyses, we might *reasonably* ask if such experiences do not have to become a systematic, essential part of the problem of intersubjectivity – also involving a more practical aspect that always might be extended, in the historically profound correlation to the life-world. And, of course, the concept of Love is in no way new in the history of philosophy;[15] in a sense it even partially constitutes the essence of philosophy itself, all from the very beginning.

University of Bergen, Norway

NOTES

[1] Edmund Husserl, *The Crisis of European Sciences and the Transcendental Phenomenology*, trans. David Carr (Evanston: Northwestern University Press, 1970). In this context I will be referring to the English versions of both the *Crisis* and the *Cartesian Meditations*, trans. Dorion Cairns (The Hague: Martinus Nijhoff, 1960).

[2] At the international phenomenological conference held in Canada in 1969, R. Ingarden asked the question, "What is new in Husserl's 'Crisis'?", and he didn't find so much "newness" of real importance to phenomenology. (See Analecta Husserliana, The Yearbook of Phenomenological Research, Vol. **II**, ed. Anna-Teresa Tymieniecka [Dordrecht: D. Reidel Publishing Company, 1972], pp. 23–48.) In my mind Ingarden is very much underestimating the importance of the *Crisis*, but this is not the issue in this paper. Our reflection will be based on the assumption that there really is something importantly "new" – especially when it comes to reflecting back into the earlier phenomenology of Husserl itself. (See also footnote 14 in this paper.)

[3] What exactly "fundamental" is meant to designate in this context, is a matter of discussion, and so is, of course, the question as to exactly at which point in Husserl's development the issue of intersubjectivity took on such a character. As might be seen from the "Zur Phänomenologie der Intersubjektivität" (*Texte aus dem Nachlass*, heraussgegeben von Iso Kern [The Hague: M. Nijhoff, 1973]), Husserl had been working on such issues already from 1905, and in the Winter semester 1910/11 he lectured on "Grundprobleme der Phänomenologie" substantially including the subject of intersubjectivity or subjects related to it. But the whole matter is rather complex, interrelating many different subjects that are also "fundamental". As this also took place before his true transcendental period, the question of what *exactly* "fundamental" designates is in a way open – and did not get its answer till some years later, in the *Ideen zu einer reinen Phänomenologie und phänomenologische Philosophie* (1913). And in that work other questions and subjects might seem to be more prominent, but in the *Cartesian Meditations* the problem seems to have been decided: intersubjectivity really is a phenomenological fundamental (transcendental) issue – if not *the* fundamental issue. Reflections of a similar kind can be made related to the concept of the Life-world also – and to the concept of historicity – both explicitly made fundamental first in the *Crisis*. In my mind (and I think also in

Husserl's) all these issues really (did) interconnect in the most fundamental way, and this constitutes a very interesting field of problems as to how they might have been working in the "underground" of what Husserl actually did in the times before the works we relate to in our reflection – it might just be mentioned in this context that Husserl (in 1921 and in 1933) made some reflections explicitly on the subject of love as a part of investigations on intersubjectivity (Kern, *op. cit.*, 1973, Zweiter Teil, pp. 165–184; Dritter Teil, pp. 597–602).

4 As a general comment it has to be said that it is, of course, not my intention to cover the quite extensive and very complicated field of problems that is present in Husserl's work from the period of his life of *Cartesian Meditations* and the *Crisis*. That would have demanded also going more specifically into the very extensive – and important – tradition of interpretation and discussion that has been (and of course still is) related to the problems Husserl was engaged in investigating at that time. My intention in this context is not a discussion on that wide tradition with its many very interesting and important representatives. It is quite unpretentious, trying only to give an example of how something really important from the life-world can be reflected on in transcendental investigations – in a way that might give life-worldly sense also to the transcendental. Therefore I will say no particular thing on this highly living tradition, although I will be working within its horizon. I assume I am not however offending anyone if I mention only one name, that of M. Merleau-Ponty.

5 The term "senses" as it is used in this context might be in need of an elucidation; I use it as something lying in the middle, between experience and meaning as we understand these two concepts in a more ordinary sense. The point is that it is a mediating expression – constituting a link between concepts usually separated and distinguished.

6 This might be seen as an unspoken point in our perspective – that the analyses also of the text of the *Meditations* itself have the life-world "leaking" into themselves, and it remains a problem with no obvious solutions how this is tackled by the reduction-procedures.

7 We are here referring to the paragraphs 50, 51 and 52 of the *Cartesian Meditations* (henceforth, *CM*).

8 *CM*, p. 113.

9 *CM*, p. 113.

10 All the quotations that follow refer to *CM*, p. 114.

11 If we go further into the text of *CM*, to paragraph 55, p. 122, after Husserl has gone through the "here-and-there-dialectic", we find him saying: "As we said once before, appresentation as such presupposes a core of presentation. It is a making present combined by associations with presentation, with perception proper, but a making present that is fused with the latter in the particular function of 'co-perception'. In other words, the two are so fused that they stand within the *functional community of one perception*, which simultaneously presents and appresents, and yet furnishes for the total object a consciousness of its being itself there". This might be a hint of the way in which "common experience" can be phenomenologically grounded.

12 In the perspective of the philosopher it might however not be impossible to study this, but not as permanently fixed "objectivity" where you have to take a "teleological" stand of the type the *Crisis* – as "historicity" in my mind, might be grounding.

13 This very brief consideration that leaves more questions than it answers, might be

explicitly related to and extended in texts of both M. Merleau-Ponty and, of course, Husserl himself. I will especially mention *Phenomenology of Perception* (London: Routledge & Kegan Paul Ltd., 1962), pp. 154–174 and the *Crisis*, pp. 106–111. Then the whole field might be laid open.

[14] As a point of view I find related to what I am indicating here, something G. Brand has said (in Tymieniecka, *op. cit.*, p. 166) is of great interest: "The *life-world* is the central theme of Husserl's thought from the beginning on in *a teleological way*. I try to show how in a certain way it is present in the motto, 'zu den Sachen selbst'. Finally it comes out very clear – this does not alter at all Husserl's other writings or his commentators because one may encompass with this concept the totality of Husserl's work. It's one way of interpreting it. . .". At the beginning of the same discussion (answering a question about whether he was interpreting Cassirer), Brand said: "I'm interpreting Husserl. You find in the by-line to the *Crisis*, I think in the appendix, the text on which my interpretation is based" (Tymieniecka, *op. cit.*, 1972, p. 160). If Brand here is referring to "The Origin of Geometry", I will be very much in agreement with him – making that text a foundation for interpreting Phenomenology.

[15] In this connection, of course, the name of Plato goes without saying, but I would say that the life-world has also had quite a decisive impact on the development of philosophical concepts and the whole of the Western scientific culture that they have been integrated with and interrelated to. The soil of growth and development has not solely been culture in the strict sense, and it should therefore be of no surprise to anyone that a concept really taking hold of the concrete historical, factual conditions for human life and understanding is made a fundamental philosophical concept. On the contrary, maybe the "big surprise" should rather be the fact that great philosophical systems and fundamental concepts have been developed without some – systematic or otherwise – reflection on this soil! That being, of course, if it really is a historical fact. As Husserl in his later years develops both the concepts of the life-world and historicity together, as intimately integrated into each other, he rightly seems to doubt this. The main tradition of Western thought has – and not only incidentally – had the life-world as a meaning-fundament, but this has at the same time been a forgotten theme. Even Kant, reflecting "transcendentally" on the most profound preconditions of science and understanding, had unexpressed "presuppositions" in taking the surrounding world of life for granted as valid. And as Husserl in his *Crisis* period understands the whole situation, it is both highly necessary and possible to do otherwise – making the life-world (also) an explicitly essential dimension of the transcendental. But of course this constitutes new problems in the analysis and understanding of transcendental phenomenology itself; reflected on in this manner, this also calls for analyses of transcendental intersubjectivity. Husserl explicitly goes into such analysis in only one paragraph (34) of the *Crisis*, but the whole issue is obviously and quite concretely working in the background of his undertaking. As to the subject of love, I do not think it is present, but according to K. Schuhmann's *Husserl-Chronik* (The Hague: M. Nijhoff, 1977), p. 456, Husserl in 1935 made some brief reflections on "Philosophie, Teleologie und Liebe, Liebe als Problem" – which however I don't know the content of.

HIPÓLITO RODRÍGUEZ PIÑEIRO

EL SER DEL SI MISMO COMO ALTERIDAD Y PARTICIPACIÓN EN PAUL RICOEUR

En este trabajo queremos explicitar el lazo que une la ontología y la ética de Ricoeur y mostrar además que el sentimiento ocupa un lugar clave en la constitución del sí mismo en cuanto revela su pertenencia al ser y sostiene la solicitud hacia los otros.

Como paso previo trataremos de explicitar el vínculo de Ricoeur con pensadores como Nabert, Marcel y Levinas, que aclaran tanto la relación de ética y ontología como el lugar del sentimiento en la constitución del sí mismo en cuanto participación en el ser y solicitud hacia los otros.

Para ello seguiremos las indicaciones de Ricoeur en los volúmenes de *Lectures* . . . en los que reseña su encuentro, su diálogo, y su desacuerdo con esos pensadores entre otros.

Luego veremos como juegan estos influjos, acuerdos y desacuerdos en la síntesis de *Soi même comme un autre*. En tercer lugar trataremos de explicitar el lugar del sentimiento en la constitución del sí mismo, tema central que consideramos necesita ser explicitado, de ese libro de Ricoeur.

Para Ricoeur el sentimiento supera, conciliándolos, deseo y razón. Es la razón interiorizada, la interiorización, en la intimidad, de la totalidad a la que se aspira.[1]

La sabiduría práctica a su vez es la visión, iluminada por la razón, de lo bueno en el marco de la autoestima, la solicitud por los otros y las instituciones justas. En esa constelación quisiéramos destacar y explicitar el lugar del sentimiento como raíz común de sensibilidad y razón, comprender encontrándose, templado, en el sentido de Heidegger.[2] O bien como deseo racional o razón deseante, en el sentido de Aristóteles.

El pensamiento de Ricoeur es, en una dimensión fundante, una hermenéutica del testimonio. El testimonio es la huella de lo infinito en lo finito, de lo absoluto en el tiempo. Huella, deuda, convicción, compromiso, son algunas de las figuras del testimonio, de lo inverificable objetivamente, del Absoluto, de lo Eterno.[3] La hermenéutica del sí mismo es en *SMA* una figura de la atestación.[4]

Nabert, Kierkegaard, Marcel, Levinas, Heidegger, prestan sus voces

A-T. Tymieniecka (ed.), Analecta Husserliana, Vol. LIV, 37–50.

a este "pensar polifónico", a esa tarea de "hacer trabajar las aporías", de Ricoeur.[5]

La experiencia y la afirmación de lo Absoluto que se da en el testimonio no debe identificarse con el ejemplo o el símbolo: "En la acción ejemplar el caso se disuelve ante la regla, la persona ante la ley" (*FF*, pp. 126–127). La experiencia de lo Absoluto dada en la convicción y en el compromiso, figuras del testimonio, no es la afirmación de la universalidad de la ley en la singularidad del caso, sino una respuesta a un sentido, una idea, con la que me identifico decidiéndome, pero que me solicita a la vez, de modo que "no podría hacer otra cosa" porque esa idea me sobrepasa, es superior a mí y de algún modo me constituye en lo que soy antes de decidirme identificándome con ella.[6]

La afirmación del sentido trascendente del testimonio requiere una dialéctica que empiece "desde abajo", dando el rodeo de la semántica a la hermenéutica propiamente filosófica a través del método dialéctico peculiar de Ricoeur, que es a la vez el método de arbitraje de las interpretaciones: mostrar las rupturas de sentido y de vida en la pluralidad de los discurso humanos y mostrar a la vez cierta continuidad de sedimentación y de innovación en cada uno de los discursos.[7] Se irá desde el uso empírico del significado del testimonio en diversos contextos de aparición (semántica), hasta una hermenéutica filosófica que rompe con el nivel anterior del discurso pero que a la vez conserva sus adquisicones.

En este primer nivel semántico el testimonio implica una relación en principio dual: el que atestigua y el que recibe el testimonio. A la vez está al servicio del juicio: "El testimonio es aquello en lo que alguien se apoya para pensar que . . . para estimar que . . . en fin, para juzgar. El testimonio quiere justificar, probar lo bien fundado de una aserción, que más allá del hecho, busca alcanzar su sentido" (*FF, ibidem*, p. 129).

En tercer lugar el testimonio se da en una situación de discurso caracterizada" presupone la Institución (*ibid.*).

Esta relación triádica: el yo, el tú, y la Institución, caracteriza todas las relaciones de despliegue del sí mismo en *SMA*: el discurso, la acción, la narración, el juicio y la acción éticos.[8] La situación institucional se señala en los rituales que hacen válido el testimonio, como se patentiza en el caso del juramento. Esos rasgos del significado común del testimonio, juntos o con otros, – como el carácter de diferendo entre partes son transpotados analógicamente a situaciones más amplias, como el testimonio histórico, entre otros. Un rasgo importante generalizable también analógicamente es la cualidad del testigo. Testimoniar no se limita a testimoniar que . . . el testimonio tiene que ver con la cualidad del testigo, con su convicción, y en el caso extremo con su sacrificio, que verifica lo que testimonia.[9] El testimonio no es del orden de lo verificable objetivamente: "lo que uno puede reconocer en un testimonio – no en el sentido de relato de un testigo que cuenta lo que ha visto, sino de una obra que atestigue – es que sea la expresión de la libertad que deseamos ser". Este movimiento "autoriza a hablar de acciones absolutas, lo que resulta insensato para los historiadores" (*FF, ibidem*, p. 155).[10]

Ricoeur toma de Nabert la ruptura ética respecto de la reflexión a la vez que la afirmación de la multiplicidad de las personas: "la esfera ética [en Nabert] es así liberada de la sutil tutela que ejerce en Kant el procedimiento trascendental de la primera crítica sobre la segunda" (*L3*, p. 90). En segundo lugar es el sentimiento de lo injustificable el que nos hace acceder a la esfera ética. En tercer lugar es en la protesta contra lo injustificable (Ricoeur, como veremos, hablará de "lo intolerable"), donde la conciencia testimonia absolutamente lo absoluto. Lo injustificable no deja medirse por la violación de las normas a las que la conciencia moral se igualaría.[11]

En Levinas el testimonio es huella del infinito. Su lugar "epistémico" es el mismo que el de la huella.[12] También aquí la revelación del sí mismo se da como pasividad absoluta, como ser afectado. Como punto de partida, la ontologia se subordina a la ética, entendida como metafísica.[13] "El uno para el otro constitutivo de mi responsabilidad es del orden de la *asignación* y no de la *manifestación*" (*L3*, p. 97).

La responsabilidad es constitutiva del sí mismo. La ipseidad es responsabilidad con y para el otro. Es anterior a la visión, a la representación del otro, ya sea que el otro aparezca a partir de la percepción sensible de su cuerpo o como el *Mit-sein* que comparte conmigo un ámbito de iluminación. Frente a la prioridad de la luz, antes que ella, está el estar-ya obligado, que se remonta a una *arqhé* que no es *arqhé* porque es anterior a todo tiempo, es impresentificable, no representable en ningún tipo de visión, de presencia. El otro, la obligación para con él, me subyuga antes de que tenga conciencia de él y antes de despertar a mi propia conciencia: Está siempre ya más acá o más allá de la conciencia y del objeto. Lo cual quiere decir también que lo "primero" si puede hablarse aquí de primero, es el estar subyugado o afectado, atado, obligado. El ser afectado es lo constitutivo de mi ipseidad, de modo que hay una prioridad del ser afectado en la constitución del sí mismo por la responsabilidad ética con el otro.

Levinas cuestiona la noción de intencionalidad. El otro que se da en la relación intencional no es otro, sino un otro cualquiera suceptible de ser representado.[14] La representación, cuando no es del dominio de la conciencia – como en Husserl –, es del dominio de la manifestación, – como en Heidegger – de la visión. Señnala Ricoeur que "la noción de proximidad . . . marca la primera ruptura, en el plano de lo sensible, con la intencionalidad husserliana, cautiva de la representación".[15]

Para Levinas la ética comienza antes que la ontología. La manifestación del mundo, si se toma como prioritaria, excluye la obligación con el otro. Por eso, para Levinas, Heidegger no sale de la relación de dominio con los entes. El mundo, el trabajo de las cosas, tienen sentido en todo caso como ofrecimiento al otro.[16]

La inteligibilidad del decir, en que me dirijo al otro, es anterior y fundante respecto de la inteligibilidad de lo dicho, en la que me refiero a las cosas o al mundo.[17]

La unidad del sí mismo está dada en Levinas por lo que lo reúne en cuanto responde a una obligación, a una asignación desde el otro. La unidad del sí mismo no es otra cosa que el mantenerse en la obligación de la respuesta: ese modo de la temporalidad tiene en la exterioridad, en la asignación, el lugar en que se constituye como el reverso de un anverso: respondo o no respondo a un llamado, una interpelación: como la circunstancia, llamado mudo o expreso, es única, la respuesta que me constituye y que constituyo es única. Las "pasividad más pasiva que la diferencia entre actividad y pasividad" se refiere a esa unidad inobjetivable, más allá de toda contraposición en que la pasividad del ser afectado es mi libertad misma. Como en Ricoeur y también en Marcel, hay una unidad o una dimensión donde el recibir no es distinto del dar, donde el recibir es crear, donde la iniciativa más propia, *es* la respuesta a un llamado, a una interpelación: esa estruetura circular es propia del testimonio, del compromiso, de la promesa, de la creación en el plano del amor de las personas y de la creación poiética.

Ese sí mismo anterior a pasividad y actividad, o actividad que se corresponde con la absoluta pasividad de ser-ya, del encontrase abierto, expuesto a lo otro, a los otros, es el sujeto, "subjectum",[18] que soporta el peso, la responsabilidad, del universo: es así que hay unidad del todo. Es así como interpretamos la diferencia del decir y lo dicho, la estrategia del exceso y de la exterioridad.[19] El si mismo no es empero una simple función o X vacía, correlato de los requerimientos empíricos que lo solicitan. Es insustituíble concreción como respuesta. Ricoeur deslinda entonces en la lectura de Levinas la diferencia entre una identiddad *idem*, la identidad dicha, y una identidad *ipse*, que correspondería al decir. El ¡héme aquí!, la constitución, si puede hablarse así, de este sujeto sujetado, impide el acceso a él desde dentro, como yo, como primera persona.

Esta constitución, en *acusativo*, del sí mismo, tiene su correspondencia en la búsqueda de Ricoeur del acceso al sí mismo desde los prononmbres reflexivos, en que el sí se designa a sí mismo reflexivamente, desde el decir, la acción, la narración, el juzgar ético: en esa desigancíón indirecta, en su multivocidad analógica, consiste el modo alético del atestar, figura del testimonio.[20]

La prioridad del otro en la constitución del sí mismo, y su manifestación en el ser-afectado, está presente en Ricoeur, en la medida en que el sentimiento de autoestima implica en sí mismo la solicitud por el otro. Pero el ser afectado por el otro, y en esto Ricoeur corrige la unilateralidad que cree leer en Levinas, tanto la autoestima como la solicitud por el otro presuponen actividad y pasividad, crear y recibir. El dar y el recibir constiyen el núcleo de la persona en Ricoeur: el don como expresión, pasiva y activa a la vez, la donación como creación poética, como descubrir, crear, y donación en la relación interpersonal como ser afectado y dirigirse al otro en el cuidado (palabra castellana que traduciría, creemos, el sentido que Ricoeur le da a la solicitud).

Al absoluto despojamiento que implica para Levinas la asignación y la responsabilidad por el otro opondrá Ricoeur la afirmación correlativa, dialógica, del yo. La reciprocidad, el reconocimiento mutuo, es la base hacia la que y desde la que se interrelacionan, en Ricoeur, la actividad del que da y la pasividad del que recibe. Actividad y pasividad intercambian sus papeles ya que dar es recibir el reconocimiento del que recibe. Recibir es dar, sacar recursos de un fondo de pasividad, de paciencia, que justifica también la actividad del poder activo de dar (*SMA* VII.1). Esta no objetivable, no verificable, convergencia da actividad y pasividad se analogiza en múltiples figuras: creación y donación, compromiso y llamado, descubrimiento y creación. Es la unidad del sentido a la que apuntan secretamente las dicotomías, unidad de pasividad y actividad en la acción, unidad de actividad y pasividad que constituye el "fondo de ser" y el "lecho de ser" en cuya participación nos constituímos como personas.

Es el fondo de ser "potente y efectivo"[21] que subyace a nuestro ser en común y también a nuestro vínculo con la naturaleza y con el mundo de la acción humana colectiva.[22] Esa síntesis postulada, a la que Ricoeur apuesta, como conciliación de los opuestos, paradójica, inobjetivable – síntesis de naturaleza y libertad,[23] de deseo y logos, es el polo ontológico de atracción hacia el que se orienta el pensamiento polifónico y conciliador de Ricoeur.

Esa referencia mutua de donación y creación, de actividad y pasividad, que diferencia a Ricoeur de la insistencia de Levinas en la pasividad como consitutiva del sí mismo, lo mantiene en cambio en la cercanía de la ontología de Marcel.

La categoria de la donación es equivalente en Marcel a la de disponibilidad. En Marcel la ontología se orienta hacia la ética, de modo que es la reflexión ética la que da sentido a la reflexión ontológica, si bien, como acotará Ricoeur, no hay en Marcel una explicitación respecto a la relación de ética y ontología.

Decíamos que la estructura circular de donación y creación, compromiso y llamado, apuntan a una unidad inobjetivable que constituiria el polo ontológico de atracción del pensamiento de Ricoeur. El postulado, la apuesta por la unidad de la verdad, del ser, del bien, esa esperanza escatológica que opera en el discurso filosófico de Ricoeur, – expresa una honda resonancia del pensamiento de Marcel.

Siguiendo el estilo marceliano de reflexión sobre experiencias nucleares concretas, Ricoeur explicita las referencias implicadas en el tema de la disponibilidad.[24] El nexo de la categoría ética de disponibilidad con la ontología de la fidelidad resuena en el nexo que Ricoeur encuentra entre la categoría ética de la promesa y sus condiciones ontológicas de posibilidad en la pertenencia al Ser que nos interpela.

Es a partir de la reflexión sobre la fidelidad con el que Marcel inicia su itinerario de pensamiento en el *Journal Metaphysique*, que se especifica el tema de la *disponibilidad* como fidelidad creadora.[25] Estos temas se desarrollan a su vez en conexión con el de la promesa, que es nuclear en el pensamiento de Ricoeur ya que ocupará el centro de la problemática de la diferencia de identidad *ipse* e identidad *idem* en *SMA*.

La promesa implica *comprometer* mi porvenir. La idea de compromiso conduce a la de *disponibilidad* (*L2*, p. 61).

La idea de compromiso plantea el interrogante de una especie de solvencia para garantizar su cumplimiento. Si nos ponemos de parte de un observador objetivo exigiremos una solvencia o garantía objetivable, es decir de la que podamos disponer en el presente, o al menos una solvencia predictible de acuerdo a a la lógica de probabilidades: "hay una alta probabilidad de que fulano pueda complir con sus compromisos" (pagar las deudas que contrae, etc.). La incertidumbre en este caso acompaña al cálculo de probabilidades. En cuanto a la actitud interna, comprometer mi futuro es afirmar que seré el mismo, en el sentido de que las contingencias o el cambio de mis estados de ánimo, de mis creencias, de mi "concepción del mundo", permanecerán inalterables. Es la posición solipsista del estoico: se trata de un dominio, en el presente, del futuro, de una decision sobre mi mismo. Pero esta decisión[26] correría el riesgo de ser un engaño para sí mismo o para aquellos con los que me comprometo: no soy dueño absoluto de mis estados de ánimo, creencias, etc., por lo tanto no puedo asegurar – a menos que mienta – que mantendré mi compromiso. O no engañaré a los otros manteniéndolo, pero podría engañarme a mi mismo fingiéndome que sigo sintiendo o pensando lo mismo en el momento del cumplimiento, cuando en realidad puedo haber cambiado. ¿En qué condiciones es posible entonces una promesa, un compromiso? ¿Qué es lo que creemos en el fondo cuando hacemos una promesa veraz, es decir cuando creemos poder cumplirla? ¿Qué es lo que sostiene mi creencia de que "a pesar de todo" podré cumplir la palabra empeñada? Y bien: no podría, ser otra cosa que una creencia en que estoy en una dimensión en la que los recursos de ser que me constituyen Sobre los que me sostengo, son del orden de la donación y no del orden económico, del orden del ser y no del tener, de la disponibilidad como creación del ser en mí y creación o donación de mí mismo en la fidelidad a los otros. La promesa implica una estructura triádica: el yo, el tú, y ese garante al que de alguna manera invoco. Ese "garante" no puede ser, no lo es en Marcel ni en Ricoeur, un fondo de ser neutro, sino un ser que se da, que es donación.[27] Ahora bien, si se trata de "condiciones de posibilidad" no hay ninguna "garantía objetiva", es decir, verificable, de su existencia.

Desde el punto de vista de lo verificable objetivamente, de lo disponible en sentido técnico-económico, el garante necesitaría a su vez de una garantía. Pero en el orden del *testimonio*, la promesa afirma un sentido desde el que vivimos *ya*. La invocación a un recurso de ser personal es *lo que da sentido* al sostenimiento de una promesa. Es decir que este acto tiene sentido, si implica la invocación a ese fondo de ser personal o "garante absoluto", a esa disponibilidad ontológica que me hace posible ser disponible, disponer de mi futuro. Ese fondo de ser proyecta sentido sobre mi acto de prometer. Sin ese fondo, recurso, Disponibilidad. Confianza, mi acto no tendría sentido, es decir no existirían promesas, es decir promesas auténticas.[28]

Decir que el compromiso se sostiene en la fidelidad a sí mismo nos lleva a la "pendiente kantiana" (*ibid.*, p. 69), de la autonomía, del formalismo y la abstracción. Es aquí donde se anuncia la idea de disponibildiad como metáfora bancaria. "¿Qué compromiso escaparía entonces a esa sospecha, no sólo de desconocimiento de sí, sino de presunión? Aquí, el salto: sólo escaparía a esa sospecha el compromiso, que, lejos de proceder a una posición de sí por sí[29] sería una respuesta a algo así como una prenda del ser sobre nostros . . . Todo compromiso es una respuesta" (*Etre et avoir*, p. 63, *ibid.*, p. 68).

Ricoeur reafirmará esa implicación ontológica del ser en la promesa, pero – consecuente con su metodología del "largo rodeo" – buscará la mediación del lenguaje como institución fiduciaria que la hace posible.

Asi como para Marcel la disponibilidad y la esperanza eran las marcas esenciales de la persona,[30] Ricoeur desarrolla la noción de compromiso, vinculada con aquellas, y en relación con el movimiento personalista y sus inquietudes sociales y políticas.

En la linea del pensamiento de Mounier, la persona la caracteriza en principio Ricoeur como soporte de una actitud, de una perspectiva, de una aspiración (*L2*, p. 194). Después de la salida del primer lugar de la escena francesa del existencialismo, la fenomenología, el marximo e incluso el personalismo, cuando no perdura ninguno de estos *ismos*, y cuando ya es extemporáneo hablar de conciencia, de yo o de sujeto, ricoeur reivindica la persistencia de la problemática de la persona y su lugar crucial en los debates de nuestro tiempo: "No podría argumentarse en la defensa de los derchos humanos, de los exiliados, de los prisioneros, si borramos de la escena el concepto de persona" (*ibid.*, p. 198). La noción de persona, vinculada por tradición al cristianismo, fue adoptado por Mounier. Esa noción se vinculaba, aún dada la amplitud no dogmática del personalismo, a un orden jerárquico de valores, influenciado por Scheler. La irrupción de la nueva ola de fondo nietzscheano en la escena francesa, cuestiona el orden de valores despues del nihilismo proclamado por Nietzsche.

Ricoeur introducirá innovaciones en el pensamiento heredado de

Mounier. Pero tenía que enfrentar al nihilismo emergente ya que éste es incompatible con una ética de compromiso con las personas. Se trataba de apuntar hacia una ontología como horizonte de sentido del compromiso con la humanidad del hombre, y es a partir de la crisis y del sentimiento de lo intolerable que nos descubrimos comprometidos sin posibilidad de evadirnos o de jugar el rol de espectadores desinteresados. El compromiso con el hombre es posible desde el horizonte de una ontología esperanzada en un sentido de la historia según el cual las conquistas del bien se acumulen y las destrucciones del mal se anulen[31] El pensamiento de Ricoeur puede interpretarse como la búsqueda de las condiciones de posibilidad, del horizonte de sentido, el horizonte ontológico, que haga comprensible una práctica ética, social y política, de compromiso con el hombre, con las personas.

El sentimiento de lo intolerable[32] nos revela nuestro compromiso con la historia, nuestra inserción en lo real, un orden preferencial de valores que decidimos nosotros, no pudiendo sin embargo no asumirlo sin traicionarnos a nosotros mismos, es decir sin dejar de ser un sí mismo. El compromiso me pone en la situación de "discernir un orden de valores capaces de requerirme en la medida en que me identifico con una causa que me sobrepasa" (*ibid.*, p. 200).[33]

NOTAS

1 *H. F.*, Libro I, cap. IV; "Le Sentiment" (en Husserl, 1859–1959, Phaenoménologica, 4, La Haya, 1959; "Simpatía y respeto" en *A l'école de la phénoménologie*, París, 1993).

Ortega, también influído en la temática del sentimiento por M. Scheler, como Ricoeur despues, en *Meditaciones del Quijote*, y en "Leyendo el Banquete de Platón" rescataba el vínculo entre la aspiración al saber de la totalidad y el sentimiento como expresión del vínculo con la totalidad.

2 *H. F. ibid.*: *SMA*, p. 22, ss; 254, ss., 308, ss.

3 "Si el testimonio debe ser un problema filosófico y no sólo, como se suele decir, jurídico, o incluso histórico, lo es en la medida en que el vocablo no se limita a designar el relato de un testigo que refiere lo que ha visto, sino que se aplica a palabras, obras, acciones, vidas que, en cuanto tales, atestiguan, en el corazón de la experiencia y de la historia, una intención, una inspiración, una idea que va más allá de la experiencia y de la historia. El problema filosófico del testimonio es el problema del testimomnio de lo absoluto, o mejor: del testimonio absoluto de lo absoluto. La cuestión sólo tiene sentido si, para la conciencia, lo absoluto crea sentido. Ahora bien, crea sentido, más allá de la crítica al argumento ontológico y a las pruebas de la existencia de Dios, más allá del derrumbe de la ontoteología, si la *reflexión*, por una ascesis tanto intelectual como moral, es capaz de elevar la conciencia de sí hasta una "afirmación absoluta de lo absoluto" (*Fe y Filosofía*, Almagesto, docencia, Bs. As., 1990).

Las figuras de la huella, la deuda, la Eternidad y el tiempo, son temas centrales del libro de Ricoeur *Temps et Récit III*. A la profunda resonancia agustiniana de estas figuras del testimonio en ese libro nos hemos referido en nuestro trabajo "Las implicaciones ontológicas de la convicción" (Vto. encuentro Nacional de Fenomenología, a publicarse en Escritos de Filosofía, Centro de Estudios Filosóficos, Academia Nacional de Ciencias, Bs. AS.) En ese libro Ricoeur limita a la relación con el pasado supuesta en el conocimiento histórico esas categorías del testimonio sin dejar de mantener el sentido de referencia a lo Absoluto como horizonte último de las categorías de deuda y de huella.

En *Soi-même comme un autre* la figura del testimonio es el hilo coductor, como testimonio de la ipseidad, hacia donde converge la "unidad analógica de la acción" en las modalidades del decir, el hacer, el narrar y la sabiduría ética.

Ver en L3, "El testimonio en Levinas, Heidegger, Nabert". Ricoeur tambiéna Marcel: (El misterio del Ser, Lección . . . : "El testimonio").

[4] En *SMA*, Prefacio, y en el *Est.* VIII. 1 Ricoeur serefiere a la frónesis como visión (acordada) de la totalidad de sentido de la vida en las dituaciones singulares y la asimila a la interpretación textual, remite a la interpretación de Gadamer de la prudencia en Aristéles.

[5] En los trabajos sobre Marcel (*L2*) insiste en ese estilo musical o pensar polifónico. También en *VH*. En *TR* se trata de "hacer trabajar las aporías" sobre el tiempo. El arbitraje en el "Conflicto de las Interpretaciones" es otra expresión de esa modalidad dialéctica y de búsquda de las síntesis, de las conciliaciones, en Ricoeur.

[6] *L2*, p. 200.

[7] *EC*, p. 149.

[8] En *L2*, pgs. 203–223 "Approches de la personne" Ricoeur hace converger la ética, la filosofía del lenguaje, de la acción y de la narración.

[9] "Todo el mundo comprende que es otra cosa distinta que un narrador exacto, incluso escrupuloso. No se limita a testimoniar que . . . , sino a testimoniar en favor, a dar testimonio de . . . Por esas expresiones nuestro lenguaje entiende que el testigo sella su vínculo a la *causa* que defiende por una profesión pública de su *convicción* (s.n.) por un celo proagador, por una consagración personal que puede ir hasta el sacrificio de la propia vida. El testigo es capaz de sufrir y morir por lo que cree. Cuando la prueba de la convicción se paga con la vida, el testigo cambia de nombre: se llama mártir. Pero ¿cambia realmente de nombre?, Mártyr, en griego, es testigo. Ciertamente, no carece de peligro evocar ese vinculo terrible entre testigo y mártir; el argumento del mártir es siempre sospechoso; una causa que tiene mártires no es, necesariamente, una causa justa. Pero, precisamente, el martirio no es un argumento, y menos aún una prueba. Es un poner a prueba una situación límite. Un hombre se convierte en mártir porque antes es testigo. Pero para que un hombre pueda llegar a ser mártir debe ser testigo hasta el fin, lo que no puede derivarse de una reflexión puramente jurídica; pues en un proceso, no es el testigo sino el acusado *el que arriesga la vida*."

". . . Nos encontramos así, hasta en el lenguaje ordinario, con expresiones diametralmente opuestas a la del 'testimonio de los sentidos', que arrastran el testimonio hacia el sentido cuasi empírico; se habla así de 'testimonio de la conciencia'. Pero, sobre todo, se denomina testimonio a una *acción*, una *obra*, al *movimiento de una vida*, en tanto constituyen la *señal*, la *prueba viviente de la convicción y consagración* de un hombre a una *causa*" (s.n.). ". . . El sentido del testimonio parece entonces invertido; el vocablo ya no designa una acción puesta en palabras, la relación oral de un testigo ocular acerca de un hecho al que ha asistido. El testimonio es *la acción misma en tanto atestigua del hombre interior, de su convicción, de su fe*, en la exterioridad". Y sin embargo no hay ruptura de sentido, al punto que los dos usos extremos podrían convertirse en puros homónimos. Del testimonio, entendido en el sentido de una narración sobre hechos, se pasa por transiciones ordenadas a la atestiguación por la acción y la muerte. *El compromiso del testigo en el testimonio es el punto fijo en torno al cual gira el abanico de sentido. Tal compromiso es el que constituve la diferencia entre el falso testigo y el testigo verídico y fiel* (*FF*, *ibid*., p. 136).

[10] *HV* "Afirmación oriiginaria": afirmar la existencia de la libertad: eso significa respetar al otro como persona.

La autoestima implica la solicitud y esta a su vez una comunidad de lenguaje y las demás instituciones sociales y políticas (*SMA*; *PE*).

[11] "En presencia de ciertos actos de crueldad, o de la degradación de ciertos hombres, o de la desigualdad extrema de sus condiciones de existencia, ¿es acaso por la idea de un desacuerdo entre estos hechos y las reglas morales que agotamos las razones implicadas en nuestra protesta? Cuando la muerte, interrumpiendo prematuramente un destino nos hiere de estupor, o que ella nos aparece como el precio de una alta ambición espiritual, ¿es acaso por la idea de lo injusto que evaluamos el acontecimiento? Todo ocurre como si, por el *sentimiento* (s.n.) de lo injustificable, se descubriera en nosotros, en ciertos casos, una contradicción más radical entre los datos de la experiencia humana y una demanda de justificación que la sola transgresión de estas normas no alcanzaría decepcionar, que la sola fidelidad a estas normas no alcanza a satisfacer" (*ibid*., ps. 90–91 *Le Désir de Dieu*, citado por Ricoeur). Veremos que en cuanto a la distancia con respecto a Kant es compartida casi en los mismos términos por Marcel y por Ricoeur.

[12] Levinas, *HOH* IX: "La huella". Seria interesante, nos parece, seguir las afinidades del pensamiento de la huella en Levinas, al menos en relación con lo transcultural, con el Heidegger de "Un diálogo sobre el lengaje" y las concomitancias con M. Ponty, al que Levinas refiere en *HOH* (*ibid*.), respecto a las relaciones "transversales" entre las lenguas y las culturas.

Ya hemos mencioado arriba el tratamiento que da Ricoeur a la idea de huella de Levinas. Ricoeur se aproxima también a Derrida (*TR* III, *passim*).

[13] Derrida discute esa posición de Levinas frente a Heidegger (*VM*).

[14] *L3*, p. 96. Ya Sartre (*El ser y la nada*) había cuestionado en el mismo sentido la experiencia del otro en Husserl. Para Sartre el otro como nóema correlativo de la nóesis es un objeto representado.

[15] Este señalamento de lo sensible como lugar de la proximidad que es a la vez distancia (Levinas: *De otro modo que* ser . . .) resuena también en la dimensión de la pasividad del "yo puedo" en la constitución del sí mismo como otro en Ricoeur: "la carne es la primera alteridad" (*SMA*, Est. X). Ricoeur distingue allí entre cuepro propio y carne, modulación que se acerca al Merleau Ponty de *Lo visible y lo invisible*.

[16] En *Totalidad e infinito* el mundo del trabajo y de la praxis histórica no arranca para Levinas al mismo de su mismidad. La historia es el reino de la totalidad, de la política, de la guerra. Es el mundo de la casa, de lo propio, de lo familiar. Es el exilio de este

mundo familiar lo que experiemento con el rostro del extrnajero. No obstante la responsabilidad con el otro implica el ofrecimiento del pan, la hospitalidad de la casa.

[17] *De otro modo que ser . . . passim.*

[18] Ricoeur, *ibid.*, p. 104.

[19] Derrida, *VM*, *op. cit.* remite esa estrategia del exceso, de la hipérbole, a la esencia del lenguaje. Lo propio del lenguaje es el exceso sobre sí mismo, el estar siempre más allá de sí mismo en cuanto dicho.

[20] Pp. 132, 145.

"Yo en sí mismo como un sonido que resonara en su propio eco" (p. 130).

"Sé muy bien que el texto oscila entre lo Desdicho de la ipseidad, en tanto que siempre ya tomada en los restos del tema, del Dicho, y una cierta reaserción de la ipseidad y de la identidad como "término en hipóstasis" (p. 135). "El sí no es el yo. Queda sin embargo que el lugar del sí es inexpugnable: "El Sí es *subjectum*: él está bajo el peso del universo: responsable de todo" (p. 147) Porque no se substituye a su ofensor más que aquel que es "uno e irreemplazable" (p. 131). Y todavía: "El no-intercambiable por excelencia, el Yo, el único, se sustituye a los otros" (p. 149). En eso consiste a título último la "recurrencia previa del sí mismo" (p. 132).

[21] *SMA*, Est. X.

[22] El espacio de experiencia horizonte de espectativa, la ideología y la utopía, la Institución y la libertad, la tradición y la crítica, son otras tantas expresiones, en el plano de la ahistoria colectiva, de esa dialéctica de pasividad y actividad.

[23] En *VI*, se refiere aes núcleo de la ontología del hombre como "libertad finita" (ver "Cuestiones de Método" y "Conclusión").

[24] Cfr. Ricoeur, "Réflexion primaire et réflexion seconde chez Gabriel Marcel" y "Entre éthique et ontologie: la disponibilité" en *L2*.

[25] "Este nacimiento quasi simultáneo de 'dos ideas musicales' es para neustra búsqueda de un gran interés, en la medida en que la idea de fidelidad tiene el mayor peso ontológico, y la idea de disponibilidad el mayor peso ético" (*ibid.*, p. 68).

[26] Ricoeur se referirá a Heidegger como comprometido en este tipo de actitud "estoica" (*TRIII*, *passim*, *SMA*, Estudio V).

[27] Cfr. en "El misterio del ser", Bs., Sudamericana.

[28] *SMA*, Est. VIII. Ricoeur interpreta del mismo modo, es decir desde una ontología del ser personal, si bien pasando por la mediación del lenguaje, la promesa. Esta presupone el carácter fiduciario de la institución del lenguaje, sobre el cual es posible la relación dual del acto discursivo de prometer, que a su vez no sería vinculante moralmente sin ese fondo o recurso ontológico que había señalado Marcel, al que Ricoeur remite. *SMA*, p. 311.

[29] Nótese el vínculo de esta posición con el cógito como autoposición, y con ello la implicación ética del problema del *cógito*, posición en la que paradójicamente estaría a su vez implicado esa especie de solipsismo estoico de Heidegger, para quién la constancia en el sí mismo, se funda en último término en la resolución para la muerte (*SMA*, Est. VIII; *TRIII*, p. 95, ss.).

La diferencia entre identidad *ipse* e identidad *idem* de Ricoeur se remonta a la distinción entre ser y haber: No puedo disponer del cambio de mis cualidades de carácter, de mis estados de ánimo, del mismo modo que puedo disponer de algo que tengo, como de una propiedad, pero puedo mantener mis promesas y en esa medida ser fiel a mi mismo en la misma medida en que soy fiel a los otros, y en eso consiste la auténtica ipseidad.

No puedo dominar el curso dol mundo ni de mis estados internos, pero sí puedo abrirme un crédito de confianza, que implica ese fondo en la confiabiidad del ser, y ese pacto humano de cofianza sin el cual el lenguaje y la sociedad no serían posibles. Solo por el sentimiento de confianza puedo abrirme ase fondo o recurso de ser, a ese Tú que me llama antes de que yo responda:

"La conjonction se fait, de l'autre côté, avec le thème de la fidelité comme prise de Dieu sur moi, par rapport à quoi ma liberté s'ordonne et se définit." "Aquello a que se opone la disponibilidad es muy exactamente la autonomía kantiana y su interpretación kantiana como *autoposición* (s.n.). Se le reprocha a Kant no haber percibido la "desconcertante alternativa", suscitada por el fenómeno de la promesa y por el estatuto de permanencia asignado al sujeto de la moralidad" (p. 70).

[30] *L2*, p. 78.

[31] *En Lectures 3, Médecins tortionnaires, Médecins résistants* (ps. 398–403), ejemplifica Ricoeur la situación del pensamiento ético y político frente a algunas de los modos de lo intolerable.

[32] "Es en este sentimiento de lo intoerable donde la crisis insinúa el discernimiento de la estructura de los valores del momento histórico" (*ibid.*, p. 200).

[33] "Aquí se descubre una relación circular entre la historicidad del compromiso y la actividad jerarquizante que revela el carácter de deuda del compromiso mismo. Esta relación circular constituye lo que puede llamarse en lenguaje hegeliano una convicción. En la convicción, yo me arriesgo y me someto. Yo eleijo, pero me digo: no puedo actuar de otro modo. Tomo posición, tomo partido y así *reonozco* lo que, más grande que yo, me constituye en deudor insolvente. La convicción es la réplica a la crisis: mi lugar me es asignado, la jerarquización de las preferenciass me obliga, lo intolerable me trasforma, de tránsfuga o de espectador desinteresado, en hombre de convicción que descubre creaando y crea descubriendo.

"Permitidme ageregar a estos dos criterios algunos corolarios. . . ." El criterio del compromiso en la crisis me autoriza a ver en la actitud persona un cierto comportamiento respecto del *tiempo*. Yo derivo esto de aquello que puedo llamar la *fidelidad a una causa*. Pienso aqui en un lealismo del viejo neohegeliano Rosiah Royce (que ha inspirado en otro tiempo a G. Marcel). El compromiso no es la virtud del instante, como lo sería la conversión o, para toda una teología salida de Barth o de Bultmann, el acontecimeitno de la palabra es la virtud de la duración. Ello en razón de la identificación del sujeto con fuerzas transsubjetivas; no es en la conciencia, en el sujeto, ni siquiera en la relación dialogal del cara-a-cara con el otro, donde encuentro este hilo de continiudad, sino en la fidelidad a una dirección elegida (p. 201).

BIBLIOGRAFIA

Ricoeur, Paul, *Le volontaire et l'involontaire*, Paris: Aubier, 1950 (*VI*).
Ricoeur, Paul, *De l'interprétation, essai sur Freud*, París: ed. du Seuil, 1965 (*EF*).
Ricoeur, Paul, *Finitud y Culpabilidad*, Madrid: Taurus, 1969 (*FC*).
Ricoeur, Paul, *El lenguaje de la fe*, Buenos Aires ed. Megápolis, 1978 (*LF*).

Ricoeur, Paul, *Temps et récit*, Paris, Vol I, 1983; Vol II, 1984; Vol III, 1985 (*TR*, *I*; *II*, *III*).
Ricoeur, Paul, *Soi-même comme un autre*, París, ed. du Seuil, 1990 (*SMA*).
Ricoeur, Paul, *Lectures 1*, Paris, ed. du Seuil, 1991 (*L1*).
Ricoeur, Paul, *Lectures 2*, Paris, ed. du Seuil, 1992 (*L2*).
Ricoeur, Paul, *Lectures 3*, Paris, ed. du Seuil, 1994 (*L3*).
Ricoeur, Paul, *Hermenéutica y Acción*, Bs. As., ed. Docencia (*HA*).
Ricoeur, Paul, Etica y Cultura, Bs. As. ed. Docencia (*EC*).
Ricoeur, Paul, *Politica y Educación*, Bs. As. ed. Docencia (*PE*).

PART TWO

HUMAN INTERPRETATION-IN-EXISTENCE: CONSTITUTIVE OSCILLATING BETWEEN THE "SELF" AND THE "OTHER"

KATHLEEN HANEY

THE DIALECTIC OF SELF

I. OTHER: SELF-CONSCIOUSNESS/SELF-RESPONSIBILITY

The human self requires reciprocal dialectical relations with others to achieve proper maturity; my interest in the essay to follow is the self that develops more or less autonomously of others, the self of the autistic child. This self albeit limited and truncated is nevertheless a seat of virtualities and habitualities. Exploring the possibilities for the self without others permits us some insight into the genetic structures of human development.

I begin with the observation that in many theological, philosophical and psychological tellings, the structure of the self is trinitarian. Plato, Aristotle, St. Thomas Aquinas, Husserl, Tymieniecka, Ballard, among others, provide various descriptions of this tripartite structure that are mutually informative for looking at aspects of the process of ontopoiesis. In order that my interest in autism in this essay be contextualized, I turn to discussions of the structures involved in the ontopoiesis of the self and the essential conditions for its possibility.

The results of this analysis shall show that the other is necessarily implicated in the mature human self. Yet, before its dialectical engagement with the other, the self must engage itself. When the self fails in this self-engagement, we "see the character of autism as a specific modification of human existence."[1] This human existence develops a kind of self; its divergencies and limitations characterize the founding level of consciousness.

In what follows, I shall focus on two philosophic analyses. First, I will be interested in the philosophical foundations of the self that is functioning well, i.e. living well. Next, I shall show the autistic moment of human existence as the genetic ground of the constitution of the other. I shall conclude with a note concerning the commerce between the happy person and his primordiality.

The late Edward G. Ballard was interested in the self in the three functions that Aristotle called making, being, and knowing (or doing). According to Ballard, these human powers exercise themselves in

A-T. Tymieniecka (ed.), Analecta Husserliana, Vol. LIV, 53–67.

"contact with other beings."[2] The exercise of these powers is the necessary although not sufficient condition for the creation of the human meaning of individuated life. The temporal unity of the self bridges: 1. the discovering self, in its protendings, the maker of meaning, 2. the acting "I," that finds its being, "am," in experience gathered up with its past, 3. the concretized "me" who made these products, these meanings.[3] The unity of a human being, the integrity that Socrates and Plato take to be essential to human happiness, depends upon the reflection that provides the meaning which can unify the person as a whole being, aware of its fecund future that combines its past with its chosen virtualities.

The discovering self for Ballard, as the creating self for Tymieniecka, has three possibilities for engagement. The self can be pointed toward its creative acts, making meanings, toward grasping the meanings it has already made or toward its being, as meaning-making. Tymieniecka describes three movements of the soul. The first, "radical examination," seeks for depth of meaning in its affections and passions; the second, recognition of finitude, "establishes its authentic existence," while the third tends toward "transcending."[4]

If the self inclines towards discovering its being through uncovering the meanings it has bestowed, the self must objectify what it has been. What the self has been might be answered (albeit inconclusively) with a report of what the self has done or made, since creating or making objects makes its creator, as well. Simply said, the carpenter develops the hands of a carpenter; the philosopher, the soul of the philosopher. The senses that Tymieniecka refers to as moral, poetic, and intellectual[5] dimensions of human life are hewn from making objects. Making them makes the creator as making music makes the person a musician. Knowing this or that about what (s)he has made subjects the maker who has been to cancellation. Oedipus cries, I have done this terrible deed, but now I disown it. It was me and it is not me. Oedipus rejects what he has made. He never wants to look upon the offspring of his union; the fruit of his body is now distasteful to him.

According to Ballard, human dramatic time, the temporality of ontopoiesis in the self-individuation of life, often displays the structure of tragedy. The protagonist chooses an identity (after first having chosen to quarrel with his fate), as the just king of Thebes, say. The protagonist takes up the burden of his decision and suffers through to the insight that he had misidentified himself. Such mistakes or misdeeds purify him so that he strips away what he is not. The proviso is that the protagonist take on the responsibility for what he has been. Until then, the plague remains on Thebes. Before responsibility can be confessed, responsibility must be taken on in the recognition of agency. Responsibility is the debt that the protagonist owes to the insight that he has chosen to act within the limits appropriate to human freedom. Oedipus participated in his fate through the self-definition that he determined within the range of his actual possibilities. Choosing this or that identity (as the maker of this or that, the painting, say, or the kingdom) determines the arena for further action including creative acts which determine the identity of the protagonist to be recovered through reflection.

Oedipus chooses the role of king of Thebes from among the possibilities for making which present themselves to him. His decision as to his identity decides what kind of making is appropriate to him. The creation the king makes rightly includes the establishment and governance of the good state. The product of the king is his rule which constructs the state. He who would undertake the role of king takes on the responsibility of creating the state. Oedipus decided by himself to become king; he could have chosen to identify himself with the role of shepherd, say. Then, the sheep which resulted from his shepherd's care would be the product he made while he was co-creating himself as the maker. We would suspect that the development of the shepherd's human excellence would demand that he had created a healthy, safe environment for his sheep to live and grow while making himself into a patient, thoughtful and dutiful man.

The shepherd would be mistaken if he were to say of himself, "I am shepherd." Indeed, the dialectic of the three functions or the three movements of the soul requires that the self, whatever it be, cannot be an "am," except during the moments it is gathering up what it has been. I am not; I become. My movements are always transcendent. Neither does Oedipus become identical with the role of the king; rather, he may develop his human excellence by creating, within the confines of his virtualities, that good state that he undertook to actualize. The king, who in his acts of bringing not being, into being makes himself just or the contrary, reckless or narcissistic, etc., is always the man, Oedipus, sometimes king, sometimes beggar, sometimes denizen of Olympus.

What had Oedipus in fact created? A sick and sorry state of barren women and mothers of dead children wept under the burden of the debt

Oedipus owed. Oedipus did not recognize his creation. He saw his initial judgment of his identity in his role reenacted in his commitment to save Thebes again. He sees himself as the same man, recreating his meaning, the just king still. Reflection through and in the other teaches Oedipus that he was never that man. Oedipus was never the just king establishing the good state. Oedipus brought the wrath of the gods on his city; the wailing, barren women told him so. Intersubjectively, Oedipus' creation did not stand the test. He was not the self he believed himself to be. The creation he believed himself responsible for was not being made. No just state arises through the efforts of the parricide until he recognizes that his creation points to a self other than the one he actualizes through his rule of Thebes.

Other people teach Oedipus who he is. After all, he is wrong about himself. In addition to his catastrophic misidentifications of his parents, Oedipus mistakenly blurs the boundaries of himself and his role. He wrongly believes himself to be fulfilling the role of just king. He is the king, but in his overly close identification of himself with his role, he forgets that knowing the past carries with it the possibility of rejecting the self whose choices and acts led to tragedy.

Here we come to the first illustration of the structure of the self which I want to explore. According to his particular tripartite analysis of the self, the self reflects on its products in order to see its history and its origin. The active self makes a meaning. The reflectively active self recognizes that it has been the maker of that meaning. We recall that the tragedy of *No Exit* is Sartre's unrelenting identification of a self with the coward that it finally was. Even for Sartre, self-knowledge requires the other.

How can one see himself without a reflection? He can judge himself by his products, but how does he judge his products? What do his values reflect? Sometimes, his best efforts at creation fail to achieve fruition; the potter's kiln may destroy his jar. Improperly created structures cannot stand. But, what about the standing structure? We would not say that its success in performing its function was the same as excellence in performing its function. What about the pot that does not fail in the fire? Must it be pleasing? When can the potter say, "I am a good potter?" The maker's values as well as their accomplishment must be evaluated.

In Jocasta, Oedipus saw his own happiness. Jocasta loved the man in her life although she did not love him well, to state the obvious. Jocasta took Oedipus to be who he said he was. She did not struggle with him to find out who he was or who they were. She believed him; she did not believe in him enough to fight for his transcendence. Peripheral seeming characters team up to engage Oedipus in what Jaspers refers to as the "loving struggle"; Tiereris loved Oedipus. Blind as he was, he was responsible to tell the truth about Oedipus as he saw it. He risks his life to persuade Oedipus to see what he is really. Finally, Oedipus listens. Too late. In her blindness, Jocasta has reconfirmed the truth about who Oedipus is. Oedipus liberates himself from what he no longer wishes to be after he hears and sees what he has been. He cannot bear the sight. To be blind is better than to see what he has done. Oedipus cancels the hold of what he has done by recognizing that his children are products of an unspeakable union. He is the self who formerly deluded himself.

The blind wanderer in *Oedipus at Colonus* lives more richly and deeply in himself than the boastful young king could have imagined. Tragedy is a kind of victory over illusion and self-deception. What is won is emancipation from the not-self. The not-self is contained in the past of the self, in its parents, in its culture. The ancestors continue in the offspring, yet they cannot fully determine the offspring. One of the layers that the self must strip away is the illusion of the self believed to be determined by the parents. Oedipus left his parents, he thought. We know that he found his parents, but harmed them because he did not recognize them as his real parents. Oedipus did not take himself to be Jocasta's son, therefore he could take his mother for his wife. Oedipus mistook his parents. To see one's real precursors is a possibility which erupts out of freedom, limited (not all precursors are possible), but, nonetheless, real freedom. Oedipus freely made of himself a protagonist who was blind to himself. Later, Oedipus freely choose to become a blind wanderer so that his outer appearance and his inner life could be more mutually reflective. Interestingly, following the tragic pattern, Oedipus learns of himself through others.

The result of tragedy is insight so that tragedy has well been regarded as a triumph in the West. Oedipus, blind, disgraced, guilty, shamed, remorseful, lonely, is happier than Oedipus, young, brave, smart king and devoted family man. By rejecting untruth, Oedipus has moved closer to the truth of who he is.

II. EMPATHY, THE POSSIBILITY FOR THE ONTOPOIESIS OF SELF

Much human preoccupation, and most modern philosophy, valorizes the encounters with the "objectively" given which the person understands or uses. Even phenomenology, foundational philosophy resting on the subject's indubitable self-awareness, has not sufficiently explored the necessity for others implied in the operative notion of objectivity. Yet, these others are themselves each more or less individuated, unique persons. If the person is significant to his encounters with the world, then the person must be already making himself as a self.

One way in which the individual makes its own life occurs in its desire for the other who can see him and hear him and appreciate his creative products. When he sees himself through the eyes of the other, he sees himself as real, as intersubjective (as Husserl would say) or "one among others" in Paul Ricoeur's phrase. He can adjudicate his own and various perspectives according to their presuppositions, their spatial location, degree of illumination, etc. (at least in principle).

Empathy, the poiesis which makes the other, makes the other as expressive, linguistic, and communicative. This making involves recognition as well as creation, knowing as well as making. Without empathy for the other, without insight into the meanings which the other intends and delivers to me symbolically, or his gestures, I cannot make sense of the other's reflection of me – I cannot fully constitute myself.

Self-reflection is the distinctly human power which issues from the virtualities of human linguisticality, but language is born in, of the other and in empathy and communication. Language is shot through with affect. Through acts of empathic intuition of the other's primordial presence to himself, revealed in his "indirect discourse," I recognize the other as he is for himself. Eventually, Oedipus sees that Teiresias is truthful. The mutual attempt to articulate the insights gleaned empathically is the becoming of each in knowing each other in shared discourse which unites the one with the many in the one world of meaning. As Jaspers understood, this kind of communication risks self-being as an identity in the loving struggle with the other. The self that makes itself in making its constitutions and its creations exposes itself and the errors of its ways to the eyes of the other. Also, this self may enjoy its visions reflected in dialogue with the other.

To summarize thus far: the making of the self, the ontopoiesis of the self, occurs in the modes of knowing, making, and doing. The being of the human self is directing its power toward becoming the self it knows itself to be, but how could the self know itself to be if it is not yet? In part, through what the other shows me about my possibilities through his reflections of me which he communicates to me in various fashions. In part, our common universe of discourse reenforces and expands my intuitions of value. Our shared enjoyment encourages me. The intersubjective acts as the test of my private world. In communication, not only do I discover what I know, but also I bring myself to birthing (Frieda Kahlo's painting of giving birth to herself exactly illustrates this expressive structure. In making, in creating the image of herself being born from herself, Kalo births herself in we who see her, as well). Also, as we have seen in *Oedipus Rex*, the other helps one rightly to evaluate what he has made.

As Plato teaches us in the *Phaedrus* (Derrida aside), the spoken word is the best means of communication. Human virtuality is inherent in language use. Language provides the space for self-critique, self-reflection, self-collection, self-recognition, self-identity, integrity, and authenticity. In Frieda Kalo's painting or in writing, the product and its reflection are not useful to the modification of the self, being of the artist who made himself in making the work. I'm not sure that we are helping Plato by reading his works now, but I am sure that Socrates helped Plato by speaking to him in a loving struggle for truth. They are children of the same god and know him in and through each other.

Words, the peculiarly human vehicle of communication, incarnate meanings. I grasp the other's meanings; I grasp the other's soul although I may grasp it in the recognition that the other who tries to deceive me is a liar. Nevertheless, the other is one, we are each an other among others. We are each one among many with similar sets of possible meanings. We learn about these meanings through empathic grasp of others and by studying the intentions of words.

The other's words are, of course, the community of many others. The native linguistic community provides values and meanings which the self chooses to take up as its own. The self can live as the agent which knows the world, produces goods, and becomes itself to the extent that it can share meanings with others. Shared incarnate meanings, common language, provide the categories, the forms, which the self requires to see itself as one, unique, among others, each unique. The self can choose to become according to its form when it can speak about itself to others and listen to how others see it. More important than living well in the world, the self cannot live in the human world at all unless it can share the meanings of incarnate words. Without language, there can be no self-reflective grasp of Logos; without language, there can be no articulated world; without language, there can be no individuated selves who possess themselves in freedom. Without the empathy with the other which presents us as sharing meanings, there can be only limited denotative language.

Merleau-Ponty distinguishes "originary language" from the pedestrian language we usually employ. Sometimes we hear ourselves saying things we have never said before or never knew we knew. This process is an analogue to the process of creation as Tymieniecka envisions it, but articulating meaning is reflecting which makes the meaning the self's to abide by or reject. In empathic communication, I recognize my own meanings as well as the others. The medium of speech transgresses the boundaries of language.

In speech, the other evaluates my meaning, by first comprehending it then finding it worthy of discussion. This kind of intersubjective, dialectical, meaning-making is thoroughly spontaneous in its utterance (although the conversation may have been planned) since to speak to the other is to embody my meanings in sounds that erupt from a bodily posture, from bodily expressions and gestures. The words and their responses are the products of this creation. The otherness of the material of sculpture, dance, music is an objective dialectic while the otherness of the other intrudes with non-objective elements. Creating objects, the subject imposes its intentions (more or less readily). Creating discourse, subjects reciprocally co-intend (with more or less overlay of meaning).

Articulating the self as it arises in dialectic which responds to the other is an attempt to cash out my reading of Husserl as a philosopher who opens pathways for understanding the absolute importance of the intersubjectivity and the empathy it rests upon.

This analysis needs to be supplemented by a discussion of the constitution of the other qua other. In outline, I recognize the otherness of the other in his opacity to me. Empathy is a perceptual intuition of dwelling in the other's body that permits me access to his physical experience of himself.[6] I can never, in principle, have direct access to the other. I can throw intentional bridges across the chiasm, however. These are well-made if empathically made. Edith Stein describes empathy as "a sui generis" sense. Empathy projects its own bodily syntheses (that it lives in) onto the other who is seen, but not felt. I cannot see my own face, but I can recognize human meanings in the face of the other. Who is this living I that functions as the possibility for empathy?

Now we turn to the subject's dialectical engagement with himself. According to Husserl, the foundationalist, the subject is prior to its engagement with others. The subject is the possibility for engagement; the subject is the transcendental condition for any experience. In its fruitful expression, the subject's process may result in what Tymieniecka calls creation, the ontopoiesis of self and culture. Without the other, the subject lacks recognition. Without experience of a subject prior to the self-with-others, there can be no other than to the subject. Or, to put the matter more bluntly, no communication without anything to say. And, there can be nothing to say without prior communication with the self.

III. THE DIALECTIC OF THE SELF

Recently, a rash of reports from autistics, often by means of facilitated communication, reveal much about their affective states and their inner lives. Some older autistics have reported, for instance, that they want to do one thing, but their bodies do not cooperate. They want to respond to commands, but are not able to get started; or their bodies react in ways which contradict their meanings. A remark made by a seventeen-year-old boy through facilitation is typical. He is purported to have typed, "my mind and my body disagree."

The grave difficulty for phenomenologists has to do with the possibility of a "pre-bodily" meaning which seems to be required for the body to be unfaithful to the mind's intentions. How are meanings constituted if not in gestures, words, and so forth? What is the possibility for intentional consciousness to differ from its meaning constitutions? Is the severely disturbed autistic child's body making its own meanings which are disconnected from conscious intentions?

Such reports are of interest to philosophy, I contend, because autistic children develop a self less modified by others than the normal child. In this paper, I take advantage of the luxury (for a philosopher) of having real examples embodied in autistic persons. My argument will be carried on as an eidetic analysis. I begin with empirical examples of autistic characteristics to subject them to eidetic analysis. The essence of autism reveals itself in the free variation of these possibilities.

The autistic self develops without recognizing the other, or recognizing the other without being able to communicate with him. Recognizing the other requires bodily activity; without these acts, the self as a member of a pair, "we," does not develop. In these cases, the self fails to develop a sense of its embodiment as an object in the world of others. In the second case, the clumsy body may be mirrored in others.

In order to justify my claim that autism sheds light on epistemological problems of interest to the phenomenologist, I turn now to methodological issues. Husserl argues that purified psychology, i.e. psychology subjected to phenomenological reduction, presents apriori structures of the ego. The purified psychological ego is freed from its involvement in concrete embodiment although a situation and history fix the concrete psychological subject as a particular enactment of apriori structures. The ontologies of essences and being, the ideal and the existential, must be related as individual embodiments of essences. Purified psychological description begins with individual instances, in order to understand essences, and recognize their instantiation. Given this proviso, autists provide access to the essence of autism. Like all essences, the essence of autism shows itself in each instance of autism, but no human child embodies the autism fully and only. Indeed, the very word "autism," from the Greek "self," suggests that autistics are more individuals than more normal persons who take their name by tending toward the regulative, the regular.

The primal self of transcendental philosophy begins in a state of consciousness that lacks differentiation between self and other. The analogue of the primal self in the realm of intentional psychology is the autistic self. All its experience is immanent to its consciousness. The distinctive feature of autism, impaired social relations, shows itself in language deficiencies. Without language, autistic development occurs without much reference to real others who guarantee a real world. If this is the case, "things" are pre-real for autistics. Their "things" are constituted "within the limits of the primordial empirical consciousness."[7] Autistics may have sensory experiences of "pre-things," "phantoms" without having objects, things in the ordinary sense.

Things are constituted through "different formations and strata." Each of these has its own character and its own unity. An example of this in the natural attitude might be the chiming bell. I hear it, locate it, but may never actually see it (inside the bell tower). The audit constitutes the bell adequately by hearing and locating it spatially. This constitution may involve memorial elements as protentions: I saw the bell once, it sounds the same, in the same location, I'll give it the same name. The doxic intention constitutes the thing, the bell, in its presence and absence.

Among the difficulties autistics report, excesses and deficiencies in sensory perceptions feature prominently. Donna Williams reports her "sensory overload" by which she means disassociated competing sensory attractions. Her eyes and ears vie for intentional regard, but the strength of attractions (even by the same "pre-object") across discrete, unassociated sensory manifolds may throw her into a confusion. She sometimes reacts by "switching off" all but the visual manifold. Dr. Temple Grandin, Ph. D., herself an autistic, takes issue with Donna Williams' statement, "Autism is not me. Autism is just an information processing problem that controls who I am." Grandin, on the other hand, believes that autism is "part of what I am."[8]

Grandin surmises that the reason that the women evaluate their autism so differently has to do with the degree that it handicaps them. Professor Grandin dedicated her latest book on autism to her mother to whom she acknowledges great gratitude since she credits her emergence from autistic mutism and sensitivity to touch to her mother's insightful interventions. She now has full possession of her linguistic powers and opportunities to enjoy herself. She believes that her autism gives her a unique visual acuity that she employs in her work in livestock handling. She can see from a cow's perspective; she can see from the perspective of the blueprints, she can see structural relations. She reports that her thinking is almost exclusively pictorial and associative. This is her autism and her talent.

Donna Williams and others report being imprisoned in their bodies. In Ms. Williams' case, her failure to synthesize sensory strata (or sometimes even to identify the strata as each separate) can still overwhelm her. She, like Temple Grandin, writes about autism. They both require self-management. Temple Grandin regulates her stress and anxiety with a "squeeze" machine of her own invention. When Donna Williams can work, she has structured her environment and her diet so that she, rather than her body, will be in control of what she does.

To the attentive reader, I have just resurrected the decayed Cartesian mind/body split. I am guilty. This topic is one which the first generation phenomenologists, Husserl and his followers, Stein, Merleau-Ponty, Pfander, Berger, Strauss, Strasser pursued. Husserl devoted the second volume of *Ideas* to the constitution of the animate self, the real psychic subject, and the soul, of the animal nature and the spiritual world.

Autistic consciousness is prior to consciousness of the world or of other persons. Is this why many autistic children are mute? Why don't they continue with infantile babbling? Even deaf babies babble until four months of age. The animate nature acts naturally. The natural possibility becomes direct activity. The normal infant continues to babble because he hears himself babbling and wants to make the noise, to enjoy his power over his voice. Later, he will recognize that his mother is imitating his babbling noises or that they are cooing together. The infant baptizes babbling as mutual meaning making and gains entry into the mystery of language. But, what if his vocal cords would not make the noise he wanted to hear? Would he forget about babbling? Would he become mute? What if he could not distinguish his mother's voice in all the other sounds?

Autistic infants are often hyper- or hypo-tonic. They may experience touch as I experience the sudden arrival of a car with a revved up boom box while I am waiting at a stoplight. Touch may be agonizing. Not only does the inhibition of the normal desire for touch jeopardize the infant's emotional development, it denies him the epistemological clout of touch as well. After all, as Husserl, Merleau-Ponty, and Sokolowski have all noted, touch teaches presence and absence. The infant can feel one of his hands touching the other hand or touching the hand of the other. The resiliency and opacity of the other is evident there incipiently in the various tactile intuitions of presence.

These insights are denied the autistic infant insofar as he avoids touch. He, therefore, misses this early clue to the nature of selfhood and others. He is blocked on this channel. He can go no further along the way to the world of the personalist, the goal of human maturation, according to Husserl. If the infant does not easily nuzzle up to the mother and suckle at her breast while he gazes at her face, he cannot see her looking at him. Not touching, not being held, or seeing himself being seen, he does not experience the mothering one as the one who holds and gazes. Untouched, her body appears thing-like to him. He has difficulty constituting her as *Leib* and pairing with her.[9] He des not practice acts that constitute sensory fields and across sensory fields into a manifold which he identifies as an object.

We now know that autism is caused by neurological abnormalities that shut the child off from normal touching and hugging. It is the baby's abnormal nervous system that rejects the mother and causes it to pull away when it is touched. There is the further possibility that secondary damage to the brain, caused by a defective nervous system, adds to the child's further retreat from normal comforting touch.[10]

The autistic infant, thus, acts on a level of consciousness beneath what Husserl describes in section 151 of volume I of *Ideas*. The most severely autistic children do not indicate that they constitute sensory manifolds as substantial-causal Thing, even if the Thing relates to only "one empirical subject." Without constituting the bodily other as a substantial-causal Thing, there is no basis in motivation for intentional pairing with the other. No other exists.

The other exists in paired meanings with the originary ego. The other is a noema that corresponds to the acts whereby the subject transfers presence: "to the seen Body there belongs psychic life, just as there does to my Body."[11] The other can be known only as a function of the originary ego's self-knowledge. This knowledge may not be self-conscious, however. The originary ego in its primal state acts. Reflection upon the objectivity of the acting agent, knowing the self, cannot be achieved unless the originary ego sees itself and, if not through the other's eyes or reciprocal gestures, or language, how? On the other hand, the more primary possibility must subsist. All egos are active as well as passive. Without activity, the ego does not live.

Aristotle was careful to draw a clear distinction between making and doing. Making involves an end product external to ourselves. Doing, though, is acting; its end is internal. It is its own end, ourselves.[12]

Autistic acts are not acts of making. The autism moment, as evidenced in the most severely autistic children, includes knowing and doing, but precludes making. The autistic infant knows to regulate his body temperature, to suckle, to cry, but it may never learn to gaze, to speak, to make eye contact, to cook, to clean, to write a letter or a poem, to live independently. The function of making must be important, indeed.

Describing autistic behavior without providing a causal account of it, phenomenologically, has led us to the signification of making, of creating, the products that make the self as maker. The self as maker can recognize himself in his products. The artist always leaves his trace. The self can see itself in its reflections. Without casting images, in the other, in the word, in the wood, in the pate, the self cannot see itself.

Children diagnosed with autism come to self-consciousness significantly later than normal children or blind children.

IV. CONCLUSION

In conclusion, then, I return to the tripartite description of the self. We have seen that autistic children fail to accommodate the necessity that the self has to incarnate its body, the experience that the autistic young man described as wanting to do something but being prevented by his body. Autism shows more than this deficiency, however. Children with cerebral palsy may share many physical difficulties, yet their handicaps are not so humanly profound.

Autistic children do not attempt to make products, objectivities outside of and independent to themselves. Without such products, without external reflection, without the activities that develop the habits of the maker, the autistic can experience himself in what he does only, not in what he knows.

The realm of self-knowledge is beyond the ken of the autistic. His tragedy is that he cannot engage in the tragic pattern which is the way of the soul in its self-knowledge. Autism shows us our dependence on our bodies. The body can make a person autistic insofar as it does not fulfil its function as instrument of communication. We can evaluate the severity of a particular child's autism along the continua that Temple Grandin suggests. She uses her peculiar "autistic" abilities to her own communicative and creative ends. Donna Williams' bodily demands are much greater. Her body imprisons her. The body of the autistic is flawed insofar as it blocks the normal fulfilment of the apriori of constitution.

The autistic is autistic to the extent that his acts are doing without making. Creation, making, is the essential condition for the possibility of self-reflection. We can create or make without recognizing ourselves as responsible. We cannot recognize ourselves without creating and making. Making brings our ownness into the intersubjective world. This making for humans is primarily linguistic or subject to linguistic expression. The self of the autist, the self that has not yet made of itself an "I," cannot ask, Who am I? Autism waits at the threshold of the ontopoiesis of self-identity. Prior to dialectic of self-reflections in others and products, Autism is the necessary ownness that finds intersubjective expression when autism is overcome.

University of Houston

NOTES

[1] Gerhard Bosch, *Infantile Autism*, Trans. Derek and Inge Jordan (New York: Springer-Verlag, 1970), p. 50.
[2] Edward G. Ballard, *Philosophy and the Liberal Arts* (Dordrecht: Kluwer Academic Publishers, 1989), p. 301.
[3] Edward G. Ballard, *Principles of Interpretation* (Athens, Ohio: Ohio University Press, 1983), p. 225.
[4] Anna-Teresa Tymieniecka, *Logos and Life*, Book 2, p. 77.
[5] Anna-Teresa Tymieniecka, *Logos and Life*, Book 2 (Dordrecht: Kluwer Academic Publishers, 1988), p. xxviii.
[6] See my discussion of these topics in *Intersubjectivity Revisited* (Athens, Ohio: Ohio University Press, 1994).
[7] Edmund Husserl, *Ideas*, Trans. W. R. Boyce Gibson (New York: The Macmillan Company, 1958), p. 419.
[8] Temple Grandin, *Thinking In Pictures* (New York: Doubleday, 1995), p. 61.
[9] See particularly chapter 3 in my *Intersubjectivity Revisited.*
[10] Temple Grandin, *Thinking In Pictures* (New York: Doubleday, 1995), p. 85.
[11] Edmund Husserl, *Ideas Pertaining to a Pure Phenomenology and to a Phenomenological Philosophy*, Second Book, Trans. Richard Rojcewicz and Andre Schuwer (Dordrecht: Kluwer Academic Publishers, 1989), p. 174.
[12] Ballard, p. 305.

ANNE-MARIE BOWERY

THE PRACTICAL SELF
A Test Case for Foucault

In this paper, I examine Bret Easton Ellis' fictional portrayal of a serial killer named Patrick Bateman. I argue that Patrick Bateman exemplifies a post-humanistic Foucauldian subject. However, his practices of self-constitution and his comportment towards them are problematic. They lead us to question the liberating possibilities of Foucault's ethics of self-care. To frame this literary analysis, I will summarize Foucault's provocative conclusions about the death of man and survey the alternative suggestions he offers for conceiving the human in the wake of this critique.

I. FOUCAULT'S CRITIQUE OF MAN

In the interview *Truth, Power, and Self*, Foucault remarks, "It is one of my targets to show people that a lot of things that are a part of their landscape – that people think are universal – are the result of some very precise historical changes."[1] Since the belief in man indelibly structures our conceptual "landscape," Foucault focuses his intellectual energies on this point. He explains that an "idea of man has become normative, self-evident, and is supposed to be universal."[2] However, if one recognizes that this idea arose through different sets of practices, "psychological, medical, penitential, educational," then one should recognize that humanism "may be quite relative to a certain situation" and not foundational at all.[3] To this end, Foucault explores the historical development of the social sciences and in doing so reveals the emergence of the modern concept, man. In *The Order of Things*, Foucault argues that the idea of man that we, at least outside the circles of philosophy, take for granted, arises out of very specific historical contingencies. Foucault avers, "As the archaeology of our thought easily shows, man is an invention of recent date." In fact, "man is only two hundred years old."[4] Not only has man recently appeared on the face of the earth, but his appearance will be fleeting at best. The claim that the era of man will not long endure intensifies the startling nature of these remarks. Indeed, *The Order of Things* ends with Foucault's famous sug-

A-T. Tymieniecka (ed.), Analecta Husserliana, Vol. LIV, 69–82.

gestion that "man will disappear erased, like a face drawn in sand at the edge of the sea."[5]

Why, though, does Foucault believe that the era of man is "perhaps one nearing its end?"[6] To be sure, the belief that the human subject is historically constituted and not universal partially explains his eschatological view. However, Foucault does not simply offer a dispassionate critique. His writing embodies more than a sober conceptual analysis. It is rhetorical, evocative, even polemical. Foucault wishes for the era of man to end. According Jon Simons, author of *Foucault and the Political*, Foucault engages in "a campaign against faith in essential human subjectivity. It is the unity of the individual, the subject, that Foucault considers the most suspicious of the truths which we hold to be self-evident."[7]

The negative consequences that arise out of this belief also cause Foucault to argue for the demise of human subjectivity. Jon Simons again explains, "Foucault's key criticism of the modern era is that the three axes of subjectification [truth, power and ethics] are so closely entangled that the only subjectivities, or modes of being a subject, available to us are oppressive."[8] According to Foucault, this oppressive situation at least partially arises out of the fact "that man, assignable in his corporeal, labouring, and speaking existence – is possible only as a figuration of finitude." He continues, "modern culture can conceive of man because it conceives of the finite on the basis of itself."[9] Hence, modern man, as opposed to concepts like "renaissance humanism" and "classical rationalism," is "the imposition of constraining limitations."[10]

This subjugation of the subject also occurs because "man appears in his ambiguous position as an object of knowledge and as a subject that knows."[11] While the average philosopher analyzing this situation might argue about problems of circularity and self-referentiality, Foucault does not. Indeed, "the crucial issue for Michel Foucault is not simply an epistemological one." Rather, as one interpreter explains, "the anthropological bias is understood to provide knowledge with an insidious potential – with a propensity for turning into a political instrument of subjection."[12] This seemingly unavoidable eventuality deeply concerns Foucault. In books like *Madness and Civilization, Birth of the Clinic* and *Discipline and Punish*, Foucault explores the multifaceted nuances of these humanistic modes of oppression.[13] Indeed, the rhetorical point of Foucault's work incites us to an existential rebellion. Ideally, "We should refuse to remain tied to the identities to which [and by which] we are subjected."[14]

II. FOUCAULDIAN PRACTICE

Foucault assumes this historical critique of the subject and the humanistic regimes of oppression it endures throughout his *oeuvre*. Like Nietzsche, Foucault's project is largely negative. By criticizing institutions and concepts we take for granted, he raises our awareness about their oppressive character. To be sure, Foucault has a positive agenda as well in that he wants us to "refuse what we are."[15] Unfortunately, it is often unclear what form this refusal should take in terms of our political commitments and individual lives. Quite understandably, most interpreters hesitate to leave the political implications of Foucault's thought in such a tenuous state. For if this "refusal" resulting from our heightened awareness is all that Foucault offers, then we are no better off than we were after Nietzsche's transvaluation of values. We know that we should not be slaves, but we are uncertain about how to "live dangerously" and "give style to our character."[16] This uncertainty becomes even more intense when faced with the proclaimed erasure of man. If no concept of man exists, how can we "venture out to face any danger?"[17]

Attempts to come to terms with the political implications of Foucault's thought generally take one of two forms. First, the minimal claim that Foucault, like Nietzsche, "maintain[s] the openness of agonistic relations" and "offers an ethic of permanent resistance."[18] Foucault himself proffers much that supports this Nietzschean orientation. For example, he believes in "challenging what exists, not proposing alternatives."[19] Foucault pessimistically remarks, "there is no point in dreaming of a time when knowledge will cease to depend on power; this is just a way of reviving humanism in a utopian guise."[20] Given this outlook, the situation seems futile. What are we to do? How are we to respond? Should we merely engage, as Foucault does, in "the constantly renewed interrogation as to how thought can reside elsewhere than here . . . in the forms of non-thinking"?[21] Questions still remain. What will these forms of non-thinking look like? How will they interact with each other?

With this problem in mind, other interpreters follow a second path. They insist that Foucault has a positive agenda as well. They insist that he offers concrete alternatives to replace the notion of essential subjectivity that he so thoroughly criticized. According to James Bernauer and Michel Nahon, Foucault "excavates a specific axis of the relationship to oneself, the ways we fashion our subjectivity."[22] In doing so, they argue that Foucault strives to envision "new forms of subjectivity."[23] Indeed, strands of Foucault support this interpretation as well. For example, when describing "the goal of my work during the last twenty years," Foucault says "my objective . . . has been to create a history of the different modes by which, in our culture, human beings are made subjects."[24] He argues that it will form a "new genealogy of how the self constituted itself as subject" and implies that this new genealogy will provide a means for self-liberation.[25] Most readers of Foucault locate the positive elements of his intellectual project in the later volumes of *The History of Sexuality* and his essay, "Technologies of the Self." In these texts, Foucault refers to "arts of the self," and explores the "care of the self," the techniques by which humans come to constitute themselves as subject.

In other words, as his career ends, Foucault shifts his field of inquiry to "what he himself called ethics."[26] To be sure, some are disappointed with Foucault's ethical *kehre.* Jana Sawicki explains, "To many of his readers, Foucault's preoccupation with subjectivity and practices of the self in his later writings have been puzzling and disappointing – even embarrassing. His turn toward an esthetics of the self appeared on the surface to fly in the face of his earlier proclamation of the death of man."[27] However, other interpreters, like Chauncey Colwell, insist that this turn does not undermine his previous work.[28] However this may be, Foucault's intellectual forays lead to these ethical issues of self-care.

In his essay, *Technologies of the Self,* for example, Foucault delineates four types of technologies: production, sign systems, power, and the self. Foucault describes this fourth category as a set of practices which "permits individuals to effect by their own means or with the help of others a certain number of operations on their own bodies and souls, thoughts, conduct, and way of being." With these self-technologies, people can "attain a certain state of happiness, purity, wisdom, perfection, or immortality." Foucault does not see these self-transforming technologies as abstracted Kantian rules for ethical action or as rigidly theoretical epistemological categories. Rather, the notion of technology becomes important because Foucault "wish[es] to discuss the subject [of these technologies] not only in theory but in relation to a set of practices."[29] Foucault's emphasis on technology is a practical concern. Whether or not it has genuine political implications remains to be seen.

The ancient Greeks referred to these practices as *epimelesthai seautou*, which Foucault translates as "to take care of yourself" or as "the concern with self."[30] Foucault elaborates these practices with respect to the Greeks in *The Use of Pleasure*. There, he reads sexual or erotic practices, as similar in nature to economics, dietetics, and gymnastics. The sexual question is not "how am I to achieve the most pleasure?", but "how am I to comport my bodily actions so that I maintain the highest level of autonomy?" Similarly, "whom shall I be with [*suneinai*]?" is the same kind of question as "what shall I eat today?" because both inquiries involve a relationship of control over oneself. Foucault explains, "the accent was placed on the relationship with the self that enabled a person to keep from being carried away . . . to achieve a mode of being that could be defined by the full enjoyment of oneself."[31] Diet, gymnastics, economics, and erotics form crucial underpinnings of self-care. By correctly practising this *epimelesthai seautou*, one exhibits "the perfect supremacy of oneself over oneself."[32] Foucault hopes that these practices will liberate the human subject from its relationship with conceptual finitude and systematic oppression.

Any thinking, feeling, politically committed person would share Foucault's laudatory hope for freedom. However, uncertainty remains about whether a desirable liberation will arise out of these Foucauldian practices. Would a self, conceived of as the product of self-care, constituted by a set of regimes and practices, truly provide a desirable *telos* for human being? Would it justify the death of humanism? I now turn to Bret Easton Ellis' *American Psycho* to provide a tentative exploratory answer to this inquiry.[33] Interestingly, the turn to fiction receives Foucault's explicit approval. Foucault embraces "all those experiences which have been rejected by our civilization or which it accepts only within literature."[34]

III. AMERICAN PSYCHO

American Psycho is a book about a serial killer. The story unfolds from the killer's point of view. Much of Bret Easton Ellis' novel describes numerous sets of practices by which the main character, Patrick Bateman, cares for and constitutes himself. Like the ancient Greeks that Foucault admires, Patrick Bateman strives for autonomy. Curiously, he displays many of the same practices that Foucault describes in *The Use of Pleasure*: economics, gymnastics, dietetics, and erotics. I will briefly describe each set of practices in turn.

Economics: Patrick Bateman has complete economic freedom which, of course, determines many of his practices. He bears a private wealth that allows him infinite trips to the cash machine, the conspicuous consumption of designer suits, high-fidelity stereo equipment, and ostentatious home furnishings. Economics is not a level on which Bateman must concern himself. True, he must replace his Platinum Card when it breaks from overuse, but his practices of self-care primarily focus on gymnastics, dietetics, and erotics.

Gymnastics: Patrick Bateman follows a rigorous fitness program "which incorporates both aerobic exercises and [weight] training." He elaborates: "On the leg machines I do five sets of ten repetitions. For the back, I also do five. Over at the free weights, I do three sets of fifteen repetitions of leg extensions, leg curls and leg presses. For the chest, I do three sets and twenty reps of incline-bench presses."[35] The novel provides list after list of infinite variation: he jumps rope, Stairmasters, attends aerobics classes, and Lifecycles vigorously. Many chapters of novel begin with his mantra: "I spent two hours at the gym today." With this set of gymnastic practices, the American Psycho cares for himself.

Dietics: "I'm drinking close to twenty liters of Evian water a day." He always drinks out of a "glass bottle, not plastic, so the mineral content is not altered."[36] He allows himself small glasses of decaffeinated cinnamon tea and rations out slices of Japanese apple pear and oat bran muffins.[37] Patrick Bateman prefers Diet Pepsi to Diet Coke because of its "fizzier and cleaner taste."[38] He meticulously takes multivitamins and potassium tablets.[39] He worries about the sodium levels in soy sauce and the grams of fat in a Dove Bar's chocolate coating. With this set of dietetic practices, the American Psycho cares for himself. However, if one considers how Patrick Bateman treats other people, these elaborate dietetic and gymnastic regimes of self-care create a disturbing dichotomy.

Erotics: The American Psycho's sexual practices are myriad and diverse. Though Patrick has a girlfriend, Evelyn, he begins an affair with her best friend, Courtney, as well. Apparently, his sexual appetite exceeds the parameters of these relationships for he regularly employs prostitutes from the streets and high-class call girls as well. He supplements these erotic activities by picking up Vassar girls on spring break and aspiring actresses at night clubs. As time permits, he masturbates and rents pornography. He also brutally destroys many of his sexual partners. With these actions, the *American Psycho* becomes a useful test case for Foucault. While Patrick Bateman spends extraordinary time taking care of himself and unquestionably demonstrates an enormous respect for his body, this respect is utterly solipsistic. It generates no kind of imperative in his actions toward others. His practices are not normative. He maims and desecrates the integrity of another body as easily as he brushes his teeth. The same man who regularly gets facials, manicures, pedicures, and shiatsu massages also kills a couple with a machete and tortures women with ice picks, nail guns, and cigar butts. How is it possible that the practices by which Patrick Bateman cares for himself also allow for, and seeming endorse, the brutal desecration of other selves? What accounts for this disparity? An exploration of the structural foundations of this dichotomy and an attempt to resolve it follow.

I argue that this dichotomy rests upon two premises. First, Patrick Bateman defines his subjectivity with these bodily practices. His body becomes himself. The second premise involves his visual orientation toward both himself and his external world. I will consider each premise and its implications for Foucauldian subjectivity in turn. The definition of the self as the body becomes apparent in the novel's earliest scenes. While a detailed delineation of his morning grooming techniques covers three full pages of text, there is little mention of his feelings. Other than the occasional reference to his irritation with rap music or his anger at an incessant busy signal, Bateman's descriptions of inwardness are almost non-existent. Before evenings out on the town, he will "take a hot shower and afterwards use a new facial scrub by Caswell-Massey and a body wash by Greune, then a body moisturizer by Lubriderm and a Neutrogena facial cream and debate between two outfits."[40] However, he does not look forward to seeing his friends nor does he reflect upon his accomplishments at work. He obsesses about hair mousse, styling gel, and a razor straight parting, but rarely offers any speculation about what might cause his fleeting feelings of nervousness and anxiety.

Nothing is internalized; Bateman defines himself as his body. He exhibits what Foucault describes as the belief that "the body is the inscribed surface of events . . . the locus of a dissociated self adopting the illusion of a substantial unity." Foucault maintains that the body "is molded by a great many distinct regimes; it is broken down by the rhythms of work, rest, and holidays, it is poisoned by food or values."[41] Bateman, on the other hand, defines the body and, hence, the self merely as what is seen, as what appears to the world. For Bateman, his body is what it looks like. Endorphin rushes are not his goal. He does not work out to feel good; he works out to look good. Similarly, what his food looks like concerns him more than its ability to assuage hunger. While his description of any given meal rivals a photograph in *Gourmet* magazine for detail, more often than not, he is "thoroughly unhungry."[42]

That the body is what it looks like merely represents the most basic example of Patrick Bateman's visual orientation. In fact, visual stimuli drive his orientation toward other people and the world. Patrick cultivates the visual. He uses a Sony Walkman to block out sound. He wears gloves and dislikes even the most accidental touch of strangers. He tunes out dinner conversation and focuses instead on a single image: a portion of "rabbit cut to look just like a star" or minuscule dots of wasabi surrounding his "lone baby prawn."[43] He makes this visual priority explicit; "I'm not really listening, watching instead."[44] He sees clothes, cars, deltoids, designer labels, and good hair cuts. Emotions, thoughts, and motivations are absent. In fact, many critics of the book point to the interminable lists of clothing, furniture, and personal hygiene products to diminish Ellis' skill as a novelist. However, these tedious catalogues exist for a reason. To his credit, Bret Easton Ellis lets us see the world as his American Psycho does.

Quite consistently, Easton Ellis' character describes people in terms of a surface veneer, how they are clothed, how they are groomed. He fixates on appearances, physical features, and flaws. Examples abound: "she looks pretty decent, a silk gazar blouse with rhinestone cuff links . . . crystal earrings . . . and gold sling-back pumps," a hardbody waitress has a "left kneecap [which] is knobblier, almost imperceptibly thicker than the right knee."[45] Someone "flashes a bright, toothy smile."[46] Another wears "a red, white and blue firework-print silk tie by Hugo Boss and plum washed-wool trousers with a quadruple-pleated front and slashed pockets by Lazo."[47] He sees a "definite model type [carrying] a pressed leather bag from Hermes."[48] For Patrick Bateman people are not even objectified because objectification would entail the intentional treatment of a person as an object. However, he merely sees visually stimulating apparitions. He knows no self other than what appears to the eye. There is nothing to objectify.

Much of the novel catalogues the American Psycho's horribly systematic destruction of bodies. Image compounds upon image. At first, he "smilingly" tells a bartender he wants to "stab [her] to death and play around with [her] blood."[49] Then, these vicious undercurrents mix with his pornographic sexual practices as he hires two prostitutes and beats them.[50] Any pretense of eroticism soon fades. His destructions become increasingly savage. There are prolonged scenes of dismemberment, decapitation and, finally, cannibalism. How can he perform such horrors? He performs them because he wants to see them.

His visual orientation enables his destruction of others. In his imagistic universe, action inevitably follows one of two choices. Either he engages in the production of images or in the destruction of images. The American Psycho destroys the images that bombard his sight. They are no different from the television screen he turns on and off every morning; they are the same videos he rents every day and frantically returns each evening. Patrick Bateman is quite honest about his visual need. "I want to watch," he explains to a victim. He lifts her up to "get a better view."[51] Later, he "stand[s] over Bethany's body, sipping a drink contemplatively, studying its condition."[52] This dependence on visualization facilitates his destruction of others. Bret Easton Ellis concretizes this visualization of violence, as he has the American Psycho videotape his dastardly deeds. He then shows these tapes to his most recent acquisition as he destroys her.[53]

Surely, such a person does not care for the body at all. Yet, this psycho is Patrick, the same person who tenderly places Clinique Cover-Up over a minuscule blemish beneath his lip. How are we to resolve this incongruity? With reference to the second outlet of action his visual orientation allows him: the production of image. Patrick Bateman is not simply a psychotic voyeur. He is not Chauncey Gardener who likes merely to watch. Other people must perceive the image that is himself. For example, he wants Helga the "skin technician" to see his totally buff abdominal muscles.[54] He is "grateful it's light in the room so [his girlfriend, Evelyn] can see how bronzed and defined his is."[55] Though irritated, he likes it when a "homosexual" asks if he's a model. So, he cares for himself to produce this perfect image. While it would be tempting to say he produces this perfect image to disguise the psychotic horror he is, such an explanation would fail. Bateman does curiously little to hide his destructed images. He kills people in restrooms, on the street, and in the back of a cab. He tells people exactly what he wishes to do to them. He wants the violence seen.

However, Patrick Bateman also realizes the danger of his incessant visualization. He senses that his produced visual image is as easily destructible as those he destroys around him. One scene makes his knowledge evident. At one of the numerous visual dinners, he "stares [at] the potpie, a yellow hexagon, strips of smoked salmon circling it, squiggles of pea-green tomatillo sauce artfully surrounding the dish." Then, Bateman notes, "my hand will not move . . . as if my hand had a mind of its own and refused to break up its design."[56] Incredibly, he refuses violence against a potpie but not against a human body. "I sigh and put the fork down, hopeless."[57] Bateman is hopeless because he recognizes that the visual image provides no real protection against destruction. His hand will not mar the visual image perhaps because he wants there to be some integrity beyond it. Bateman wants to create something like a "mind of its own" for his body. Therefore, he must provide it. The methodical and scrupulous care for the body is, in fact, a desperate attempt to create a self. He attempts to create something other than a visual body, something that will provide a mechanism against destruction. Unfortunately, he fails.

He has no answer for "why I should have done this instead of that."[58] For him, "there is no catharsis," no escape in sight.[59] Though he ends his narrative wondering "what being Patrick Bateman means to me," he can only see a sign in a movie theatre whose "colors match the red velvet drapes," a sign flashing, "this is not an exit."[60] For the American Psycho, the rigid sets of practices he adopts to care for himself imprison his notion of selfhood as a visually inscribed body. He does not know himself. Therefore, he cannot answer who he is. More importantly, without self-knowledge he cannot achieve autonomy. Nonetheless, he desperately clings to these practices. They determine who he is, what others are, and his interactions with them. He is caught just as surely as if he were in "the trap of humanism, which binds us ever more tightly to our subjectivities through our efforts to liberate ourselves."[61]

But what about Foucault? Why do he and his interpreters focus repeatedly on the notion of praxis, on sets of practices, on interwoven practices of thought and action, mind and body? Foucault himself articulates a partial answer. He focuses on practice to question our fundamental insistence on the unquestionable goodness of humanism. Foucault fears that humanism "presents a certain form of our ethics as a universal model for any kind of freedom."[62] In other words, humanism can be oppressive and imprisoning for those who do not share this "certain form of our ethics." Foucault and his interpreters hope for liberation. James Bernauer, for instance, speaks of "Foucault's passion for freedom in his last works."[63] He describes Foucault's "impatience for liberty, for a freedom that does not surrender to the pursuit of some messianic future but is an engagement with the numberless potential transgressions of those forces that war against our self-creation."[64] As early as *The Order of Things*, Foucault refers to "autonomous subjects who are free . . . in so far as they have been able to transgress the limits that constituted them as what they are."[65] Later in life, Foucault himself articulates a desire "to show people that they are much freer than they feel . . . and show which space of freedom we can still enjoy and how many changes can still be made."[66] These hopes for something that "enables one to get free of oneself" hinge, at least partially, upon the practices of self-care.[67]

While Foucault points out the atrocities that occur under the auspices of humanism, he ignores the possibility that practices, even when not tethered to an ideal concept, may oppress both self and other. The American Psycho reminds us of this possibility. While some might argue that Patrick Bateman's extreme visual orientation indicates a world-view more in keeping with enlightened modernity than postmodern Foucauldian praxis, such an argument would fail. His practices of self-care produce this orientation toward his reality. They are not a product of it. Patrick Bateman is no humanist. His notion of self is not universally defined; it does not depend on an immortal soul or a rational cogito. He has no notion of progress nor of a metaphysical reality undergirding the visual representations bombarding his life. He is, rather, a practical man. The very Foucauldian practices by which he cares for himself define the American Psycho's psyche.[68] However, he practices nothing that even remotely resembles a desirable ethics. He has no place in what Richard Rorty refers to as "the liberal utopia."[69] He, like Foucault, does "not reconcile with a notion of community or polity."[70] He does not "inaugurate new experiences of self and human solidarity."[71] Perhaps we should not so easily accept the implied normativity of techniques of the self as an all-encompassing answer to the imprisoning nature of metaphysical humanism. In fact, it is just possible that Foucault constructs a new prison of oppressively practical regimes from which there is, as Patrick Bateman suggests, no exit.

Baylor University

NOTES

[1] Michel Foucault, "Truth, Power, Self," *Technologies of the Self: A Seminar with Michel Foucault*. Eds. Luther H. Martin, Huck Gutman, and Patrick Hutton (Amherst: University of Massachusetts Press, 1988), p. 11.
[2] *Ibid.*, p. 15.
[3] *Ibid.*, p. 15.
[4] Michel Foucault, *The Order of Things: An Archaeology of the Human Sciences*. Trans. Alan Sheridan (New York: Pantheon, 1994), p. 387.
[5] *Ibid.*, p. 387.
[6] *Ibid.*, p. 387.
[7] Jon Simons, *Foucault and the Political* (London and New York: Routledge, 1995), p. 2.
[8] *Ibid.*, p. 2.
[9] Michel Foucault, *The Order of Things: An Archaeology of the Human Sciences*. Trans. Alan Sheridan (New York: Vintage Books, 1994), p. 318.
[10] *Ibid.*, Jon Simons, *Foucault and the Political* (London and New York: Routledge, 1995), p. 6.
[11] Michel Foucault, *The Order of Things: An Archaeology of the Human Sciences*. Trans. Alan Sheridan (New York: Vintage Books, 1994), p. 312.
[12] Karlis Racevski, *Michel Foucault and the Subversion of Intellect* (Ithaca and London: Cornell University Press, 1983), p. 27.
[13] Michel Foucault, *Madness and Civilization*. Trans. Richard Howard (New York: Pantheon, 1965). Michel Foucault, *Birth of the Clinic: An Archaeology of Medical Perception*. Trans. Alan Sheridan (New York: Pantheon, 1973). Michel Foucault, *Discipline and Punish: The Birth of the Prison*. Trans. Alan Sheridan (New York: Vintage, 1979).
[14] Jon Simons, *Foucault and the Political* (London and New York: Routledge, 1995), p. 2.
[15] Michel Foucault, *The Subject and Power*, Afterword. *Michel Foucault: Beyond Structuralism and Hermeneutics*. By Hubert L. Dreyfus and Paul Rabinow (Brighton: Harvester Press, 1982), p. 216.
[16] Friedrich Nietzsche, *Gay Science*. Trans. Walter Kaufman (New York: Vintage, 1974), p. 228. *Ibid.*, p. 232.
[17] *Ibid.*, p. 280.
[18] Jon Simons, *Foucault and the Political* (London and New York: Routledge, 1995), p. 4, *ibid.*, p. 6.
[19] Michel Foucault, "Questions of Method. An Interview with Michel Foucault." Trans. Colin Gordon. *I & C* 8 (Spring), p. 13.
[20] *Power/Knowledge. Selected Interviews and Other Writings 1972–1977*. Ed. Colin Gordon. Trans. Colin Gordon, Leo Marshall, John Mepham, Kate Soper (New York: Pantheon, 1980), p. 52.
[21] Michel Foucault. *The Order of Things: An Archaeology of the Human Sciences*. Trans. Alan Sheridan (New York: Vintage Books, 1994), p. 324.
[22] James Bernauer and Michael Mahon, "The Ethics of Michel Foucault," in *The Cambridge Companion to Foucault*. Ed. Gary Gutting (New York: Cambridge University Press, 1994), p. 143.

[23] Michel Foucault, *The Subject and Power*, Afterword. *Michel Foucault: Beyond Structuralism and Hermeneutics*. Hubert L. Dreyfus and Paul Rabinow (Brighton: Harvester Press, 1982), p. 212.
[24] *Ibid.*, p. 208.
[25] Rux Martin, "Introduction to *Technologies of the Self.*" *Technologies of the Self: A Seminar with Michel Foucault*. Eds. Luther H. Martin, Huck Gutman, and Patrick Hutton (Amherst: University of Massachusetts Press, 1988), p. 4.
[26] Arnold Davidson, "Ethics As Ascetics: Foucault, the History of Ethics, and Ancient Thought," *The Cambridge Companion to Foucault*. Ed. Gary Gutting (New York: Cambridge University Press, 1994), p. 115.
[27] Jana Sawicki, "Foucault, Feminism, and Questions of Identity," *The Cambridge Companion to Foucault*. Ed. Gary Gutting (New York: Cambridge University Press, 1994), p. 287.
[28] Chauncey Colwell, "The Retreat of the Subject in the Late Foucault," *Philosophy Today* **38** (1994): 56–69.
[29] Michel Foucault, "Technologies of the Self," *Technologies of the Self: A Seminar with Michel Foucault*. Eds. Luther H. Martin, Huck Gutman, and Patrick Hutton (Amherst: University of Massachusetts Press, 1988), p. 18.
[30] *Ibid.*, pp. 18–19.
[31] Michel Foucault, *The Use of Pleasure*. Trans. Robert Hurley (New York: Vintage, 1990), p. 31.
[32] *Ibid.*, p. 31.
[33] Bret Easton Ellis, *American Psycho* (New York: Vintage, 1991).
[34] Michel Foucault, "Revolutionary Action: Until Now," *Language, Counter-Memory, Practice: Selected Essays and Interviews*. Trans. Donald Bouchard and Sherry Simon, Ed. Donald Bouchard (Ithaca: Cornell University Press, 1997), p. 22.
[35] Bret Easton Ellis, *American Psycho* (New York: Vintage, 1991), p. 69.
[36] *Ibid.*, p. 106.
[37] *Ibid.*, p. 29.
[38] *Ibid.*, p. 97.
[39] *Ibid.*, p. 28.
[40] *Ibid.*, p. 76.
[41] "Nietzsche, Genealogy, and History," *The Foucault Reader*. Trans. Donald F. Bouchard and Sherry Simon. Ed. Paul Rabinow (New York: 1984), pp. 85–87.
[42] Bret Easton Ellis, *American Psycho* (New York: Vintage, 1991), p. 123.
[43] *Ibid.*, p. 123, *ibid.*, p. 335.
[44] *Ibid.*, p. 398.
[45] *Ibid.*, p. 77, *ibid.*, p. 48.
[46] *Ibid.*, p. 41.
[47] *Ibid.*, p. 42.
[48] *Ibid.*, p. 42.
[49] *Ibid.*, p. 59.
[50] *Ibid.*, pp. 170–176.
[51] *Ibid.*, p. 170.
[52] *Ibid.*, p. 252.
[53] *Ibid.*, p. 328.
[54] *Ibid.*, p. 114.

[55] *Ibid.*, p. 23.
[56] *Ibid.*, p. 49.
[57] *Ibid.*, p. 49.
[58] *Ibid.*, p. 399.
[59] *Ibid.*, p. 377.
[60] *Ibid.*, p. 399.
[61] Jon Simon, *Foucault and the Political* (London and New York: Routledge, 1995), p. 47.
[62] Michel Foucault, "Truth, Power, Self," *Technologies of the Self: A Seminar with Michel Foucault*. Eds. Luther H. Martin, Huck Gutman, and Patrick Hutton (Amherst: University of Massachusetts Press, 1988), p. 15.
[63] James Bernauer and Michael Mahon, "The Ethics of Michel Foucault," *The Cambridge Companion to Foucault*. Ed. Gary Gutting (New York: Cambridge University Press, 1994), p. 143.
[64] *Ibid.*, p. 155.
[65] Michel Foucault, *The Order of Things: An Archaeology of the Human Sciences*. Trans. Alan Sheridan (New York: Vintage Books, 1994), p. 43.
[66] Michel Foucault, "Truth, Power, Self," *Technologies of the Self: A Seminar with Michel Foucault*. Eds. Luther H. Martin, Huck Gutman, and Patrick Hutton (Amherst: University of Massachusetts Press, 1988), pp. 10–11.
[67] Michel Foucault, *The Use of Pleasure*. Trans. Robert Hurley (New York: Vintage, 1990), p. 8.
[68] This situation is particularly ironic if James Bernauer is right to suggest that Foucault identifies with history's victims. James Bernauer, *Michel Foucault's Force of Flight* (Atlantic Highlands: Humanities Press, 1990), p. 1.
[69] Richard Rorty, *Contingency, Irony, and Solidarity* (Cambridge: Cambridge University Press, 1989), p. xvi.
[70] David Hiley, "Foucault and the Question of Enlightenment," *Philosophy and Social Criticism* **11** (1985): 77–80.
[71] James Bernauer and Michael Mahon, "The Ethics of Michel Foucault," *The Cambridge Companion to Foucault*. Ed. Gary Gutting (New York: Cambridge University Press, 1994), pp. 155–156.

RAYMOND LANGLEY

MIRROR, MIRROR . . . WHO IS THE REAL SELF?

The doubled mirror phrase is stolen from Disney's film, *Snow White*. In this movie, mirrors prattle on, making comparative judgments about physical beauty, but ignore substantive issues like the Oedipal struggles of the heroine and conflicts between old and young, ineffectual fathers, evil stepmothers and seven psychologically arrested dwarfs. The mirror "on the wall" comforts the evil queen by declaring her the fairest of all, but the mirror is silent about the extent of reference. A later mirror "in the hand" has no mental reservations and responds with a bit of taunting that Snow White is "lovelier far o queen than thee."[1]

The human imagination has been captivated forever by the physical properties, figurative uses and symbolic powers of mirrors. All objects reflect light but polished surfaces retain enough undiffused light to capture images. An ordinary mirror image is identical to the object except that it is laterally reversed so print appears backwards and our own image mirrors the way that others see us. The horizontal but not vertical distortion of mirror images is puzzling, for as Alice noted in peering through the looking glass, the room seems the same "only the things go the other way round."[2] Curved surfaces reflect either concave, larger images like shaving mirrors or convex, smaller ones like rear view mirrors. A mixture of flat and curved surfaces creates the distorting effects of fun house mirrors, and the bending of light rays in abnormal air density creates inferior and superior mirages of wet pavement and heat waves. By the second century B.C., "burning mirrors" of bronze or silver were used to focus the sun and ignite ships off shore. Human and cultural evolution are mirrored in physical artifacts; for, "No matter the technology, and no matter the ideology, society cannot escape, even in extinction from impressing, from forming a mirror of their ways."[3]

The figurative uses of mirror imagery in conventional associations of picturing and reflecting are inexhaustible. Thus, there are mirrors *for* greatness, man, Americans, and there are mirrors *of* art, fashion, life and death, and mirrors *to* all sorts of places. Mirrors play a symbolic role in intentional resemblance. Narcissus falls in love with his mirror reflection. The *Ode to Solomon* bears the title "Our Mirror is the Lord."[4]

A-T. Tymieniecka (ed.), Analecta Husserliana, Vol. LIV, 83–99.

The Apostle Paul describes our image of heaven: "For now we see through a glass, darkly." In eastern thought, the renowned sutra of "dust wiping" uses mirror images. Shen-hsiu states: "Body is the Bodhi tree./Mind is like bright mirror-stand,/Take care to wipe it continuously/And allow no dust to cling." To this, Hung-Jan responded: "There never was a Bodhi tree/Nor bright mirror-stand/Originally, not one thing exists/So where is the dust to cling?"[5] A more literal-minded Hamlet instructed actors "to hold, as 'twere, the mirror up to nature; to show virtue her own feature, scorn her own image, and the very age and body of the time his form and presence." In philosophy, Wittgenstein tried to get us to understand propositions as pictures of facts, and Heidegger directed us to reflect on "The Age of the World Picture."

There is, to adopt a modern phrase about an old dilemma, a "hermeneutic of suspicion" about the mirror image, the object and the self. Mirrors are open to all sorts of distortions and illusions. We are afraid that mirrors reflect more about ourselves than objects. Thus Francis Bacon warned, "The human understanding is like a false mirror, which, receiving rays irregularly, distorts and discolors the nature of things by mingling its own nature with it."[6] Conversely, Sartre dramatized hell as a place where there are no mirrors, and our subjectivity is objectified by the look or gaze in the eyes of the other. In Borges' short story, "The Library of Babel," all possible combinations of letters are stored in galleries and they are reflected infinitely in a mirror that "faithfully duplicates appearances."[7]

The attractiveness, fecundity and elusiveness of mirror images conflicts with imagination's other, or logos. The suspicion of reason is that mirrors make only simulacra or illusions. Plato's contempt for the visual arts, including speaking and writing, is expressed in a *reductio ad absurdum* of mirrors. The *poietes* make things appear but not "as they truly are." Even you and I, says Plato, can make anything in the universe: "You could do it quickly and in many places. The quickest way is to carry a mirror with you everywhere; you will then quickly make the sun and things in the heavens, the earth as quickly, yourself and other living creatures, manufactured articles, plants, and all that was mentioned just now."[8] René Magritte captures the mirror ambiguity between appearance and reality in his painting of a smoking pipe which is enigmatically entitled, "This is not a pipe."[9]

This essay will examine three mirror images of the self. The experiential and reflective basis for these mirror selves is archetypal. Our interior life is a stream of images mixing memory, imagination, desire and thought; we reproduce the image of our being in our children, we stamp our office by our labor and leave mirror traces of our collective selves in art and culture. The three images of the self share a common metaphor in that the relations of self and nature, self and other selves, and self to itself are ones of representation, imaging, picturing or, in short, mirroring. Each categorization also contains a dialectic of immediacy and mediation which culminates either in strict antithesis or in synthesis.

This metaphor of the self as a mirror is philosophically useful because it is no easy task to explain what and who is a self and what selves do. Ontologically the self has been credited and praised as indubitable existence and also denied, decentered and erased as a no-thingness without content. As to who is a self, the mirror presents a whole carload of candidates: I, subject, person, individual, ego, etc. and these images stand for those conglomerate, hyphenated and contradictory selves that roam the OED from self-abuse to self-will. From these mirrored images a consensus emerges. There is a minimal agreement that I/we selves have experiences, bodies and minds. Provisionally, then, the self is the continuing existence of that indivisible entity I call myself and that makes me acknowledge your originativeness and individuality or, as Cole Porter put it so well, "it's the way you wear your hat, it's the way you sip your tea."

The purpose of this tortuous plea bargaining is to fill in the notion of the self as subject of consciousness or self-consciousness. But this I/it, person/thing, ego/non-ego, subject/object set of dichotomies and their counterparts of individual/social, particular/universal, presence/absence are attempts to translate the experiences of I see, feel, think, and dream, doubt and love into the language of objectification as the referent of subjectivity. All the richest metaphysical definitions and explanations of self and subject involve analogical and metaphorical thinking as they are attempts to think of one thing in terms of another. Theory itself is a way of seeing, and the mirror metaphors are attempts to spy out or catch sight of ourselves as in a *specula* or glass or mirror.

The three mirror self metaphors are seductive because of their explanatory power, and they dialectically overlap traditional definitions of human reality.[10] The elusiveness of these images results from the shifting metaphoric between subject, object and the mirroring representation. Metaphors carry presuppositions. In this instance the relation between the object and its mirror image, the represented and its representation is assumed to be mimetic. And these imitations are assumed further as relations of identity, correspondence or coherence. But will the simile of inimitable and imitation as a mirror relation hold still when it is extended to the signified and sign, the bond between reality and appearance, and mirror of law and society, egoism and altruism, private and public, individual and community and *Eros* and civilization?

One has to attend to the shifting metaphors of mirrored selves like a shell game. First the self mirrors the exterior world, and then exterior society mirrors me and the images are redoubled or one is caught in the mirror as an in-betweenness reflecting only itself. So one must be reminded that something or someone is being represented or reflected as an image or idea to someone or something else, and that the relations between self, nature, and others are being thought in mirroring metaphors.

I.

The epistemological mirror of the self originated with the representative theory of perception in the seventeenth century. The theory contended that the self was a pure subject or knower, a disembodied mind, and that its own existential certainty entailed incorrigible, internal mental pictures or images of external material objects. Understanding or reason was the process of subsuming the assemblage of representations under formal concepts. Such *a priori* categories were representations of the possible ways of understanding reality. Hence philosophy became epistemology, a theory of knowledge about how the mind mirrored nature as it appeared and as it was understood. In its practical dimension, reason transcended the phenomenal world to constitute or make the self an autonomous agent. This view of reason led to a conception of the self and subject as the noetic mirror of images and concepts. Philosophy was the "queen of the sciences" because it had "privileged representations" which were "automatically and intrinsically accurate", and these served as the foundation of knowledge and the standard of adjudication which discovered the nature and the limits of all other disciplines.[11]

Richard Rorty argues that the self as the mirror of nature is a metaphor gone wrong, and that this simile is something close to an original sin of Western thought. Interpreted as a mirror, the mind functions like an instamatic camera, and in the transition from perception to conception we have a quasi mechanical account of how our immaterial wax tablet gets dented by the material world. In Rorty's deconstruction of the mirroring self, a series of cumulative mistakes are exposed. First Plato invented *Nous* or immaterial mind on the assumption that the certitude of universal and necessary knowledge required parallel differences in objects. His fascination with mathematical and geometrical structures, which can be visually seen in the flux of nature, led Plato and us to remember unchanging, universal and eternal forms as the archetypes of perishing sensibles. Thus the claim to unchangeable knowledge of changing appearances presupposes eternal essences as intuitable by the inner eye of our immaterial souls.

This mistake was codified by philosophers from Descartes to Locke. Our "glassy essence" or mirror-like mind assumed that knowledge was a relation between persons and objects in which the accuracy of the latter was guaranteed by the former.[12] Then Kant tried to show how we get from the certainty of inner space to the doubtful representations of outer space by arguing ingeniously that the outer intuitions were constructed out of prior, inner ones. Kant, to Rorty, confused synthesis with predication. Thus the history of philosophy founders on the metaphor of the self as a mirror since it presupposes that the ultimate relation of minds to objects is visual, and epistemology is the foundational discipline which tries to fix those true images which correspond to reality.

These objections are familiar ones. Hume had argued that it was impossible to get outside the mind to see if inner pictures corresponded to anything outside. Kant concluded that it was equally impossible to get outside of the mind's *a priori* categorical scheme to see if things were the same or different from the ways in which concepts represented them. A subtext of Hegel's *Phenomenology of Mind* was that if truth was the totality then certitude was nowhere. Thus the mind or the self as a mirror of nature fails because we cannot do more than assume that images or ideas form a mirror of reality similar to the way a picture represents a landscape.

Rorty claims that philosophy's last gasp to resuscitate the mirror self was language. But propositions do not divide themselves into upper and lower divisions between sentences corresponding to something and those statements which are true by convention. The search for a transcendental truth beyond space and time and the search for the truth of this world both founder on a notion of correspondence. Just as there is no way of getting outside impressions or ideas, there is no way of breaking out of language without using language, and attempts to hook language up to reality fail on the same grounds as representative images and concepts. We cannot step outside our mirrors to see if either sensations, concepts or words hook up with reality. Thus the problematic of the history of philosophy moves in stages from analyzing the structure of the real world; then, to analyzing the structure of the mind; and, in the final stage, to analyzing the structure of language. But the *Vorstellung* or representations also move from descriptions and referrals to depictions and stipulations to mere talk or words, words, words.

II.

What cracked this mirror of the self was the philosophic inability to resolve the oppositions between sense and reference, thought and object, mind and matter. But the fault line of this metaphor ran beyond the failure to demonstrate a necessary correspondence between subjective mirrors and the objective world. The mirror of nature also reflects contradictory images of the self. On one side, the mirror reflects objects through images or concepts and this self is passively egocentric and empty of content. But standing behind the empty mirror is a theoretical being, Kant's non-cognizable but self-unifying and world-constituting power as the necessary condition for the empirical self and phenomenal world. So the mirror self does not reflect the subject.

The second mirror image of the self dialectically turns from looking at objects out there to reflecting on others like myself. The mirror image of the social self adjusts its metaphor from passive reflection of objects and blindness to itself, to finding self-identity mirrored in the minds of others. This second mirror pays heed to the actual continuity and discontinuity of experience.

The *cogito*, for example, is a working hypothesis only during instants of self-reflexive thinking. Hume describes these isolated impressions of self as a "bundle of perceptions" loosely tied by associations of memory and imagination. And the strongest glue for a self-identical person is the history of my passions and instincts, which provides the frame of continuity for my fleeting impressions and ideas. The way to catch a glimpse of the self which mirrors nature is to look at other selves. For as Hume claims, "the minds of men are mirrors to one another, not only because they reflect each other's emotions, but also because those rays of passions, sentiments and opinions may often be reverberated."[13]

The mirror of the social self shifts its focus from the transcendental subject as a pure mind or spectator to a richer self image mirrored by other persons. Here the mirror self is not a mere theoretical spectator, but a witness to an experienced world in which others have bodies, emotions and passions like ourselves. The mirror of the social self expands the catalogue of I see, think, doubt, and feel to include I give witness. This metaphor of the social self mirror explains the objective socialization of subjective beings.

The shift from spectator to witness is the root of the moral sense theory of Hutchinson, Hume and Smith which founds ethics on reciprocal feelings of praise and blame. Smith even characterized the subject as a spectator who must utilize a looking glass or mirror to measure his intentions and actions through the eyes of others. He writes: "Were it possible that a human creature could grow up to manhood in some solitary place, without any communication with his own species, he could no more think of his own character, of the propriety or demerit of his own sentiments and conduct, the beauty or deformity of his own mind, than of the beauty or deformity of his own face. All these are objects which he cannot easily see, which naturally he does not look at, and with regard to which he is provided with no mirror which can present them to his view. Bring him into society, and he is immediately provided with the mirror which he wanted before. It is laced in the countenance and behavior of those he lives with, which always mark when they enter into, and when they disapprove of his sentiments, and it is here that he first views the propriety and impropriety of his own passion, the beauty and deformity of his own mind."[14] Smith did acknowledge the limitations of the "moral looking glass" into which I look to see another's impartial regard of my own conduct by pointing out that the imagination was the "great smoother of wrinkles" with respect to "personal blemishes of character."[15]

This mirror in which I see myself in others fits Mandeville's great conceit of "private Vices, public Benefits" and Alexander Pope's poetic notion that "partial ill is universal good."[16] The explanatory power of this mirror metaphor is to show how one gets from egoism to altruism. Self-love, self-preservation, ambition, greed, avarice and fear are recognized motives of conduct. How, then, does the self arrive at moderation, prudence, concern for others and for the good of the whole? The relations between my empty, egoistic self-love and my fear and sympathy for others produces those mirrored rays of passions and sentiments which

Hume described. And the ambivalence of these two mirrors of the self, one of which thinks itself as an empty reflection apart from the world, and the other mirror which sees itself in the behaviour of others, flows throughout the great chain of being.

These reverberations and echoes of the self as mirrored in others become the "invisible Hand" which explains natural liberty and *laissez faire* economics. These twin concepts of the free man and the free market are the premises for capitalism as a radical theory of social organization. That is, *The Wealth of Nations* offers the metaphor of the "invisible hand" which guarantees that where each pursues his own interest, the total welfare is promoted. The invisible hand transforms the pursuit of private gain into public welfare by increasing supply to meet demand in commodity exchange, and generates both economic expansion and development which, in turn, increases and circulates the wealth of the nation in the same way that blood circulates through the body.

The mirror image of the social self is very powerful. As above, the metaphor of "I do unto others as I see others behave" was an ethical perspective. But the moral looking glass became the premise for a series of sweeping analogies and inferences. The theoretical I, as contentless master of the phenomenal world, finds his politicized cousin in the theory/practice dualism of Rousseau's "man" who "is born free and everywhere is in chains." And just as the knowing subject is free from determinations of the object, so bourgeois liberty is free from all external restraint, just as the market economy is free from government intervention.

One can see how the passive ego and contentless subject became the transcendental self. And this self obtained objective socialization in ethics, politics and economics. For Marx the theoretic ego and the autonomous ethical self are fabrications, ideological distortions of dehumanizing capitalism mystified by notions of empty liberty and social atomism. Marx attempted to show dialectically how the ahistorical, privileged I was transformed into empty liberty, and how capitalist social relations were viewed as laws of nature. He attacked the mirror image of the social self. The social mirror, like a *camera obscura*, reverses its own image. Under capitalism the human subject becomes an objective commodity; namely, the labor power which he sells for wages. And the surplus value of the product of his labor acquires the status of private property and the rights of a person. Capitalism is that fantastical process of transformation in which a man becomes a thing and a thing becomes a legal person. Thus the moral looking glass produces commodity fetishism and ideological false consciousness. For Marx, the social mirror distorts the true economic base of the political, ethical and legal superstructure since it rests on the assumption that fetishized things are valuable, and that the human beings who produced the commodities are not valuable. Marx writes: "Thus the mystery of the commodity form is simply this, that it mirrors for men the social character of their own labor, mirrors it as an objective character attaching to the labor products themselves, mirrors it as a social natural property of these things . . . we are concerned with a definite social relation between human beings which in their eyes has here assumed the semblance of a relation between things."[17]

In sum, the metaphor of the social self mirror provides an account of ethical, economic and even political socialization. This self is a passive spectator of the objective world and subjective witness of his moral autonomy within the world.[18] These mirror selves are mediated by those twin realities Kant described as "the starry skies above and the moral law within." I think the politics of representation throughout the seventeenth to nineteenth centuries is derived from these mirror metaphors. Thus, the *Leviathan*, the various social contract theories, the notion of checks and balances are theorizations of bourgeois liberalism and capitalism. The mirror images apply equally well to representative politics where sovereign and subject, authority and body politic, the republic and its constituents, maintain social solidarity of the general over private interests in the name of elected representatives who serve intermediary bodies which represent citizen to state and state to citizen.[19] Marx's critical analysis of capitalism historicized those intra-subjective ethical rays and reverberations of passions and opinions into the concrete, objective and material relations of production and reproduction. I note that in this century the emerging disciplines of psychology and sociology largely ignored the mirror of the epistemological self in favor of a soft determinism called "the looking glass self" derived from the social self mirror.[20]

III.

Both mirror reflections of the self are Janus-faced. The egocentric mirror shows nature but not the self. Yet when we turn the mirror to see ourselves, we see only objective images of how we appear to others. In neither mirror metaphor does the self grasp its experienced reality. In the third mirror of the self, one turns from looking out there at objects and other selves to peer inside the mirror. Can we steal inside these reflections, between the mirror and its *tain*, to see how subjectivity constructs the world mirrored in nature; and how, in turn, the constituting mirror is itself constituted as a self through its objectification in society? This look inside is accomplished by a hyper-modern metaphor of the self as a mirror of reflections which substitutes false images of our other mirrored selves. The conceptualization of this mirror self moves away from philosophy and socio-economics to borrow categories from biology, psychoanalysis and linguistics.

In the twentieth century existentialism and psychoanalysis are two intellectual movements which have tried to reach inside the mirror. Existentialism has focused directly on the *Lebenswelt* as the unique experience of the human self. Phenomenological ontology, according to Heidegger, addresses the question ignored in the quest for noetic and socio-economic objectivity; namely, "the subjectivity of the subject." The human subject is fallenness, a *Dasein*, literally "being there," in "genuine inauthenticity," thrown in a world of "average everydayness" and "idle chatter." Sartre"s *vécu* or "lived experience" is a dialectic of separation between the *pour soi* and the *en soi*, the subject's struggle to maintain its free project against the objectifying mirroring "look" or "gaze" of the other and the envelopment of the practico-inert world of materiality.

But it is to Freud that we must turn for an attempt to get behind and inside the mirror. Before one is an *ego cogito* reflecting the world and before one sees oneself mirrored through social classes, the self is already there as that primordial organic vegetative, animalistic evolutionary *bios* whose constitution is written in genes, neurons and synapses. Freud expanded the concept of matter to include the psychically real. In contrast to the logocentric metaphysics of presence, which is the framework of both egoistic and social self mirrors, Freud offers a phenomenology of absences: things forgotten, slips of the tongue, jokes, daydreams, sexual fantasies, and dreams themselves "as the royal road to the unconscious," and their attendant behaviors of free-floating anxiety, hysteria, megalomania, compulsion and obsession; in short, the repertoire of disorders that are attributable only to persons.

Psychoanalysis is a radical departure from the mirror selves of the knowing ego and socially formed individual. Freud's depth analysis of the psyche argues that ego and social self mirrors are manifest by-products of the latent reality of the unconscious. The unconscious is the originary cause of the distorted mirror selves. The depiction of the unconscious is like a Greek drama of biological/psychical drives striving toward discharge and encountering external resistance from the world, from others and internal opposition from our mirroring egos and social selves. This biologically driven psychical warfare between unconscious desire, conscious ego and the social repression of the superego is a repetitive, compulsive life and death struggle to maintain homeostasis between the positive and negative cathodes of pleasure and pain, individual and society, between animal drives and civilized repressions.

Freud circumscribed the failure of this dialectic with the elegiac phrase "the economics of libidinal gratification."[21] The symbiotic unity of infant and mother is the complete identity in difference between the subject and object of desire and this experience is the archetype of the pleasure principle. This fleeting experience is soon gone from consciousness but never forgotten. This infantile wish for identity with the other is not capable of realization; and thus, desire is the unconscious origin of both the ego and the social self as the bio/psychical drives to find a symbolic substitute for "the hand that fed us." Freud's account bears a striking parallel with Hegel's dialectic of self-consciousness. For Hegel also, life itself is the immediacy of desire which dialectically produces both self-reflection and the demand for recognition by an other self as an identity in difference in the antithetical relations of master/slave, stoic/skeptic and Hegel's definitive expression of modern selfhood, "the unhappy self-consciousness." For Freud, the science of psychoanalysis is the "talking cure" which promotes ego strengthening and social adjustment to those who are "suffering from memories."

It is this last art that brings us back to mirrors and metaphors. Therapy, which aims at making the unconscious conscious by talking about images and memories, is a linguistic and symbolic cure. Freud famously used the simile of the iceberg to explain the relation between the unconscious and conscious psychical realms. In the physics of displacement, two thirds of an iceberg are under water; hence, by analogy conscious beings do not understand the unconscious origins of what they feel or think two thirds of the time. The unconscious decenters the subject in that the thinking mirror and the social self mirror are distorted images of unconscious processes.

Lacan draws on empirical data and experiments to describe what he calls the "Mirror Stage" as a crucial turning point in the constitution of the self. Between six and eighteen months, the infant sees itself and others in a mirror. The infant's reaction proceeds through stages which Lacan elaborates. At first the infant identifies its mother in the mirror and briefly regards itself "soberly." Then the infant notices other babies and is sociable with its playmates. Then the infant connects the mirror image with itself, and there is pointing and conversational testing. This is followed by coyness in which the baby will turn away from and then quickly turn back to catch its image. The point of all this is that the infant, – who cannot stand upright without support – suddenly recognizes itself with what Lacan calls "jubilant interest."

This is a crucial moment for the constitution of the self on several levels. The emotional joy at self recognition of the *imago* or image is that actually the infant feels its body as fragmented "in bits and pieces." Thus what is seen in the mirror is a proto self of what the child will become, namely, a full self. On a social level the totalizing image of myself that comes from outside is illusionary. It is the mis/representation and mis/recognition that allows the baby to assume "jubilantly" the upright posture as "already there," when, in fact, such a state will "only later become." On the level of reflection, one finds the following hypotheses: the basic function of the ego is misrecognition. That is, the ideal formed in the mirror stage situates the activity of the ego in the line of fiction. The self is constituted by anticipating what it will become and then this totalizing and anticipatory model is used to gauge retroactively what went before. This mirror stage is for Lacan the "root stock" or formation of the "I" which will later have "my" history. Lacan transforms the resolution of the oedipal conflict into a purely linguistic affair in which the symbolic replaces the imaginary. The mirror image reflects an imaginary me who answers to my name which is replaced by the symbolic subject "I." Thus, the "I" expresses itself as a signifier "in the name of the father," and the imaginary level of desire for mother/other becomes inserted into the chain of signification or social order and history and into the discourse of meaning.

Lacan's provocative statement that "the unconscious is structured like a language" seems to mean that we are blind to the real self because we cannot say our true desire. In the context of our mirror metaphors of the self, the decentered ego and social self reflects false mirror images of an imaginary me and a fictional I. Inside or behind the mirror is the unconscious structure of language which makes the "I" into a floating signifier wandering between imaginary and symbolic reflections. If Lacan literally means that "the unconscious speaks us," then our mirror metaphors of the self imply either a transcendent infinite self which all "I" 's reflect, or the absolute non-existence of self and the transcendental illusion of mirror images. Lacan prefers enigma to clarity as to whether he is thinking about being or language, and so he leaves us with a puzzle or *koan*: "The mirror would do well to reflect a little more before returning our image to us."[22]

IV.

We asked the mirror to reflect the real self, and we have examined three mirror metaphors of knowing subject, social self and imaginary and symbolic fiction. Hegel's *Phenomenology* tracked the contradictory appearances of absolute reality or *Geist*. Often Hegel paused in the text to ask us readers what "we" were learning from these dialectical mediations. Similarly I ask, what we can learn about the self from these mirror images.

I think we suspect correctly that self realization cannot be done with mirrors or metaphors. Lacan's well founded injunction against mirror reflections, "How do we know that we are not impostors?" is worth thinking about. On the other hand, we may learn that a skeptical attitude is the basis for a dialectically critical method. From this we can come to the perspective that if we do not know everything about the self, it does not follow that we do not know anything. If we cannot find a fixed human nature in the images of the epistemological, social and psychoanalytical selves, this does not logically entail that there is no subject. If a critical and dialectical examination of self images leads to the realization that mirrors distort and produce mirages and illusions, then we have learned a lot. For example, we know that many distorting effects can be corrected. To follow Hegel a little further, absolute knowledge meant that we knew the limitations of all particular viewpoints. Thus, the certitude of every immediate form of consciousness has to be mediated or negated in order to preserve its naive, but real truth.

Let us apply the stages of critical dialectic to the mirror images of the knowing subject, the social self and the psychoanalytic construct. All three reflections of the self were derived from a mirror metaphor. On the level of immediacy, each mirror self as knower, as social being, as existential patienthood appears, appeals and accounts for wide and rich dimensions of experience. But the self-certainty of each mirror self is naive in assuming demonstrable identity between knower and world, sociability and society, care and cure. On the level of mediation or negation we have seen internal and external opposition to the reality and truth of each of the three mirror selves. Consciousness and self-consciousness are hopelessly divided between the empty and contentless empirical self and the active, unifying and world creating subject of reason. The social self is also a social atom, and both selves are subject to objective reification in which persons are things as mere bearers or by-products of vast impersonal forces. And what real difference is there between self-renunciation and acquiescence to social conformity if the mirror self is a linguistic construct? Arguments in support of each mirror self tend to refute alternative mirror perspectives. On the level of synthesis, or the negation of the negation, each mirror image has made a contribution to knowledge and technology, to theories of ethics and politics and to a humanistic dimension in the social sciences and history. In conclusion, all three mirror selves stand and fall only in dialectical apprehension and criticism and thus, the method discloses the content.

The transcendental implications of this examination suggest that just as the mirror images turned reflection from things to persons and reflected both back into the language of desire and self-realization, so likewise we can dialectically see from this that the self obtains objective knowledge in technology and simultaneously that the self is a socio/political and historical product. The subject is not just a product of language since the interpreter is free within the boundaries of discourse and meaning. So at the end we are left with metaphors and the healthy suspicion that they are not exhaustive, for as the poet Coleridge said, "no simile runs on all four legs."[23] But my modest claim is that knowledge of limitations is a dialectical transcendence and that we are sure/unsure that ideas transcend ordinary experience and that some experiences transcend understanding. We can choose among mirrors of self as spectators, witnesses and participants who claim to possess or own meaning individually. That meaning is owned by no one. That meaning is shared by the dialectic of the self, the other and representations.[24]

Manhattanville College,
Purchase, New York

NOTES

[1] Cf. Bettleheim, Bruno, *The Uses of Enchantment, The Meaning and Importance of Fairy Tales* (New York: Alfred A. Knopf), pp. 194–215.
[2] Cf. "Mirror Reversal" in *The Oxford Companion to the Mind*. Edited by Richard L. Gregory (Oxford: Oxford University Press, 1987), pp. 491–493.
[3] *Human Mirror, Material and Spiritual Images of Man*. Edited by Miles A. Richardson (Baton Rouge: Louisiana State University Press, 1974), p. 6.
[4] Pulver, Max, "Jesus' Round Dance and Crucifixion According to the Acts of St. John," *The Mysteries, Eranos 2* (Princeton: Princeton University Press, 1955), pp. 169–193.
[5] Yampolsky, Philip B., *The Platform Sutra of the Sixth Patriarch*. The text of the Tun-Huang Manuscript with translation, introduction and notes (New York: Columbia University Press), pp. 130 & 132.
[6] Bacon, Francis, *Novium Organum, and Related Writings*. Edited with an introduction by Fulton H. Anderson (Indianapolis: The Bobbs-Merrill Co., Inc., 1960). Aphorism 41, p. 48.
[7] Borges, Luis, "The Library of Babel" in *Labyrinths, Selected Stories and Other Writings*. Edited by Donald H. Yates and James E. Irby, with an introduction by André Maurois (New York: Modern Library, 1988).
[8] Plato, *Republic*, Book H. 596e, translated by G. M. A. Grube (Indianapolis: Hackett Publishing Company, 1985 (1972)). This passage is cited by M. H. Abrams, *The Mirror and the Lamp: Romantic Theory and the Critical Tradition* (New York: Oxford University Press, 1974), p. 30.
[9] Foucault, Michel, *This is Not a Pipe*, with illustrations and letters by René Magritte. Translated with an introduction by James Harkness (Berkeley: University of California Press, n.d.).
[10] For traditional definitions and citations concerning the self, cf. John MacMurray, *The Self As Agent* (London: Faber and Faber, 1957) for the notion of the empty theoretic ego; Karl R. Popper and John C. Eccles, *The Self and Its Brain* (Berlin: Springer International, 1977) for the footnote citing Adam Smith; Max Scheler, *Man's Place in Nature*, translated with an introduction by Hans Meyerhoff (Boston: Beacon Press, 1961).
[11] Rorty, Richard, *Philosophy and the Mirror of Nature* (Princeton: Princeton University Press, 1979), Chapters 3–6, pp. 131–313.
[12] Shakespeare, William, *Measure for Measure*, Act II, Scene ii, line 117.
[13] Hume, David, *A Treatise on Human Nature* (Oxford: Oxford University Press, 1962), II, ii, 5, p. 365.
[14] Smith, Adam, *The Theory of Moral Sentiments* (Oxford: Oxford University Press, 1976), Part III, Ch. 1, Sec. 3, p. 110.
[15] *Ibid.*, p. 112.
[16] Mandeville, Bernard, *The Fable of the Bees or Private Vices, Publick Benefits*. With a commentary by F. B. Kaye (Oxford: The Clarendon Press, 1924). Alexander Pope, *Essay on Man*, Book II, line 110.
[17] Marx, Karl, *Capital*, Vol. I. Translation by Eden and Cedar Paul of *Der Produktionsprozess des Kapitals*. Introduction by G. D. H. Cole, addendum and bibliographical note by Murray Wolfson (New York: Dutton Everyman's Library, 1978), p. 45. Other editions substantially alter this passage by eliminating the mirror and visual analogies. The

passage then reads as follows: "Consequently the social relation of the producers to the sum total of their own labour, presents itself to them as a social relation, not between themselves but between the products of their labour. Thanks to this transference of qualities, the labor products become commodities, transcendental or social things which are at the same time perceptible by our senses. In like manner, the impression which the light reflected from an object makes upon the retina is perceived not as a subjective stimulation of that organ, but in the form of a concrete object existing outside the eye. But in vision, light actually passes from one thing, the external object, to another thing, the eye. We are dealing with a physical relation between physical actualities. On the other hand, the commodity form, and the value relation between the labour products which finds expression in the commodity form, have nothing whatever to do with the physical properties of the commodities or with material relations that arise out of these physical properties. We are concerned only with a definite social relation between human beings which in their eyes has here assumed the semblance of a relation between things. To find an analogy, we must enter the nebulous world of religion. In that world, the products of the human mind become independent shapes, endowed with lives of their own, and able to enter into relations with men and women. The products of the human mind do the same thing in the world of commodities. I speak of this as the *fetishistic character*. . . ."

[18] The terms "spectator, witness, participant" were borrowed from an unpublished paper of Havier O. Monasterio, "On Sartre's Trans-Cartesianism" presented at the inaugural meeting of the Sartre Society of North America held at the New School, New York City on October 6–8, 1985.

[19] *International Encyclopedia of the Social Sciences*, vol. 13. David L. Sils, editor (New York: The Macmillan Company and Free Press), p. 461ff.

[20] Largely ignoring the mirror of the epistemological self, social theorists developed a distinction between mental states that apply solely to the individual and the acquired or learned sets of beliefs, values and practices which express the social reality of the group or social class rather than of the person. For example, Charles Horton Cooley writes: "The looking-glass self, the social self, arises reflectively in terms of reaction to the opinions of others on the self." This 1902 text is cited in T. R. Young, *New Sources of Self* (New York: Pergamon Press, Inc., 1972), p. 16. Young also discusses the work of George Herbert Mead, Talcot Parsons, Erving Goffman and Heinz Kohut, who separately develop and expand the notion of mirroring and mirror transference. The consensus of the "looking glass" theory is that the self is derived from the social order rather than the social order being derived from the self.

[21] Cf. C. Brenner, *An Elementary Textbook of Psychoanalysis* (New York: Doubleday Anchor, 1957). The author summarizes Freud's four meta-theoretical conceptions of the human psyche: the topological view of the relations between unconscious/conscious; the structural view of id, ego and superego; the dynamic view of the conflict between the pleasure and reality principles; and the dialectical view of "the economics of libidinal gratification."

[22] Lacan, Jacques. *Ecrits, A Selection* (1977) which contains "The Mirror Stage as Formative of the Function of the I," pp. 1–8 and *The Four Fundamental Concepts of Psycho-Analysis* (1978), both books translated by Alan Sheridan (New York: W. W. Norton & Company). Cf. Lacan's *The Language of the Self*, translated with notes and commentary by Anthony Wilden (Baltimore: The Johns Hopkins Press, 1968). Important secondary

sources include: "Interpreting Lacan" edited by J. H. Smith & W. Kerrigan, *Psychiatry and the Humanities*, Vol. **6** (New Haven: Yale University Press, 1983); "French Freud," *Yale French Studies*, 1976; "Literature and Psychoanalysis," *Yale French Studies*, 1977; Jane Gallop, *Reading Lacan* (Ithaca: Cornell University Press, 1985).

[23] Cited by M. H. Abrams, *The Mirror and the Lamp, op. cit.*, p. 35.

[24] Distinctions concerning the ownership of meaning are suggested by an unpublished paper of Michael Holquist, Slavic Department, Indiana University, entitled, "The Politics of Representation."

LOUISE SUNDARARAJAN

THE "MADMAN" AND THE DREAMER

Towards an Aesthetic Approach to the Other

INTRODUCTION

My thesis, put simply and straightforwardly, is twofold: a. the ethical, problem-solving orientation, which is characteristic of the medical approach to the other, does not do justice to the other as Other, as "the absolutely other" (Levinas, 1969, p. 39); and b. the aesthetic perspective constitutes a less violent approach to the Other, an approach which deserves serious consideration. This view is diametrically opposed to Levinas' privileging the ethical over the aesthetic (1969). To facilitate contrast and comparison between these two approaches to the Other, ethical versus aesthetic, I examine the role of the other in two different settings: as patient and "madman" in the clinical setting; and as a dreamer in a Chinese poem. The following discussion is divided into five sections: section one examines "madman" as the prototypical Other; section two considers the dreamer as the Other; section three compares the problem-solving approach to the "madman" with the aesthetic approach to the dreamer; section four further elaborates on the aesthetic approach to the Other as "relation without relation"; section five draws some tentative conclusions.

"MADMAN" AS THE OTHER

Levinas finds the face of the Other in that of "the poor, the stranger, the widow, and the orphan" (*ibid.*, p. 251). This paper intends to add to this list another manifestation of the Other, namely the (absent) face of the "madman," who, according to Foucault, is the "perfect stranger" (1965, p. 249). However, before we approach the "madman," we need to examine the larger context, the patient as the Other.

"Patient as the Other" is any oxymoron, for the simple reason that there is no room in the healing tradition for the patient to be the Other. The patient role requires that he or she be a compromised other, not the "absolutely other." This point may be illustrated by the model patient

A-T. Tymieniecka (ed.), Analecta Husserliana, Vol. LIV, 101–114.

of all times, "Lazarus saved," to borrow a metaphor from Blanchot (1981, p. 46). The story of "Lazarus saved" is a familiar one:

> Now a certain man was ill, Lazarus of Bethany. . . . Now when Jesus came, he found that Lazarus had already been in the tomb four days . . . [H]e cried with a loud voice, "Lazarus, come out." The dead man came out . . . (John 11: 1, 17, 44).

What transpires in the above story is this, that so far as his "otherness" is concerned, "Lazarus saved," the patient who always responds readily to the call of the (divine) healer, is in the same predicament as the Heideggerian hammer, which works so well that its "thingness" has disappeared into its function and use. As Heidegger argues persuasively, the hammer needs to break down for its "thingness" to be noticed (1962, see also Macomber, 1967). By the same token, if the "otherness" of "Lazarus saved" is *not* to be overlooked, he needs to become "Lazarus lost," the hypothetical patient who is lost to language, according to Blanchot (1981, p. 46), hence remains unresponsive to the call of his healer. It is in the same vein that Derrida claims that "for the face of the Other not to be overlooked, he must present himself as absence, and must appear as nonphenomenal . . . I must only *reach* it [the face of the Other] as the inaccessible, the invisible, the intangible" (1978, p. 103).

But "Lazarus lost" does not fare any better than "Lazarus saved" in the medical system – his "otherness" still ends up being "overlooked." the story of "Lazarus lost" may be told by that of the "madman." In fact, the "madman" is the Other *par excellence*, for he is basically, as we shall see, silence and absence. His story will be told in the following, with a special focus on two themes, language and glance, which, according to Levinas, constitute the two components of the face-to-face encounter with the Other (1969).

The Language of Monologue

The "madman's" speech is a "language without response" (Foucault, 1965, p. 251), a monologue, which is more akin to silence than language, as is described by Foucault so eloquently:

> All those words deprived of language whose muffled rumbling, for an attentive ear, rises up from the depths of history, the obstinate murmur of a language which speaks by itself, uttered by no one and answered by no one, a language which stifles itself, sticks in the throat, collapses before having attained formulation, and returns without

incident to the silence from which it had never been freed – the charred root of meaning (1965a, p. 285).

The "madman's" monologue is matched by that of psychiatry, which, according to Foucault, is "a monologue of reason about madness" (1965, pp. x–xi). Consequently, the "madman's" speech is "a monologue in a language which exhausted itself in the silence of others" (Foucault, 1965, p. 261).

The Gaze of "Observation"

The "madman" lives in a fish bowl, which places every move of his under the vigilant gaze of "observation," a surveillance which Foucault calls "a sort of invisible tribunal in permanent session" (*ibid.*, p. 265). One of the consequences of this "observation" is that the "madman" can never "present" himself in his "nudity," namely, as Levinas puts it, in his "destitution and hunger" (1969, p. 200). Every time the "madman" tries to "impose" himself on the "staff," he will invariably get his diagnosis confirmed or revised. With increasing sophistication in the science and technology of "mental illness," the "madman's" face is likely to be forever "overlooked," or, as Foucault puts it, "forgotten" by the professionals: "In our era, the experience of madness remains silent in the composure of a knowledge which, knowing too much about madness, forgets it" (1965, p. xii).

Furthermore, as Foucault points out rightly, such an "observation" reduces everything to the surface:

> . . . for under observation madness is constantly required, at the surface of itself, to deny its dissimulation. . . . Madness is responsible only for that part of itself which is visible. All the rest is reduced to silence. Madness no longer exists except as *seen* (*ibid.*, p. 250).

Thus "drawn to the surface of himself" (*ibid.*, p. 249), the "madman" is betrayed by his appearance. Levinas makes the similar point about behavior or works as betrayal: "To approach someone from works is to enter into his interiority as though by burglary" (1969, pp. 66–67); and again, "In the work someone's intention has been divined, but he has been judged in absentia. . . . One has penetrated into his interior, but in his absence" (*ibid.*, p. 181).

Summing up the story of the "madman," we may say that his silence is not heard, his absence overlooked, and his transcendence as the Other not respected in the clinical setting. Now the stage is set for a contrasting scenario, a situation in which the Other is heard in her silence and accepted in her absence. This we find in a Chinese poem about a dreamer.

THE HIDDEN FACE OF THE DREAMER

In many respects, the dreamer is the counterpart of the "madman." For instance, Derrida points out that Descartes considers the dreamer "madder than the madman" (1978, p. 51), for the dream perception is completely devoid of objectivity. To facilitate the comparison between the "madman" and the dreamer, the latter's situation will be explored also in terms of language and glance. First, let us meet the dreamer in a poem, written by a ninth-century Chinese poet, Wen T'ing-yün, entitled "Song of Water Clock at Night":

A

The tendrils of willow are long
And spring rains fine.
Beyond the flower bed, the water clock drip-drops.
Startling cry – wild geese at the frontier;
Stirring – crows on the city wall;
On the painted screen, gold partridges.

B

The incense smoke is thin
and seeps through the curtains.
Melancholy is the Hsieh's poolside pavilion!
In the red taper's shadow [back]
A brocade curtain hangs closed
A long dream: you wouldn't know.
(Adapted from Liu and Lo, 1975, p. 250.)

The poem unfolds along a simple story line: someone (presumably a lady, as is indicated by "the Hsieh's," which is a classical allusion to courtesans) is dreaming on a rainy spring night. What is striking about this poem is the fact that its rich and detailed description of the lady's chamber seems to function as a "cover up" for the paucity of information concerning personal aspects of the lady. For instance, her face is never revealed, except that it is hinted at by the mention of its antithesis: the "back" (of the red taper) in line B4. Similar to the "madman's," the dreamer's (hidden) face is underlined by two motifs: monologue and surface phenomena.

Monologue of the Dreamer

Similar to the "madman," the dreamer also speaks a "language without response." We discern two types of monologue in the poem: a. the monologue in the dream state is referred to as "a long dream," which is unknowable, ineffable, and is more akin to silence than language; and, b. the monologue in the waking state, in which the dreamer reflects upon her dream. This reflection upon the dream is a monologue about monologue, a monologue which comments on the impossibility of having a dialogue about the dream: "A long dream: you wouldn't know." This concluding line of the poem is open to multiple interpretations:

A. The dreamer dreams of her lover, "you," and is probably engaged in a dialogue with him. But upon awakening, she realizes that her dream dialogue is ultimately a monologue, a "relation without relation," to borrow an expression from Blanchot (quoted by Wood, 1990, p. 126), for the dream dialogue never reaches the interlocutor "you."

B. "You wouldn't know, and you never will." The dreamer's waking monologue may be tinged with a sense of defiance. This note of defiance is captured by Blanchot, whose description of human remains reminds us of the dream state: "The remains appear in the strangeness of their solitude, as something disdainfully withdrawn from us . . . the sense of an interhuman relationship is broken . . ." (1981, p. 82).

C. "You wouldn't know, regardless of how much I wish you would!" Here we discern a note of resentment, lamenting the impossibility of dialogue and self-disclosure. To hear this note, we need to take a closer look at the last three lines of the poem.

First of all, consider line B4, "In the red taper's shadow [back]." According to professor Zheng Qian (1952, p. 3), the term "back" refers to the dimming of light as an attempt to prevent it from shining forth. Thus with its negative spelt out, this line would read: "the red taper is *not* showing its light fully." The next line, "a brocade curtain hangs closed," also contains an implicit negative, which, if spelt out, would read: "a brocade curtain is *not* open." The third and last line already has its negative spelt out: "a long dream: you wouldn't know." If we go one step further and put all the negatives "under erasure" (Derrida, 1978) by means of a "crossing out" practice initiated by Heidegger (1959), we get the following transcription:

> The red taper is showing its light fully,
> A brocade curtain is open,
> A long dream: you would know.

All the three terms under erasure, e.g., "show," "open," and "know," are terms of presence and unconcealment. As we dwell upon the erasure/trace of these terms of presence, we begin to notice an undercurrent beneath the dominant images of concealment (the dimmed light of the candle, the closed curtain and the undisclosed dream): it is an undercurrent of nostalgia for presence and unconcealment, and of resentment over the impossibility of this possibility.

In conclusion, then, the dreamer is situated at the intersection between monologue and dialogue. Her monologue has the intentionality of dialogue, as it is intended for the interlocutor "you." As a dialogue in the absence of an interlocutor, her monologue is full of ambiguities: it is tinged with the defiance of concealment and withdrawal, on the one hand; and on the other, it laments the impossibility of unconcealment. In the final analysis, hers is a monologue which is meant to be a dialogue, and a dialogue which is destined to be a monologue; it is also a monologue about a monologue which is lost to language. All these nuances in the dreamer's discourse can be found in the "madman's" monologue.

The "madman" is also engaged in two types of monologue: a. the monologue of the psychotic state, which is, as Foucault points out, more akin to silence than language; and b. the monologue about monologue as he reflects upon his psychosis. Furthermore, the poem has also shed some light on the "madman's" seclusiveness and withdrawal. Behind the "madman's" non-communicativeness, we may discern two conflicting sentiments: There may be a note of defiance: "you wouldn't know, and you never will," as well as resentment over the success of his concealment, a sentiment which is succinctly summed up by a wise psychiatrist, who once said, "It is nice to hide, but a disaster never to be found."

The Phenomena of Surface

Not unlike the "madman," the dreamer does not reveal herself through her glance. Taking a hint from the "back" of the red taper, we may infer that the dreamer's glance is averted, as her face is turned away from the spectator. Furthermore, similar to the "madman," the dreamer is the object of a glance which focuses on surface phenomena: the willow, the flowers, the rain, the curtains, the incense smoke, and so on. However, these surface phenomena do not betray the dreamer, but rather serve as a veil, which protects her in her absence and concealment.

In fact, one may discern numerous veils or barriers throughout the poem: the willow, which hangs like a curtain, the frontier, the city wall, the painted screen are all barriers of one form or another. In fact, the "curtain" is explicitly mentioned twice (lines B2 and B5). Furthermore, there is one more barrier between the dreamer and the "you," namely the dream behind which the dreamer lies hidden (one is tempted to say the "dream screen," although this is not an indigenous term). The motif of concealment finds an eloquent expression in the gold partridges (line A6). Their silence and immobility seem to be more disturbing than the noise and commotion of the wild geese and city crows (lines A4 and A5), for in the dissimulating appearances of these painted birds is manifest the utter incommensurability of the Other.

So much for the dreamer as the Other. The affinities she shares with the "madman" may be assessed by the following case vignette.

The "Madman" and the Dreamer

Mr. X was a psychiatric patient in a locked ward. For several weeks on end, he had been socially withdrawn, constantly crying, and delusional, until one day his mother passed away. Then all of a sudden, Mr. X's "symptoms" cleared up dramatically: his speech was no longer delusional, and he stopped weeping. He also denied suicidal thoughts when questioned by a psychiatrist. Three days later, Mr. X made a nearly fatal suicide attempt, when he left the ward for a walk.

Not unlike the dreamer, Mr. X's face seems to be hidden behind layers and layers of concealment – there were the "dissimulations" of denying suicidal thoughts to the psychiatrist, and of using the excuse of a walk to get off the ward; there were "concealing denials" (Heidegger, 1971, p. 55) such as the sudden remission of "symptoms." Concealed also was Mr. X's mourning and grief over the death of his mother. Furthermore, concealed behind all these concealments was Mr. X's anguish over the fact that his self-concealment was all too successful. Mr. X remarked after his suicide attempt that he wished that the "staff" had (seen through his concealing denials and consequently) turned down his request to leave the ward.

Having mapped out the overlaps and affinities between the "madman" and the dreamer, we are now ready to examine the different approaches to the other that these two scenarios represent.

TWO APPROACHES TO THE OTHER

Mr. X and the dreamer bring to light two very different approaches to the Other: one is the ethical, problem-solving orientation as represented by the clinical field; the other is the aesthetic approach as embodied by the Chinese poem. These two approaches will be examined in terms of, again, speech and glance.

Rejection of Monologue

As Heidegger points out so rightly, the social sciences, of which psychiatry and psychology are prime examples, share with metaphysics a preoccupation with presence, and a corresponding difficulty in "thinking absence" (Bernasconi, 1985, p. 83). Consequently, the monologue of the "madman," with its implied absence of an interlocutor, presents a problem to the clinician. A solution to this problem is usually sought in two alternatives: either a. to reject the monologue of the "madman," by disqualifying it as speech; or, b. to create a pseudo-dialogue by "teaching" the "madman" to speak the language of the professionals. While (a) is the standard procedure of the "asylum," (b) describes the practice of various forms of "psychotherapy."

Acceptance of Monologue

In contrast to the problem-solving orientation, the aesthetic approach corrects nothing and resolves no problems. It "lets be," as Heidegger would say (1971). Thus the poem not only listens to the dreamer's monologue, but also celebrates its problems, such as the absence of the interlocutor, the impossibility of dialogue, and so on. Furthermore, this aesthetic listening differs from the psychoanalytic listening, which Foucault hails as restoring "in medical thought, the possibility of a dialogue with unreason" (1965, p. 198). The aesthetic listening is prudent and reverential: it does not convert the silence of dream into speech, for it is willing to stop at "not knowing," thus preserving the dreamer's monologue in its mystery and silence. In contrast, psychoanalysis betrays the "madman" by translating his silence into the language of reason; thus Derrida writes, "The misfortune of the mad, the interminable misfortune of their silence, is that their best spokesmen are those that betray them best" (1978, p. 36).

The Glance that Denies Absence

The "observation" that keeps vigil over the "madman" is guided by a problem-solving thrust, which approaches surface phenomena with either denial or rejection.

The mechanism of denial seems to be operative in the glance of behaviorism. Confident that whatever is observable is knowable, behaviorism denies the absence of depth and substance in surface phenomena. In other words, it tends to read presence into absence, such that it approaches behavior as signifier, as a sign which refers to something of substance and truth, such as "personality," "mental status," or the "agent" of behavior.

This myopia about absence is manifest in the professional observation of Mr. X: his denial of suicidal thoughts was taken at its "face" value, and his remission of "symptoms" was viewed as "improvement." To wit, upon inquiry as to why they gave permission for Mr. X to leave the ward, thereby creating ample opportunity for him to make the suicide attempt, staff said almost unanimously that Mr. X was "improving" after his mother's death. This particular "reading" of Mr. X seems to be the logical conclusion of the view that whatever cannot be observed does not exist. Thus excluding the alternative understanding of absence as concealment, the absence of "symptoms" was interpreted as "improvement."

Psychoanalysts would not have made such a mistake. They would have recognized the concealing, dissimulating aspects of surface phenomena right away. They would have made an attempt to penetrate the surface so as to reveal the hidden, unconscious meaning of things. The violence of this approach, however, lies in its rejection of surface phenomena as obstacles to "truth."

The Aesthetic Glance

The aesthetic glance entails neither denial nor rejection of surface phenomena. It is, on the contrary, a welcoming acceptance of the facade. Levinas makes the significant point that "It is art that endows things with something like a facade" (1969, p. 192). The gaze of "observation" sees the behavior, but not facade; the insight of psychoanalysis sees defenses and denials, but not facade. Only aesthetics sees the facade.

What is facade? Levinas' description of it reminds us of our experience with the dreamer in the poem:

> But in it [the facade] is constituted the beautiful, whose essence is indifference, cold splendor, and silence. By the facade the thing which keeps its secret is [not] exposed [but] enclosed in its monumental essence and in its myth, in which it gleams like a splendor but does not deliver itself. It captivates by its grace as by magic, but does not reveal itself (1969, p. 193).

The facade, therefore, reveals nothing. It is a signifier which refers to nothing but itself. It betrays nothing but its own emptiness, its lack of depth, ground, and truth. In other words, the facade is surface accepted as such, accepted in its absence of depth and truth.

This welcoming acceptance of absence may be considered a form of respect, which, according to Caputo, has to do with feelings in response to the absence and concealment of Being: "It is *lethe* at play in *a-letheia* which inspires the mix of awe, fear, and admiration which Kant calls respect – and which Heidegger calls letting-be" (1987, p. 227). This respect, however, is not guaranteed by the glance, for according to Levinas, it is made possible only by "metaphysical desire." Derrida explains, "According to Levinas the glance *by itself* . . . does not *respect* the other. Respect . . . can be only as [metaphysical] desire" (1978, p. 99). In the final analysis, then, the aesthetic approach to the Other is based on metaphysical desire. To this topic we now turn.

Need versus Desire

Levinas makes the important distinction between need and metaphysical desire: the former is problem-solving oriented; it seeks satisfaction by eliminating the lack. The latter, in contrast, is "a desire without satisfaction" (1969, p. 34), for, as Derrida puts it, "metaphysical desire does not seek to consume" (1978, p. 99). Thus while need thrives on presence, metaphysical desire thrives on absence. In light of this distinction, we may conclude that the problem-solving approach to the "madman" is subtended by need, whereas the aesthetic approach to the dreamer is fuelled by desire. Thus while clinicians strive to "overcome" obstacles to dialogue and "truth" in their treatment of the "madman," the Chinese poem celebrates the impossibility of dialogue, the impossibility of knowing the dreamer, and the impossibility of depth or truth. What can the clinician learn from the poem? To answer this question, we explore further implications of the "metaphysical desire".

A RELATION WITHOUT RELATION

From the point of view of a desire, which "desires for the absolutely other" (Levinas, 1969, p. 34), and which "understands the remoteness, the alterity, and the exteriority of the other" (*ibid.*), monologue has its advantages. The monologue of the "madman" or the dreamer, for instance, helps metaphysical desire to hear, if we may borrow a Zen metaphor here, the sound of one hand clapping (Suzuki, 1964, p. 105).

The Zen metaphor is used here to refer to a relation which relates us to what is, like the missing hand in the clap, "relationally unavailable" (Wood, 1990, p. 126). Such a relation without reciprocity or mutuality is referred to by Blanchot as "a relation without relation" (*ibid.*), and by Derrida as "the relation of interruption":

> . . . a crazy relation, a relation without relation, which understands the other as other in a certain relation of incomprehension. . . . [I]t is necessary that at a given moment the other remain as other, and that if it is the other it is other: at this moment the relation to the other as such is also the relation of interruption (Derrida and Labarrière, 1986, p. 82).

Two observations may be made concerning this "relation of interruption": first of all, it describes the ideal discourse with the Other, a discourse punctuated or interrupted by the impossibility of dialogue, disclosure, truth and depth. From the point of view of metaphysical desire, which is "desire for the invisible" (Levinas, 1969, p. 33), such an interrupted discourse is tribute to the other as Other.

Secondly, Derrida points out rightly that "interruption here does not interrupt the relation to the other, it opens the relation to the other" (Derrida and Labarrière, 1986, p. 82). Extending this insight in another direction, I would claim that such interruptions open up an important dimension of our being as *dasein*. I refer in particular to the being-with dimension of *dasein*, which may be brought into sharp relief by an interrupted discourse, just as the Zen clap is rendered sonorous by the missing hand.

To borrow once again the parable of the Heideggerian hammer, which works so well that its "thingness" gets overlooked, the being-with dimension of the self is usually absorbed in and masked by our social interactions. But when such interactions are interrupted, the self in its being-with will find itself out on a limb, as it were, without any justifications. Thus with the interruption of social discourse, the being-with dimension of the self is not thereby cancelled, but instead made doubly conspicuous and questionable. And if we follow Heidegger, who maintains that it is the questionableness of our being that defines our authenticity as *dasein* (1962), we may conclude with a Buberian note (Buber, 1970) that it is the Other qua Other that makes the self authentic as self.

We experienced something of a simultaneous anguish and serenity, when we met with the dreamer, whose averted glance and withheld speech inaugurated a "relation without relation," a relation which makes possible not only our relation to her as the Other, but also our relation to ourselves as *dasein*. I believe that a similar experience is possible in our encounter with the "madman," if he is to be approached from the perspective of metaphysical desire.

SUMMARY AND CONCLUSION

In conclusion, the story of "Lazarus lost" reveals an intrinsic flaw in the ethical approach to the Other, namely that the problem solving zest of the ethical orientation leaves little room for the Other to assert his or her autonomy as absence, or in Derrida's words, as "the inaccessible, the invisible, the intangible." Our comparative study of the "madman" and the dreamer further shows that the contrast between the problem-solving and the aesthetic approaches to the Other is ultimately that between rejection and acceptance of absence.

To the spectator, the "madman" and the dreamer both appear as monologue, which is marked by absence of the interlocutor, and as surface phenomena, which are marked by absence of depth or "truth." In the clinical setting, which is guided by the problem-solving orientation, the "madman's" silence and monologue are rejected as speech, and his appearance becomes, under the gaze of "observation," "behavior," which betrays him in his absence. In the Chinese poem, on the contrary, the dreamer's silence and monologue are heard, and her appearance accepted as such, namely as surface devoid of depth or "truth," in other words, as facade, which protects her in her absence.

In the final analysis, the contrast between the two approaches to the other seems to coincide with the distinction in-between need, which is preoccupied with presence, and metaphysical desire, which thrives on absence. Thus clinicians, guided by the ethical, problem-solving orientation, are invested in overcoming the absence of dialogue and concealment in their encounter with the "madman." In contrast, the Chinese poem, which shares with metaphysical desire the predilection for absence, celebrates the impossibility of dialogue, of knowing, and of "truth" in the dreamer's silence and dissimulating appearance. As a discourse punctuated by the impossibility of dialogue and "truth," this aesthetic approach to the Other seems to be a "relation without relation," a relation which keeps open not only our relation to the other as Other, but also our relation to ourselves as *dasein*.

While this paper argues for the superiority of the aesthetic over the ethical, it does not deal with the practical considerations of whether and how to negotiate between these two approaches to the Other in the clinical setting. Readers interested in these practical questions may consult Sundararajan (1995).

Rochester Psychiatric Centre, N.Y.

REFERENCES

Bernasconi, R., *The Question of Language in Heidegger's History of Being* (New York: Humanities, 1985).

Blanchot, M., *The Gaze of Orpheus and Other Literary Essays* (L. Davis, trans.) (Barryton, NY: Station Hill, 1981).

Buber, M., *I and Thou* (W. Kaufmann, trans.) (New York: Charles Scribner's Sons, 1970).

Caputo, J. D., *Radical Hermeneutics: Repetition, Deconstruction, and the Hermeneutic Project* (Bloomington & Indianapolis: Indiana University, 1987).

Derrida, J., *Writing and Difference* (A. Bass, trans.) (Chicago: University of Chicago, 1978).
Derrida, J. and Labarrière, P., *Altérités* (Paris: Osiris, 1986).
Foucault, M., *Madness and Civilization* (R. Howard, trans.) (New York: Random House, 1965).
Foucault, M., *Madness and Civilization* (New York: Vintage Books, 1965a).
Heidegger, M., *The Question of Being* (W. Klubah and J. Wild, trans.) (London: Vision, 1959).
Heidegger, M., *Being and Time* (J. Macquarrie and E. Robinson. trans.) (New York: Harper and Row, 1962).
Heidegger, M., *Poetry, Language, Thought* (A. Hofstadter, trans.) (New York: Harper & Row, 1971).
Levinas, E., *Totality and Infinity* (A. Lingis, trans.) (Pittsburgh: Duquesne University, 1969).
Liu, W. C. and Lo, I. Y. (eds.), *Sunflower Splendor: Three Thousand Years of Chinese Poetry* (Garden City, NY: Anchor, 1975).
Macomber, W. B., *The Anatomy of Disillusion* (Evanston: Northwestern University, 1967).
Sundararajan, L., "Patient as the Other," *The Humanistic Psychologist* **23**(1) (1995, Spring): 63–70.
Suzuki, D. T., *An Introduction to Zen Buddhism* (New York: Random House, 1964).
Wood, D., *Philosophy at the Limit* (London: Unwin Hyman, 1990).
Zheng, Qian. (ed.), *Ci Xuan* [selected lyrics] (Taipei: Zhong Hua, 1952).

PART THREE

THE TRANS-ACTIONAL WEAVING OF THE HUMAN WORLD OF EXISTENCE

Richard Kulis and Ella Buceniece.

KELLY OLIVER

THE PHENOMENOLOGY OF INTERSUBJECTIVITY

Contemporary continental philosophy has moved away from phenomenology, especially Husserlian phenomenology with its transcendental subject. Currents in contemporary theory that have been called "post-structural" and "postmodern" share the rejection of the Husserlian theory of subjectivity. "Post-structuralism" and "postmodernism" are also post-phenomenological. Levinas, Derrida, Lyotard and Kristeva, for example, criticize Husserl's conception of the subject. Of course, Husserl's immediate successors, Heidegger, Sartre, and Merleau-Ponty also rejected his conception of the transcendental subject. Returning to different versions of Hegel's dialectical conception of subjectivity, some contemporary theorists challenge the Husserlian notion of a unified transcendental subject by insisting that the subject is not only fragmented or decentered and embodied, but also that it is the result, even the effect, of relationships to something or someone other than itself.

Coming under the direct and indirect influence of Emmanuel Levinas, much contemporary Continental philosophy attempts to come to terms with the "other." Discussions have moved away from a philosophy of subjectivity towards a philosophy of alterity. But, if we look back through the phenomenological tradition, we can already read a dehiscence in the notion of the subject in Husserl, through Heidegger, Merleau-Ponty, and Sartre, that sets up post-phenomenological theories of subjectivity as fundamentally intersubjective. From the beginnings of phenomenology, the other has been at the core of the self. In this essay, I return to Husserl's notion of intersubjectivity to point out how even his transcendental subject is dependent both upon its embodiment and upon its relationship to something other than itself. Within Husserl's philosophy, there is already a split within the subject whereby the very act of conscious ownership necessitates the activity of disowning. I trace the development of this split within subjectivity, this disowning at the heart of owning, through Heidegger's phenomenological ontology which rejects the metaphysics of subjectivity. At the heart of Heidegger's notion of ownness or the proper (*Eigen*), I find the necessary activity of disowning or impropriety. I analyze the importance of intersubjectivity for Heidegger's notion

A-T. Tymieniecka (ed.), Analecta Husserliana, Vol. LIV, 117–139.

of authenticity (*Eigenlichkeit*) or ownness. I move to Merleau-Ponty and discover not only the embodiment of the Husserlian subject but also the suggestion of a dissolution of borders between the self and the other to the point where one person can almost experience the other as he or she experiences him or herself. Next I take up Sartre's notion that the ego is the product of the look of the other, who cannot be reduced to an object even though it is still a corollary of intentionality. Finally, I read this history of phenomenology through the contemporary phenomenology of Levinas in order to offer a reading of the trajectory of otherness and intersubjectivity within the phenomenological tradition. By so doing, I situate Levinas *vis a vis* the phenomenological tradition and I provide a way of reading one of Levinas' most troubling notions in *Otherwise Than Being*, the notion of responsibility.

HUSSERL

Although there are three volumes in the *Husserliana* entitled *On the Phenomenology of Intersubjectivity* (*Zur Phänomemologie der Intersubjektivität*), Husserl's theory of intersubjectivity is best known from his discussion in the fifth *Meditation* of the *Cartesian Meditations*. There, Husserl founds his theory of intersubjectivity on the ownness of consciousness. He maintains that ownness is the eidetic structure of consciousness (CM 93). What I do not experience as part of my primordial ownness – what I can image as other than myself – is other or alien. The other is constituted through my relation with the ownness of my consciousness and the connection between that ownness and my body. Husserl describes the other as a mirror of my ownness, an analog of me, my alter ego (CM 94). The other's subjectivity or consciousness cannot be directly perceived or presented to me, so it is apperceived or appresented. It is only indirectly perceived or presented as an analog of my own consciousness.

While the other is known through analogy, Husserl insists that the process of constitution is not one of analogical reasoning, but of analogical transfer (CM 111). Upon perceiving the body of another we immediately transfer the sense of an animated organism with its own subjectivity onto that perception. The perception of the other is constituted as a perception of an animate conscious organism. And, although the other is my alter ego or my analog, Husserl maintains that I do not apperceive the other as a duplicate of myself; rather, the process is a type of reproduction: "I do not apperceive the other ego simply as a duplicate of myself and accordingly as having my original sphere or one completely like mine. . . . Its manner of appearance does not become paired in a direct association with the manner of appearance actually belonging at the time to my animate organism (in the mode Here); rather it awakens **reproductively** *another*, an immediately similar appearance included in the system constituted of my animate organism as a body in space" (CM 117–118; bold is my emphasis).

Husserl describes a type of association that is indirect and reproductive. What is an indirect reproductive association? The ego and alter ego are always paired through association, which we discovered in the fourth *Meditation* is the result of a passive synthesis (CM 112). Because the association is between my body and the body of another, pairing is always a pairing of my concrete or personal ego with a concrete alter ego (CM 123). Wherever the alter ego appears, I am its companion. It never appears without me. This is because all sense is constituted through my transcendental ego (CM 99, 115, 136). And all intersubjectivity is founded on egology (CM 155). That is to say, the other is always an other *for me*.

What troubles Husserl is how I can distinguish my own ego from that of another. Yet, if I can separate the two, then how is empathy or communion with another possible? What guarantees that my ego is mine and not yours is that my ego is presented to me and yours is always only appresented to me (CM 119). In the fourth *Meditation* Husserl concludes that through the epoché and then the phenomenological reduction, which is performed through imaginative or fictive variation, I can bracket out my personal ego and intuit the eidetic structure of the transcendental ego. The transcendental ego is whatever is left over, what cannot be doubted or imagined away; it is what I cannot imagine to be different than it is. This indubitable remainder is the sphere of my ownness; it is what I cannot doubt is my ego. While I have direct access to my ownness, I have only indirect access to your ownness.

In spite of the radical separation of my own ego from yours, communion is possible through empathy. Husserl maintains that I can put myself in your place and view the world from your perspective. This operation is possible through the same kind of imaginative variation that allowed me to perform the transcendental reduction. Leaving aside the obvious problems with the notion that I can empathize with someone by occupying their space, or imagining occupying their space, leaving aside the fact that as Husserl sets it out, the other is no threat to my sense of ownness because it is always *my* other and therefore never truly other, I want to suggest that not only is ownness the result of the other, but also that the alter ego is first *within* my own ego. In other words, it follows from Husserl's analysis that ownness is only possible through disowning.

First, my experience of ownness is only possible through an experience of otherness. As Husserl describes it, it is through the process of imaginative variation that I intuit my ego by excluding all that could belong to some other. What remains after I identify what is not my own or other than me, and exclude it, is undoubtedly mine. It seems that I must first have this imaginative relation with what is other than me before I can intuit my ownness. Derrida makes an analogous argument in *Speech and Phenomena* when he points out that ownness is possible because of internal time consciousness, and yet the movement of time undermines the self-presence of ownness:

> . . . does not everything that is announced already in this reduction to "solitary mental life" (the transcendental reduction in all its stages, and notably the reduction to the monadic sphere of "ownness" – *Eigenheit* – etc.) appear to be stricken in its very possibility by what we are calling time? . . . Is not the concept of pure solitude – of monad in the phenomenological sense – *undermined* by its own origin, by the very condition of its self-presence, that is, by "time," to be conceived anew on the basis now of difference within auto-affection, on the basis of identifying identity and nonidentity within the "sameness" of the *im selben Augenblick*? (SP 68)

Husserl's notion of the alter ego is another appearance of "difference within auto-affection" or nonidentity within the ownness of the ego. Husserl maintains that "in a free variation, I can phantasy *first of all myself*, this apodictic de facto ego, as otherwise and can thus acquire the *system of possible variants of myself*, each of which, however, is annulled by each of the others and by the ego who I actually am. It is *a system of a priori incompossibility*. Furthermore the fact, 'I am,' prescribes *whether* other monads are other for me and *what* they are for me. . . . If I phantasy myself as a pure possibility different from what I actually am, that possibility in turn prescribes what monads exist for him as others" (CM 141). I have a sense of the other by freely varying myself and discovering that such variations are not compossible with who I am. The other is who I am not and cannot be. Yet, as Husserl describes the process of finding the other as it is founded in, and constituted by, my own ego, the rupture in my own ego becomes more pronounced.

With his distinction between the personal, psychological, concrete, or empirical ego and the transcendental ego, Husserl already splits the ego. He insists that the ownness of my transcendental ego is primary and the ownness of my psychological ego is secondary (CM 100). Still, it is only on the basis of a pairing of **concrete** egos (as they are embodied) that I can apperceive the other (CM 123). In what sense, then, is my experience of the other constituted by the transcendental ego? It must be constituted in the same way that my experience of my own concrete ego is constituted by the transcendental ego. The other, as Husserl says, "is therefore conceivable only as an analog of something included in my peculiar ownness" (CM 115). But what is there present in my ownness of which the other is an analog? Indeed, how is the ownness of my own embodied concrete ego constituted from within the transcendental reduction except through an analogical transfer of sense from the ownness of my transcendental ego to the ownness of my empirical ego? What I am suggesting is that in the Fifth *Meditation*, the other can appear as alter ego only because I find the alter ego already within my sphere of ownness.

My alter ego is all of those accidental empirical qualities that I bracket off in order to uncover the eidetic structure of my consciousness. In addition, my alter ego is all of those possibilities that I imagine for myself that allow me to delineate my sphere of ownness. I am who I am by virtue of my relation with my own alter ego. It is through a fundamental split in my ego that I am. It is through a process of disowning the alter ego that I intuit my ownness. In order to apperceive an other I merely reproduce this experience of disowning to redefine myself, my ownness, in relation to my other alter ego. The alter ego is not only "demonstrated precisely within the experiencing intentionality of my ego" in the sense that it is constituted through a relationship with my ego, but also in the sense that it is experienced in the intuition of my ego itself (CM 148).

The way in which my ego **reproduces** the otherness of the other from the otherness of myself takes on new connotations, when in *On the Phenomenology of Intersubjectivity* Husserl uses the mother-child relation as an illustration of the relation between the other and the I:

(There) is an excellent (example in the case) where the other is interpreted as (intentionally) related to my "I" and to that which pertains to my "I" (*mein Ichliches*) and (where) I actually experience this. Thus, the unity within the multiplicity of interpretational experience here finds a point of fulfillment in my specific self-experience. . . . In any case, if we think of the most primordial genetic continuity between **mother and child** and of the importance of the social I-Thou-life (*Ich-Du-Leben*) (it becomes clear that) this mode of fulfillment plays a special role (quoted in Bernet *et al.*, p. 163; PI 504; my bold).

The mother-child relation is used to illustrate the way in which the other is related to the I. The mother gives birth to the child; they are genetically contiguous. So too, the transcendental ego gives birth to, reproduces, the other. In fact, Husserl figures the realm of pure consciousness out of which all objectivity originates as the realm of Mothers. The constitutive activity of consciousness is likened to procreation:

. . . to express it in the myth of Goethe, find the way to the Mothers of knowledge, to discover their realm of pure consciousness, in which all being originates constitutively and out of which all knowledge as knowledge of what is must draw its ultimate intelligible explanation . . . one is dealing with nothing other than the science of the formations of the essence of consciousness as such, as the science of the Motherly origins (Hua XII, 233; quoted in James Hart, *The Person and the Common Life*, Boston: Kluwer Academic Publishers, 1992, p. 3).

In Goethe's *Faust*, "the Mothers" is the name given to the goddesses who inhabit the realm of the absolute, where there is no finished creature or world, but where the generation of all creatures and everything in the world begins. The path to the realm of the Mothers is lonely and desolate. In his unpublished notes, Husserl asks "Do we not shudder in the presence of these depths? Who has seriously made them a theme in the millennia of the past? Who, in connection with the first reflections of Augustine, has dared to risk his life on the way to the 'mother'?"[1] James Hart suggests that the transcendental reduction, or the way to the Mothers, requires a form of violence to oneself, even death of oneself. The life that is forfeited, is the life of the natural attitude.[2] The Mothers require the death of the natural life, but they promise a new life both before and beyond that forfeited life. The body and empirical ego must be left behind in this Husserlian reunion with the Mothers of knowledge.

HEIDEGGER

Against Husserl, Heidegger insists that the ego is in the world. In *Being and Time*, he reverses the order of Husserlian epistemology. The transcendental attitude does not reveal the essential structures of the natural attitude. Rather, in Heidegger's terminology, the structure of reflective consciousness is already implicit in preflective understanding. Although he rejects the Husserlian transcendental ego, Heidegger does retain a version of Husserlian intentionality with his notion of care. Like Husserl, Heidegger uses a maternal metaphor for the intentional character of consciousness. In *Being and Time*, the centrality of care to Dasein is illustrated using the "ancient fable" of the birth of man, shaped by Care from the body of Earth, given spirit by Jupiter, and named by Saturn, who arbitrates a debate between the other three: "Since you, Jupiter, have given its spirit, you shall receive that spirit at its death; and since you, Earth, have given its body, you shall receive its body. But since 'Care' first shaped this creature, she shall possess it as long as it lives. And because there is now a dispute among you as to its name, let it be called '*homo*,' for it is made out of *humus* (earth)" (BT 242).

In *Being and Time*, Heidegger describes care as Dasein's fundamental relation to the world. On a preflective level, Dasein is engaged with the world; Dasein is always oriented in relation to the world. This is to say that Dasein is intentional. Yet, this intentionality is not given to reflective consciousness as an essential structure of a transcendental *I*. Heidegger maintains that if we start with the givennesses of the *I*, then we soon discover that I am not that *I*, that what belongs to me, what is my own, is also not mine: "Dasein is in each case mine, and this is its constitution; but what if this should be the very reason why, proximally and for the most part, Dasein *is not itself*?" (BT 151).

The notion of ownness or own (*Eigen*) permeates Heidegger's writings from the earliest, *Being and Time* (1927), to the latest, *On Time and Being* (1964).[3] In *Being and Time*, the most renowned use of *Eigen* is in the derivative *Eigentlichkeit*, authenticity. Heidegger's notion of authenticity should *not* be reduced to the pop-existentialist notion of being true to yourself. Rather, as Heidegger describes *Eigentlichkeit*, and other forms of *Eigen* in his later works, it becomes clear that authenticity not only involves ownness, or what properly belongs to oneself, but also authenticity involves owning up to the ways in which one is inauthentic. Modes of revealing proper to human beings are those which reveal the process of concealing. In other words, human beings' own, proper, or authentic, relationship to their own existence is to recognize the various ways in which they conceal that their relationship to their own existence is one of concealment.

Like Husserl's notion of ownness, there is a disruption at the center of Heidegger's notion of ownness. The way in which Heidegger's notion of ownness turns in on itself, against itself, in order to open onto something other than itself, becomes more apparent in his later writings. In

Being and Time, the structure of ownness is set up, but the essence of ownness, still not made explicit, is the recognition of disowning as the operation that defines *Dasein*. In later works, the essence of ownness is the operation of disowning itself (in the full ambiguity of "itself" in this context).

In *Being and Time*, *Dasein* designates the existential structure, or position, of a being who takes its own being as an issue. Dasein is unique among beings because it experiences anxiety in the face of its ownness and anticipation of its own death or impending nothingness (see BT 232, 308). The special position, both ontically and ontologically, of Dasein in relation to Being and other beings is clear from the first chapter of *Being and Time*. What is more ambiguous is the relation between authentic and inauthentic Dasein and the relation between Dasein's own and what Heidegger calls "the They."

Heidegger concludes that "*authentic Being-one's-Self* does not rest upon an exceptional condition of the subject, a condition that has been detached from the 'they'; *it is rather an existentiell modification of the 'they' – of the 'they' as an essential existential*" (BT 168). This is to say that authentic Dasein is a modification of the actual particular existence (an existentiell) of the They as an essential structure of existence (an existential) or as a mode of understanding Dasein. The They is an essential part of the constitutive structure of Dasein (BT 167). Yet, in the same section Heidegger distinguishes authentic Dasein from the They: "The Self of everyday Dasein is the *they-self*, which has been taken hold of in the *authentic Self* – that is, from the Self which has been taken hold of in its own way (*eigens ergriffenen*). As they-self, the particular Dasein has been *dispersed* into the 'they,' and must first find itself. . . . If Dasein discovers the world in its own way (*eigens*) and brings it close, if it discloses to itself its own authentic Being, then this discovery of the 'world' and this disclosure of Dasein are always accomplished as a clearing away of concealments and obscurities, as a breaking up of the disguises with which Dasein bars its own way" (BT 167). Is authentic Dasein the breaking up of "the They"? Is it born out of the They? Concealed within the They? What is the relation between authentic Dasein and the They?

At the beginning of Part I of *Being and Time*, Heidegger identifies both authentic and inauthentic as modes of Being for Dasein. Dasein has different modes of Being because it is, in a sense, possibility: "And because Dasein is in each case essentially its own possibility, it *can*, in its very Being, 'choose' itself and win itself; it can also lose itself and never win itself; or only 'seem' to do so. But only in so far as it is essentially something which can be *authentic* – that is, something of its own – can it have lost itself and not yet won itself. As modes of Being, *authenticity* and *inauthenticity* (these expressions have been chosen terminologically in a strict sense) are both grounded in the fact that any Dasein whatsoever is characterized by mineness (*Jemeinigkeit*)" (BT 68). If both authentic and inauthentic modes of Being belong to the structure or essence of Dasein, then how can Heidegger identify authenticity as what is appropriate to, or what belongs to Dasein's own? Unless Heidegger is suggesting that to be authentic is simply to own up to Dasein's possibility for different modes of Being, including inauthenticity. On this reading, authenticity is a mode of Being in which Dasein admits its possibilities, including the possibility for inauthenticity, which is the concealing of its possibilities from itself in the they-self.

Own and ownness as the admission of disowning is taken one step further in Heidegger's later writings. In most of his texts, Heidegger is concerned with the question of the proper. On closer examination, however, what is proper turns out to be what opens onto its other. What brings something into its own is not what limits or confines, but rather what opens out onto multiplicity. I will give several examples. In *What is Called Thinking*, Heidegger makes a distinction between the common meaning of the word "call," to name, and a rare meaning of the word "call," to command. To command, in turn, in its original meaning is to commend or entrust (WCT 387). Although Heidegger insists that "the current meaning of the word cannot simply be pushed aside in favor of the rare one, even though the rare signification may still be the proper one" (WCT 389), what does it mean for Heidegger to identify one meaning as proper or original? How do we determine the proper meaning?

On Heidegger's analysis, the proper sense of call is command, entrust, or call something forth, because this meaning gives rise to its other meanings, including the common meaning, to name. Things come into their own insofar as they are entrusted to words, names. Things become present (to us) when they are given names. The name calls forth in the sense of command. Thinking is the response to this same sort of call (cf. WCT 391; "The Thing" 183). For Heidegger, in some sense, what is proper to thinking is determined by something other than thinking; what is thinking's own does not belong to it. Or, as Heidegger says "only the peculiar quality of that which demands of us above all else to be admitted (to thinking) can decide about that (what is proper)" (End of Phil. 72). If what is proper to thinking is to respond, if responding and recalling are its own, then its own is precisely *not* its own because its own is a response to something other. A response is always a response to something prior.

Heidegger's analysis in "The Question Concerning Technology" shows the point more forcefully. There Heidegger identifies enframing – taking everything as ordered, standing, reserved for our use – and the supreme danger because, among other things, it leads to "one final delusion: it seems as though man everywhere and always encounters only himself" (QCT 332). The final delusion of the supreme danger is that we will insulate ourselves completely from any other; we will see only ourselves everywhere.

In this essay, Heidegger once again identifies a proper meaning which he relates to a more common meaning. The common meaning of technology, which he calls the instrumental definition, is that technology is a means to an end and a human activity. Heidegger traces the etymology of "technology" back to the Greek *techne*, which he interprets as a bringing forth or revealing. He suggests that this prior sense of *techne* makes even the current sense of technology as a means or instrument possible. *Techne* as a bringing forth can bring forth a multitude of possibilities. While this proper or original sense of *techne* opens onto multiple possibilities, the mode of technology as enframing limits possibilities. In fact, the danger of enframing is that it will become the only possible relationship to the world; it will become the only mode of revealing. It can do this only by concealing the fact that it is a mode of revealing. Once it reveals that it is a mode of revealing, a bringing forth, then it is one mode among many. It cannot dominate.

This same dynamic holds true for the other examples that I have analyzed where Heidegger identifies the proper or authentic. The proper, authentic, or own most, is that which opens onto possibilities, even possibilities that disown. As Heidegger says in one of his latest writings, translated as "Time and Being":

> Appropriating makes manifest its peculiar property, that Appropriation (*Aneignung*) withdraws what is most fully its own from boundless unconcealment. Thought in terms of Appropriating, this means: in that sense it expropriates itself of itself. Expropriation belongs to Appropriation as such. By this expropriation, Appropriation does not abandon itself – rather, it preserves what is its own (TB 22–23).

In this lecture Heidegger defines Appropriation as that which gives both Being and time and makes possible the relationship between them. He calls *Ereignis* (event, occurrence) the "event of Appropriation (*Aneignung*)" (TB 19). Like Heidegger's central notion from the first part of *Being and Time*, authenticity (*Eigentlichkeit*), both Appropriation (*Aneignung*) and Event (*Ereignis*) have their root in own or proper (*Eigen*). The activity of appropriation (*sich aneignene*), of making something its/one's own, determines both Being and time. What is, is determined as its ownself – it is present (in the sense of Being and time) – through the activity of appropriation. The strange thing about appropriation, however, is that what is proper to it is the activity of expropriation. Furthermore, it is through expropriation, disowning, that appropriation preserves its own. This is to say that things become present to us through a process of revealing and concealing. As some things become present or revealed, others recede or become concealed. In order for something to appear, to become present, other things have to disappear. Appropriating, making something its/one's own, necessarily involves expropriating.

Extending his analysis from "The Question Concerning Technology" in "The Principle of Identity," Heidegger claims that enframing or the instrumental view of technology reduces everything down to man (PI 34). Yet, as in "The Question Concerning Technology" there exists the hope of what he calls there a "saving power." In "The Question Concerning Technology" Heidegger maintains that the supreme danger of enframing brings with it the saving power. In that essay the relationship between the supreme danger and saving power remains for the most part imbedded in Heidegger's dense language:

> The granting that sends one way or another into revealing is as such the saving power. For the saving power lets man see and enter into the highest dignity of his essence. This dignity lies in keeping watch over the unconcealment – and with it, from the first, the concealment – of all essential unfolding on this earth. It is precisely in enframing, which threatens to sweep man away into ordering as the ostensibly sole way of revealing, and so thrusts man into the danger of the surrender of his free essence – it is precisely in this extreme danger that the innermost indestructible belongingness of man within granting may come to light, provided that we, for our part, begin to pay heed to the essence of technology (QCT 337).

"The Principle of Identity" may shed light on the relationship between the supreme danger and the saving power. There, Heidegger again maintains that enframing may make it possible to "overcome the mere dominance of the frame to turn it into a more original appropriating" (PI 37). How can the frame that dominates turn into a frame that frees us up for our essence? By showing itself as a frame. "Assuming we could look forward to the possibility that the frame – the mutual challenge of man and Being to enter the calculation of what is calculable – were to address itself to us as the event of appropriation which first surrenders man and Being to their own being; then a path would be open for man to experience beings in a more originary way . . ." (PI 40). If the process of enframing – how a frame takes over our way of seeing the world and becomes naturalized – displayed itself, then that frame could no longer dominate, because it would be seen within a larger frame. The process of framing is the result of the activity of appropriation, of the world becoming present to us in a certain way. If we see that it can become present to us in various ways, through various frames, then no one frame can dominate. The frame that borders this way of seeing; the activity or process of appropriation, opens onto the outside rather than holding the interior firmly in place. The self-same, or one's own, necessarily opens onto its other.

In Heidegger's writings, however, it is questionable how the openness to the other carries over to intersubjective or interpersonal experience. Part of the reason may be that Heidegger is attempting to work outside of a phenomenology of subjectivity, and as his writings evolve he moves further away from any notion of subjectivity. Even so, earlier in *Being and Time*, he did address relations with other people. There, the positive relation with other people is ambiguous. On the one hand he maintains that *Dasein* is essentially Being-with (*mit-Sein*) others, and on the other hand he discusses the isolation inherent in the being towards death which authenticates *Dasein* (see BT 155–157; 307–308). In addition, the relationship between *Dasein* and the general other levels down and deprives *Dasein* of its answerability – its ability to respond to the other (BT 165). Later, in *What Is Called Thinking*, Heidegger suggests the importance of interpersonal relations when he uses the example of hands: "The hand does not only grasp and catch, or push and pull. The hand reaches and extends, receives and welcomes – and not just things: the hand extends itself, and receives its own welcome in the hands of others" (WCT 381).

Heidegger rejects Husserl's account of intersubjectivity. In *Being and Time*, he alludes to Husserl when he claims that empathy does not provide the first bridge to the other. In fact, empathy is the result of being-with and not visa versa (BT 162). In addition, the other is not merely a duplicate of myself; for this disregards the otherness of the other (BT 162). Against Husserl, Heidegger argues that being with others is not a matter of a collection of subjects intending each other (BT 163). Rather, "so far as Dasein *is* at all, it has Being-with-one-another as its kind of Being" (BT 163). Yet, after this brief discussion of being-with in *Being and Time*, Heidegger rarely directly discusses interpersonal relations.

MERLEAU-PONTY

Although like Heidegger, Merleau-Ponty rejects Husserl's notion of the transcendental ego, in his early writings he describes intersubjective relations in terms familiar from Husserl. In "The Primacy of Perception," Merleau-Ponty introduces the notion of the lived body into phenomenology. All experience is embodied. Although one might think that this radical change in the emphasis of phenomenology would have drastic effects for the discussion of intersubjectivity, in "The Primacy of Perception," Merleau-Ponty claims that the body of another is perceived as an object like any other object except that this object has the ability to communicate (PP 18). Reminiscent of Husserl, he claims that the relation to the other is a relation to another myself (PP 17). Morality is possible as a result of the interaction between isolated (now embodied) subjects who appear as my alter egos: "Just as the perception of a thing opens me up to being, by realizing the paradoxical synthesis of an infinity of perceptual aspects, in the same way the perception of the other founds morality by realizing the paradox of an alter ego, of a common situation, by placing any perspectives and my incommunicable solitude in the visual field of another and of all the others" (PP 26). Although, unlike Husserl, Merleau-Ponty recognizes the experience of an alter ego as paradoxical, in this early work he still asserts that the subject is primary and the other is constituted in my lived experience (see PP 31).

As Merleau-Ponty's writing evolves, however, a radically new notion of intersubjectivity begins to appear. This evolution is apparent even within his last and unfinished work, *The Visible and the Invisible*. There, while admitting in the beginning that my perception of the other is in each case mine, and that the world of the other is private, he begins to open up the boundaries between self and other (VI 11, 82). Rather than reject any philosophy of subjectivity, Merleau-Ponty rethinks subjectivity through the body. Like Heidegger's hand that both welcomes and receives welcome, Merleau-Ponty's flesh makes communication and communion possible. Both the thickness of the flesh and the permeability of the skin make communication with the world and other possible (VI 135).

The body itself "solves" the philosophical problem of the relationship between subject and object, self and other. The body is both subject and object "because a sort of dehiscence opens my body in two, and because between my body looked at and my body looking, my body touched and my body touching, there is overlapping encroachment, so that we must say that the things pass into us as well as we into the things" (VI 123). The thickness of flesh and the permeability of skin makes "intercorporiety" possible (VI 141). Merleau-Ponty asks "if my left hand can touch my right hand while it palpates the tangibles, can touch it touching, can turn its palpation back upon it, why, when touching the hand of another, would I not touch in it the same power to espouse the things that I have touched in my own?" (VI 141). This raises once again Husserl's problem that if I cannot distinguish my own experience from the other's, then either everything becomes part of me or my experience becomes lost to the other.

Merleau-Ponty addresses this problem by distinguishing between Husserl's conception of my own and the other experienced through consciousness, and my own and the other experienced through the body. The problem only arises when consciousness is seen as the source of experience. Consciousness relates to objects; and to make either the self or the other an object makes them identical in this regard. For Merleau-Ponty, the body is not an object, but a synergy; and "consciousness" is inherent in the body's experience. "My consciousness" is not a Husserlian synthetic unity of "consciousnesses of" (VI 141). Rather, it is the unity of the body that makes the unity of consciousness possible (VI 141–142). So, my experience of the other is not a consciousness of another object; rather, my experience of the other is embodied.

As Merleau-Ponty describes it, the unity of my experience of my body is possible through my body's relationship to other bodies. All my experiences are bound together within my body "in such a way as to make up with them the experience of one sole body before one sole world, through a possibility for reversion, reconversion of its language into theirs, transfer, and reversal, according to which the little private world of each is not juxtaposed to the world of all others, but surrounded by it, levied off from it, and all together are a Sentient in general before a

Sensible in general" (VI 142). Just as all of my singular experiences are linked to each other, so too are my experiences linked to the experiences of others. Merleau-Ponty asks "now why would this generally, which constitutes the unity of my body, not open it to other bodies? . . . Why would not the synergy exist among different organisms, if it is possible within each?" (VI 142). He maintains that it is possible to think about this kind of synergy between bodies only if we give up the notion of "belongingness to one same 'consciousness,' the primordial definition of sensibility, and as soon as we rather understand it as the return of the visible upon itself, a carnal adherence of the sentient to the sensed and of the sensed to the sentient" (VI 142).

Merleau-Ponty describes the reversibility of the flesh through which flesh folds back onto itself and makes all communication with the world and others possible. We are not consciousnesses cut off from the world trying to find an adequate bridge to our objects. Rather, we are part of the flesh of the world; and we are one body by virtue of our relations with others. The reversibility of the tangible opens up an "intercorporeal being" which extends further than any one individual and founds the "transitivity from one body to another" (VI 143). There is no longer a problem of an alter ego because egos are merely by-products of fleshy relations. The thickness of the flesh guarantees relations, while the skin insures that we can distinguish our experience from the other's. Yet, since the flesh and skin are not objects, but synergies, we are never cut off from the other. The skin is a boundary, but a permeable boundary.

Merleau-Ponty suggests that I can almost experience something of the other's embodiment in the same way that I experience my own embodiment. He gives the example of the difference between hearing my own voice and hearing the voice of another. I have a different relation to my own voice because it emanates from my body and I am affected by the vibrations of my body as I speak. "But if I am close enough to the other who speaks to hear his breath and feel his effervescence and his fatigue, I almost witness, in him as in myself, the awesome birth of vociferation" (VI 144). At this point, Merleau-Ponty suggests the radical notion that I can feel the other's pain. There can be an exchange of synergy between bodies and if I am close enough to another person, I can experience the movements of her body in the same way that I experience the movements of my own. More than this, I can feel her effervescence and fatigue. The boundary between self and other has been opened. It is through the reversibility of the flesh that visible meets invisible (see VI 152).

SARTRE

Influenced by German philosophy, particularly the work of Hegel, Husserl and Heidegger, Sartre devoted a large section of *Being and Nothingness* to a phenomenology of intersubjectivity. Going beyond Heidegger's suggestions of a fundamental being-with others, Sartre rejects Husserl's notion that the other is known by analogical reasoning from my own experience: ". . . the Other is given to me as a concrete presence which I can in no way derive from myself and which can in no way be placed in doubt nor made the object of a phenomenological reduction or of any other epoché" (BN 362–363). Contra Husserl, Sartre denies that the other is ever given in my experience (BN 309). Sartre describes a dialectical movement between self and other that constitutes my self awareness through the look of the other. He maintains that I see myself only when I see another person looking at me; once I am captured by the look of another person, I am revealed to myself. Yet, as he describes it, this dialectical movement is not teleological or linear, but circular; neither the experience of self nor the experience of other are primary, but they are co-constitutive.

For Sartre, the primary experience that brings me to myself is the experience of shame that results from being caught in the act by another person. Although he insists that both pride and shame are the result of the look of the other, he does not analyze pride. Instead, he spends much of his discussion describing shame in front of the other. In fact, as it turns out pride is only the antithesis of shame; even pride is derived from the primary feeling of shame: "Pride does not exclude original shame. In fact it is on the ground of fundamental shame or shame of being an object that pride is built" (BN 386). In his analysis of concrete relationships with others, Sartre devotes most of his discussion to feelings of conflict; even love is ultimately the experience of conflict (BN 477).

Like the Hegelian theory of the onset of self-consciousness through the master-slave relationship, Sartre's theory is premised on the claim that even in concrete relations each person is attempting to enslave the other: "while I seek to enslave the Other, the Other seeks to enslave me. We are by no means dealing with unilateral relations with an object-in-itself, but with reciprocal and moving relations. The following descriptions of concrete behavior must therefore be envisaged within the perspective of conflict. Conflict is the original meaning of being-for-others" (BN 475). As Sartre describes the conflict, it is a conflict over freedom. The other both takes my freedom, possesses it, and is the catalyst for my recognition of my freedom (BN 473). I am imprisoned by the look of the other, yet through the look of the other I am aware of myself as a subject; by being an object for another, I become aware of myself as a subject for myself; I become aware of myself as a subject who escapes objectification, even my own attempts at objectifying myself.

The other can exist in different capacities, but in so far as it is dialectically engaged with my own consciousness of myself, it exists primarily as the bearer of the look. In fact, it exists as the look itself since in the look the other recedes; the eyes of another person give way to the look. When I am caught in the act, I don't see the eyes of the other: "The other's look hides his eyes. He seems to go *in front of them*" (BN 346). I only see the accusing look. This look is totalizing. Although the other is not given to me as an object of my experience, my experience of the other is a total experience, an experience of the world. Sartre says that "the world announces the other to me in his totality and as a totality . . . the other's presence to me as the one who produces my object-state is experienced as a subject-totality" (BN 390). I am overwhelmed by this other, who within this experience is all there is.

For Sartre, in spite of the totalizing presence of the other, the experience of the look, the experience which totalizes the other, is always *for me*. The look of the other always refers back to me: "The look which the *eyes* manifest, no matter what kind of eyes they are, is a pure reference to myself" (BN 347). The other's look is always directed *at me*. In this regard, Sartre follows Husserl's notion of the intentionality of intersubjective relations. Unlike Husserl, Sartre gives up the notion of the transcendental ego but attempts to retain the notion of intentionality free from the transcendental ego (TE 38–40).

In *Transcendence of the Ego*, he suggests that he replaces the Husserlian egological structure with the notion of consciousness as a nothing (TE 93). Consciousness is not a type of being determined in advance. Consciousness aims at non-being; it aims at what has yet to be, it projects itself into a time yet to come. This is the essence of human freedom:

> Human reality is free because it is *not enough*. It is free because it is perpetually wrenched away from itself and because it has been separated by a nothingness from what it is and what it will be. It is free, finally because its present being is itself a nothingness in the form of the "reflection-reflecting." Man is free because he is not himself but presence to himself. The being which is what it is cannot be free. Freedom is precisely the nothingness which is *made-to-be* at the heart of man and which forces human-reality *to make itself* instead of *to be*. As we have seen, for human reality, to be is to *choose oneself* (BN 568–569).

Sartre proposes a radical notion of freedom in which each human being is responsible not only for his or her actions but also for his or her emotions, beliefs, and desires. Our freedom and our responsibility are not limited to voluntary acts (BN 572). We are responsible not only for what we do but also who we are, what it is to be human: "In choosing myself, I choose man" (EH 37). Indeed, in choosing myself, I choose the world, since for Sartre all truth, meaning, and value have their source in human subjectivity (see e.g., EH 32, 34, 61).

Although in his discussion of freedom in *Being and Nothingness*, Sartre stresses an individual's freedom and responsibility, we know from earlier sections that the other plays a crucial role in my freedom. It is the look of the other which gives rise to my own recognition of my freedom. The look of the other turns me back on myself so that I can see the way in which I always escape myself; I see my possibilities. The look of the other confronts me with the nothingness at the core of my being when I see myself as an object for the other and realize that who I am is constituted in that look at the same time that I refuse to be reduced to that look. I both am and am not the object seen by the other. So begins my struggle with the other which supports Garcin's conclusion in *No Exit* that "there's no need for red-hot pokers. Hell is other people" (NE 47).

LEVINAS

In an important sense, the work of Emmanuel Levinas is motivated by the Husserlian problematic which insists on a sharp distinction between my experience and the experience of the other in order to guard against either everything becoming part of me or my experience becoming lost to the other. If Husserl tends toward the pole of the *I* over the other, Levinas tends toward the pole of the other over the *I*. Levinas insists on the radical alterity of the other such that I cannot presume anything about the other from my experience of myself; indeed, I cannot know the other at all. Recall that for Husserl, although my experience of the other is based on my experience of myself, I do not know the other through reason; rather, I experience an analogical transference to the place of the other. This analogical transference is unique amongst my cognitive experiences. Merleau-Ponty criticizes the Husserlian transference because it is disembodied. He insists on an embodied experience of the other which, like Husserl's transference, is based on a relationship to the place of the other. For Merleau-Ponty, however, the place of the other is the location of the other in space, and not the location of the ego of the other. Physical proximity, especially intimacy, allows for my experience of the other.

Like Merleau-Ponty, Levinas develops a philosophy of intersubjectivity based on physical proximity. Unlike Merleau-Ponty or Sartre, Levinas maintains that this relationship is not a matter of intentionality. Husserl's notion of intersubjectivity and consciousness itself is centered on the notion of intentionality. Whereas Heidegger takes over the Husserlian notion of intentionality in *Being and Time* with his notion of care, and Merleau-Ponty and Sartre embrace Husserl's notion of intentionality even while challenging his notion of the transcendental ego, Levinas rejects the notion of intentionality as auto-affection and self-possession (OTB 111). Levinas responds that our relationship to an other, to our neighbor for example, is not a Husserlian "consciousness of." To describe relations between people in terms of consciousness is to reduce them to knowledge; another person becomes an object of knowledge for me.

Levinas objects that "the hypothesis that the relationship with an interlocutor would still be a knowing reduces speech to the solitary or impersonal exercise of a thought, whereas already the kerygma which bears its ideality is, in addition, a *proximity* between me and the interlocutor, and not our participation in a transparent universality. Whatever may be the message transmitted by speech, the speaking is contact. . . . Speech and its logical work would then unfold not in knowledge of the interlocutor, but in his proximity" (LP 115). Like Merleau-Ponty, Levinas insists on proximity; but unlike Merleau-Ponty, Levinas also insists that proximity is not intentional. Proximity cannot be reduced to "a noetico-neomatic structure" (LP 119). Proximity engenders sensibility, tenderness, caresses, none of which are reducible to subjects knowing or intending objects.

Like Sartre, Levinas insists that my experience of the other as other is never an experience of an object. But, whereas Sartre describes the experience of the other as an experience of a subject-totality, Levinas describes the experience of the other as a nontotalizable experience of infinity. Sartre emphasizes a radical subjectivity at the center of the universe, while Levinas emphasizes a radical alterity at the center of the universe. For Sartre, all meaning is the product of human subjectivity, while for Levinas all meaning comes from the other. For Levinas, subjectivity is a relational product of the other, who cannot be experienced as an object of my consciousness.

Although Husserl does not maintain that another person or consciousness is an object of rational knowledge, he does posit that another consciousness becomes an intentional object of my consciousness through analogical transfer. My consciousness is the active agent in the intentional relation to the other. In "Language and Proximity," Levinas points to the passivity upon which the seemingly active Husserlian subject operates. The fact that consciousness can intend objects is the result of what Husserl calls "passive synthesis" which insures that various moments of consciousness are synthesized in time; the result is what Husserl calls internal time consciousness. What Levinas argues is that if consciousness is dependent upon a passive synthesis, then it cannot be intentional: "Consciousness as the passive work of time which no one activates cannot be described by the categories proper to a consciousness that aims at an object" (LP 114). Active, intentional consciousness is dependent on this passive nonthematic consciousness.

In *Otherwise Than Being*, Levinas points to another nonthematic element of Husserl's thematic intentional consciousness. There, he maintains that within Husserl's notion of empty intentions and the possible fulfillment of intentions there is a desire in excess of intentionality itself. If intentionality is merely the thematic by which everything is a consciousness of, then what is it that motivates the fulfillment of intentions? What motivates thematization? Levinas identifies the Husserlian movement from empty to full intentions as desire, tendency, hunger, which are nonethematic and nonintentional (OTB 66).

Levinas suggests that the Husserlian notion of intentionality cannot account for sensation and that the relationship with others begins in sensation (see for example OTB 67). The relationship with an other begins not with intentional consciousness, but in the proximity of two bodies. As Levinas says, "only a subject that eats can be for-the-other, or can signify" (OTB 74). Embodiment is necessary for any thematization or signification, not just in the obvious factual sense that we need eyes to see and mouths to speak, but in the ontological sense that signification is a response to "an infinite which commands in the face of the other" (OTB 97). This infinite which calls for a response cannot be aimed at; it is not an intentional object for consciousness. Thematization by intentional consciousness "undoes" this approach of the other which calls for a response (OTB 94). This is why Levinas insists that ethics, rather than epistemology or ontology, is the first philosophy. The relation that makes signification possible can only be described as an ethical relation (OTB 94).

This relation which makes signification possible, whose necessary consequence is signification, is a relation of responsibility. Subjectivity and signification – these two cannot be separated – are the result of responsibility to the other: "The knot of subjectivity consists in going to the other without concerning oneself with his movement toward me. Or, more exactly, it consists in approaching in such a way that, over and beyond all the reciprocal relations that do not fail to get set up between me and the neighbor, I have always taken one step more toward him – which is possible only if this step is responsibility. In the responsibility which we have for one another, I have always one response more to give, so I have to answer for his very responsibility (***Moi** j'ai toujours une réponse de plus à tenir, à répondre de sa responsabilité* (AE 134)) (OTB 84).

It might be helpful to note that the term responsibility (*responsabilité*) is from the Latin *respondere*, to respond, to answer, to answer to one's name, appear to be present, and *responso*, to keep answering. Subjectivity is fundamentally intersubjective if I *am* by virtue of my response to the other. I am by virtue of my responsibility to answer, even for the other's very answerability. This responsibility to answer is not a moral responsibility and it cannot be described in terms of ethics (OTB 120). Rather, this responsibility is an ontological responsibility that makes the ethical relation with an other the foundation of ontology.

My subjectivity is continually constituted through my response to an other. This response is not a Sartrian defense of myself against the other nor a Hegelian struggle for recognition. Neither is it Husserl's or Merleau-Ponty's exchange of places with the other.

> But communication would be impossible if it should have to begin in the ego, a free subject, to whom every other would be only a limitation that invited war, domination, precaution and information. To communicate is indeed to open oneself, but the openness is not complete if it is on the watch for recognition. It is complete not in the opening to the spectacle of or the recognition of the other, but in becoming a responsibility for him. The overemphasis of openness is responsibility for the other to the point of substitution, where the for-the-other proper to disclosure, to monstration to the other, turns into the for-the-other proper to responsibility. The openness of communication is not a simple change of place, so as to situate a truth outside instead of keeping it inside oneself . . . [I]s not the ego a substitution in its solidarity as something identical, a solidarity that begins by bearing witness of itself to the other? (OTB 119)

The response commanded by the other is to answer for the appearance of myself as a solitary ego. The response "here I am" is not only to answer for one's presence but also to turn oneself over to the other. The biblical "here I am" not only responds with the location of the *I* but also turns the *I* over to the other, God (cf. OTB 199, fn 11). The *I* is by virtue of becoming a hostage to the other: "Subjectivity is being hostage" (OTB 127). The responsibility that calls the I to the other in this way is not an action taken or chosen by the subject. Rather, subjectivity itself – the possibility of action or choice – is constituted through this responsibility that brings the *I* into being by taking it hostage, calling it forth (see OTB 117). "The unconditionality of being hostage is not the limit case of solidarity, but the condition for all solidarity" (OTB 117).

The University of Texas, Austin

NOTES

[1] This quotation is from unpublished manuscripts in the Husserl Archives in Louvain BI 14 XIII, 27; quoted in James Hart, *The Person and the Common Life* (Boston: Kluwer Academic Publishers, 1992), pp. 3–4.

[2] James Hart, *The Person and the Common Life* (Boston: Kluwer Academic Publishers, 1992), p. 33.

[3] It is interesting to note that Husserl's *Cartesian Meditations* were delivered after he had read Heidegger's *Being and Time*. In fact, passages in *Cartesian Meditations* read like a response to Heidegger's criticism of phenomenology in *Being and Time*. For example, on a page that Husserl inserted after he had delivered the lectures, as a response to Heidegger's criticism of the transcendental attitude, he says: "The all-embracing constitution of the world within the ego is outlined as a problem only/as far as the theory of clues – as far as the consideration of the world (that is to say: the ontological consider-

ation thereof, as transformed into a constitutional-ontological consideratoin). Somewhere in that context, naturally, the problem man must present itself? But what is the proper order?

The first procedure in Meditations I–IV is to awaken the guiding thought: The world is a meaning, an accepted sense. When we go back to the ego, we can explicate the founding and founded strata with which that sense is built up, we can reach the absolute being and process in which the being of the world shows its ultimate truth and in which the ultimate problems of being reveal themselves – bringing into the thematic field all the disguises that unphilosophical naiveté cannot penetrate" (CM 52 footnote 1).

DANIELA VERDUCCI

LIFE, SPIRIT, WORK

Notes on Erkenntnis und Arbeit *by Max Scheler*

1. METAPHYSICAL ROOTS OF THE TRANSITION TO MODERNITY, ACCORDING TO SCHELER

From the first pages of *Erkenntnis und Arbeit*, dedicated to examining the possibility of interwoven links between cognition and work,[1] Max Scheler affirms almost without hesitation the presence in the modern era of a model of "intimate connection between science and the technical-mechanized form of production of goods," or "a systematic and internal connection between work and cognition within our society."

By this he means that "the specific nature of modern Western experimental science, sustained by mathematical theories" and "the nature of modern organized entrepreneurial capitalism and of its economic system" have become reciprocally congruent and integrated, so that modern Western knowledge and economics are separated and divided "from the entire past and contemporary world." On the one hand, he judges, our modern science separates itself sharply from "all forms of contemplative 'science'," including Chinese, Indian and Greek sapiental teachings, while on the other hand, the capitalistic economy clearly supersedes every system of pure satisfaction of needs and of craftsmanship's direct tradition, introducing, vice versa, a methodical study of the work experience on things, which can then be applied systematically to other fields and in its turn can evoke new needs and produce an incessant knowledge-experience-work dynamic.[2]

Thus for Scheler, modernity is characterized by a unitary plexus in which the two anthropological dimensions of theory and practice are closely connected, while other *Weltanschauungen* regard these dimensions as clearly distinct.[3] In addition, the link between theory and praxis is so intimate here that one can hardly identify the *primum movens*; it is unclear if at first "the modern technique of work and production of goods consists exclusively in the practical use . . . of a knowledge of nature . . . which is purely theoretical," or whether, on the contrary, the start of the new epoch is due to the emergence of a "consciously (or unconsciously) held instinct" (*Trieb*) and of "a will to control nature"

A-T. Tymieniecka (ed.), Analecta Husserliana, Vol. LIV, 141–165.

which only later would have produced modern science, as a systematic container for the experiences accomplished "acting on nature and shaping it" (*im Handeln und Bilden an der Natur*).[4]

Naturally, the dynamics in question are not individual or subjective, as if the concomitant or even collective will of some researchers had unleashed an epoch-making transformation.[5] Scheler relates the observation of Liebig, that "precisely those knowledges of the laws of nature which lend themselves to the most fruitful technical use were not sought and discovered with this utilization in mind, but resulted exclusively from the very logic of the method in its purely theoretical aspect." Confirmation of this singular connection between theoretical intent and technical usefulness is found in Liebig's own chemical knowledge, in Weber and Gauss' telegraphic experience, in Moritz Schlick's *Allgemeine Erkenntnislehre* notes and in Poincaré's maxims.

Besides, Scheler continues, even Herbert Spencer taught in his *Etica* that "man's *maximum fortunae* constitutes the objective criterion of 'good' and 'evil'," provided that the single man does not render it unnatural, taking it on subjectively as the motivation and objective of his action; rather, he recognizes it in "those rules of justice" which are "innate and immediately evident" to him. Only in their objectivity can these rules guide human individuals to attain the objective of the *maximum fortunae*, because only as such do they truly represent "the sediment of humankind's accumulated experiences of success and failure, of pleasurable and unpleasurable results of his activity."[6]

If the cognition-work synergism typifying modernity is reflected in the psychology of the researcher, who provides a useful confirmation of a general situation, it is also true however that this individual and subjective level in and of itself does not offer elements which explain the conditions of possibility which enable theoretical-spiritual acquisitions to translate themselves systematically into practical uses, and, vice versa, which constantly cause practical experiences to modify ideal systems. Nor does the recent speculative texture, levelled on the particular sciences, seem to be able to open adequate horizons of sense.[7]

Scheler finds a resolutive opening precisely in the difficulty just discussed, that is, the difficulty of finding in the conceptual context inherited by modernity the *raison d'être* of the new dynamic that incessantly transforms work into cognition and cognition into work. In the circle of theory and praxis, in whose closed environment after Hegel and Marx it seemed that every possible interpretation of the world had to run dry, Scheler discerned no longer an ultimate character, but rather a value of adjournment:[8] in fact, in the events of western modernity, compared to those of other regions or epochs, a strange inversion of sense has occurred, so that "the maximum objective utility of knowledge in a practical sense" was achieved, even though "knowledge as a value in itself" was exclusively sought.[9] In other words, the more rigorously the theoretical stance was followed, the more adequately the practical achievement was obtained, as if the difference between the two dispositions of the human being were only the appearance of a much deeper substantial unity.

This fall into circularity by the evolutionary linearity of theory and practice remarked upon by Scheler, on the other hand, could force recourse to explicative models based on dialectic: to idealistic models, to historicist models or to those of philosophy of praxis which can all integrate the differences into a unity of procedure. But Scheler sees another possibility: there may be a structural and original link between cognition and work which goes beyond the anthropological environment in which both activities take place. At the same time, it is not thinkable that the essential relationship between cognition and work of which Scheler has just a hazy notion can permit him to use as an exhaustive interpretive reference any ideal context with monistic-immanentistic tendencies, even if its form is dialectical, such as that of Hegel or of Marx. In other words, Scheler cannot use any ideal context in which differentiation presses toward overcoming and systematic synthesis rather than opening research for a common root which, whilst gathering into unity that which is multiple, at the same time preserves the different individualities.[10]

Therefore, the problem posed by the modern configuration of cognition and work in a unitarian and synergetic plexus, seems because of its radicality to stretch the limits of psychological, social and physical sciences and to demand a metaphysical study of the question. Perhaps the moment has come to give work activity, which in appearance is so closely calibrated to the existential limitations of human beings, the rank of universal and absolute device which continually readies the *trait d'union* between the two metaphysical forces of spirit and of life – or, that is, between the two attributes known to us of the one ground of the world (*Weltgrund*), as Scheler expresses it.[11] In this way, work activity, placing eidetic totality materially in being, transforms it into reality. Thus Scheler can propose an itinerary of completion for previous metaphysics, be they classic, rationalistic or idealistic, which limited themselves to contemplate the ideal work, or at best, to reproduce it theoretically.[12]

2. SCHELER'S PHENOMENOLOGICAL METHOD

Scheler owes to the rigor of the phenomenological vision his ability to trace philosophical features and values of a phenomenon such as work, traditionally relegated to the particular realm of satisfaction of needs and of the related socio-economic sciences.[13] However, he succeeded above all because of those methodological elements that clearly distinguish his way of proceeding from that of Husserl, and which drew him to get involved in the essential description of work activity even before the publication of Husserl's *Logische Untersuchungen.*

Scheler favors the ethical-anthropological realm as the field of application for the theory of intentionality drafted by Brentano, Meinong and Ch. von Ehrenfehls and borrows from the early Husserlian *epoché* procedure, which enables the phenomenologist to establish a bridgehead in the ideal world without excluding continued nourishment from the lymph of the real world.[14]

In fact, Scheler wants to construct realistic metaphysics, following the path of ethics[15] rather than the traditional naturalistic road. For this reason, he intends to forge for himself a method which not only leads to the reign of the essences, but above all guarantees the continual ebb and flow of living experience from the ideal world to the real world, and vice versa. As expressed in the categorical affirmation of *Vom Ewigen im Menschen. Probleme der Religion,* "our spirit possesses neither innate nor inborn ideas. . . . Even the idea of *ens a se* presupposes the cognition of some contingent entity and becomes comprehensible only in terms of such an example, namely as evident condition of any contingent being."[16] Because of this underlying need Scheler did not follow Husserl in that worsening of eidetic reduction into transcendental reduction, which the latter's so-called "idealistic turnaround" ensued from.[17]

In particular, according to Scheler, the Husserlian technique of reduction needs to be radicalized. In fact, Scheler thinks that Husserl, to gain access to the essential dimension of experience, would have been content to use *epoché* on the judgment of the world's existence, forgetting all that his own phenomenology had taught; that is, that the logical moment is not the final moment, because every judgment is founded on an intuitive living experience which, for its own part, continues to be active even after having placed in parenthesis the judgement which expresses it.[18]

On the other hand, Scheler's own speculative interests, directed more towards the moral world of liberty than to that of cognition, draw him to return to Husserl's phenomenological reduction and to go to the bottom of it, not so much to achieve a further decantation of the eidetic datum, as to vigorously undertake the topic of all the living experiences, which in the very same reduction give themselves, in order to acquire phenomenologically their dimensions of sense.

Applying this orientation to the judgement of existence that belongs to the natural and scientific vision of the world, Scheler reveals its intuitive fulfillment in the living-experience and employs, on the latter, phenomenological *epoché*.[19] His phenomenological *epoché*, however, can no more reduce itself thus to a pure thought process, which only places a judgement in parenthesis: it must have the value of a true *Techné*, which conquers forces that oppose it and, inhibiting the acts sustaining the moment of reality, promotes an effective abrogation of this moment, subtended to the judgement of existence.[20] Thus Scheler's phenomenological reduction is entitled to open two worlds and not just one, like Husserl. In fact, on the one hand this phenomenological reduction, putting off the instinctive forces of vital reality that pulse below judgement of existence, attains the true de-realization of the natural world and leads to the imperturbable ideal reign of the essences which is the object of phenomenological description. However, on the other hand, and precisely by means of the same inhibitory acts which compose the Schelerian *epoché* process, this total reduction also leads to manifestating a world which is "other" than the preceding world, that is, the world of life, in its tumultuous quality of resistant opposition to the spirit.[21]

The metaphysical gain from such a phenomenological proceeding seems at first dualistic: in fact, two forces are led to manifestation as those controlling the space of the living experiences and these forces have shown that they constitute two worlds. However, the situation thus described phenomenologically is no longer a dualistic situation-of-being. This situation shows itself more exactly as a transitional situation-of-being which utilizes the "work" factor, as a dynamizer which assures the continual self-production of the reign of the essences from the material of the instinctive living experiences, as much as it assures, vice-versa, the incessant reflux of ideal forms on the indeterminate flow of vital reality.

Therefore, Scheler can shed light on the metaphysical quality of work, because he carried out the integral phenomenological reduction where the work activity shows itself to secure the continuous transit between the world of essences and that of life. At the same time, he can denounce the insufficiency and the unilaterality of every static metaphysic, such as traditional theism, which considers the unification of spirit and life not only as already having happened, but also as being already philosophically earned; in addition, he can plausibly propose his own metaphysical reform, that is, the introduction of a theism *in fieri*.[22]

3. NEW RELATIONSHIPS BETWEEN COGNITION AND WORK

a) *The Pragmatistic Knowledge-Work*

The phenomenological acquisition of a double field of cognition, that is, the ideal world and the real world, offers Scheler an opportunity to further explore the meaning of that closed and apparently self-founding circularity that he believes the theory-praxis relationship forms in modernity.[23] This is even more necessary if one considers that recent philosophical pragmatism has persuasively extended the practical-empiricist reduction of the ideal moment of every cognition and action to the anthropological field, proposing the understanding of *homo rationalis* as *homo faber*.[24]

In fact, in order to make our ideas more clear, Peirce hypothesized a pragmatic simplification of cognition, according to which the sense and meaning of a thought can be found by simply "identifying the procedure capable of forming that thought."[25]

James further simplified and amplified this beginning, affirming:

> The concrete reality which underlies all our ideal distinctions, no matter how subtle, is the following: none of these distinctions is so subtle that it can consist in something other than a difference which is capable of influencing action. Thus, in order to make perfectly clear our idea of an object, we must only consider what practical effects this object contains in itself, that is, which perceptions we must wait for and which reactions prepare. Then, our representation of these effects, be they mediate or immediate, constitutes for us the whole representation of the object, in the measure in which this idea has in general a positive meaning.[26]

In James' radical version, therefore, the sense and meaning of a thought lie in its having practical consequences, so that the proposition "Material matter with its movement is the origin of all things" is equivalent in terms of sense and meaning to the proposition "God is the origin of all things," since notwithstanding the terminological diversity, the two propositions entail no diversity of consequences; that is, James concludes in the texts reported by Scheler, that the same kind of life and action can descend from both thoughts.[27]

Scheler observes that James' affirmation of the practical principle of the distinction of ideas expresses in pragmatist terms a cognitive criterion already present in Aristotle, Locke and Berkeley, and called by Leibniz the "*principe d'observabilité*."[28] But here the matter at hand is not a principle among other principles, but the one criterion for cognitive distinction.

In fact, Stuart Mill defines this same material matter in the limits of the "totality of all ordered perception possibilities" and excludes the meaningful use of the notion of substance, considered a superfluous concept: "Posit or do not posit the existence of a material substance, independently of our possible perceptions or sensations – in both cases nothing will modify itself in our experience. Thus the assumption of one such substance is senseless, being superfluous, that is, not modifying in anything the context of the possible experience."

Scheler continues that Mach also "was used to liquidating with the same principle all 'metaphysical' questions, in as much as he considered them 'senseless,' repeatedly noting that all questions either should be proven solvable by a possible observation or should be abandoned as senseless questions."[29]

The peculiar totalitarian and radicalizing pragmatic version of the principle of observability reveals itself above all in its extension to the laws of logic, since here the necessary observation of the consequences is not immediate, but its realization requires the preparation of an adequate genetic theory of the laws of thought. In fact, only by casting light on how the latter are in turn the consequences of an acting on the world, is it possible to return them from the logical plane, eccentric in comparison to the pragmatist verifiability, to the practical plane, the only place in which they can give themselves sense and meaning understood pragmatistically. In the pragmatistic genetic interpretation, sustained with the motorial theory of perception, the laws regulating thought processes have taken form "gradually, according to the criterion of 'success and failure,' during action on the world, beginning with an experience by attempts – not with the things themselves, but with signs in the place of things –; at the same time, with the accumulation of results, fruitful roads of thought have slowly consolidated themselves and, either by social tradition (James' opinion) or by blood heredity (Spencer's opinion), these paths have transferred themselves, according to the principle of heredity, to functionally acquired psychic qualities as well."[30]

In conclusion, according to Scheler, "the pragmatist thesis attributes to all types of theoretical acts, thus in equal measure to intuition, perception, remembering and thinking, its 'principle' by which every knowledge, genetically, is only the result of a kind of interior acting and preparation for the transformation of the world": therefore, according to this principle "from the point of view of teleology and of the theory of values as well, every knowledge must serve action."[31]

Returning the constitution of the same logical laws of thought to the practical plane of their formation process, pragmatism has achieved its original intent, directed at promoting a knowledge-of-work, that is, a knowledge that totally adheres to facts, and directed to expunging from cognition every empirical reference, both transcendental and ontological: in fact, by now sense and meaning of every thought derive exclusively from its "having consequences," that is, from its being an action-of-work that consists, as origin and as development, in the manipulation of the world. Neither a sense of judgement different from the act of judgement, nor "facts" pre-existing cognition can any longer present themselves since, in the absence of an extra-process causal horizon, the "sense" as much as the "facts" must necessarily be held to be produced by the same knowledge-work dynamic, in which they manifest themselves and come to the light. Therefore, according to Scheler, in the pragmatistic vision "the world, in as much as it is unknown, is similar to a 'plastic mass,' to an entirely indeterminate 'Hyle,' that man forms for himself, into a world at the same time of sense and of facts (like a cosmos)" in this inseparable dynamic process of ideal project/action-sensation-observation/new action.[32]

Thus, according to Scheler, philosophical pragmatism fulfills modernity's own vision of an immanentistic world. Modernity, with the pick-lock of voluntarism, first released the processes of human rationality from contiguity with the divine creator logos and then, introducing the mechanical interpretation of nature and extending it to the maximum, made scientific inquiry and technical intervention ever more autonomous and absolute, to the point that the very human essence, operator of these activities, appears transformed from thinking to maker.[33]

Nevertheless, precisely from the results of such inquiries, in virtue of the radicality with which pragmatism pursues and conducts the practical-immanentistic reduction of knowledge, Scheler uncovers some metaphysical truths: these agree particularly with his phenomenological acquisitions, but they are stubbornly denied by other philosophical positions seemingly closer to his way of proceeding than the pragmatistic positions.

Above all, against every spiritualistic purism, pragmatism demonstrates energetically that "man's – and every organism's – primary stance in terms of the world is not theoretical at all, but practical; therefore every natural vision of the world is guided and sustained by practical motives."[34] Thus, Scheler continues, ". . . with regard to the technique for the production of a purely theoretical relation of man with the world, all these motives, efficacious in the area of the formative law of the natural vision of the world, must be eliminated piece by piece with their roots, by means of a particular spiritual technique – and this first of all considering a 'philosophical' treatment of things."

In addition, for pragmatism, "also the highest cognitive ends (*Erkenntnisziele*) of the positive science of nature and of descriptive psychology are practically co-conditioned – even though these sciences must strive to exclude every specific anthropomorphism."[35] This means that if the ends of science must be sought inside the process constituting the sciences, not even pragmatistically configured knowledge can do otherwise. Only by seeking and placing terms of reference or ends, in the fluid process of experiential Hyle, in fact, can pragmatistic cognition produce the world of sense and of facts, without needing to resort to any transcendence.

However, as long as finalization remains in anthropological terms, the full identification of thought with its practical consequences cannot be achieved; therefore it is also necessary to untie the idea of man and the connected gnosiological anthropomorphism, leaving at being only one factual element that permits the construction of knowledge in the natural and human sciences practically, but not anthropomorphically: in this way pragmatistic reductivism attains its own irreducible and extreme term, or in other words, must recognize as its ultimate, inevitable reference, biological essence or life, to which consequently each phenomenon, both act and fact, must relate itself, since now ultimate meaning consists in being a consequence which can be experienced vitally.

Thus Scheler argues:

since no thinkable living thing – no psycho-physical organism – can practically and directly change its nature, that is, by its own amechanized spontaneous motion, positive science takes interest in and has affective preference for that which can be set in motion by a 'living thing' in general – set in movement directly or indirectly – therefore for that which in nature can be guided, commanded and impeded.[36]

In other words, according to Scheler, pragmatism, in seeking to make cognition adhere perfectly to facts and to eliminate every metaphysical obscurity, has reached the opposite, a gnoseologic dimension of metaphysical depth. Even while pragmatistically oriented inquiries employ knowledge of facts and a science of work, they end up identifying a very general and irreducible level of being: that of life, indispensable also for the constitution of a knowledge that wants nothing to do with metaphysics. In such extreme and ultimate being, termed by Scheler "biological essence" or "living being," pragmatistic knowledge finds, in fact, the foundation for achieving its own originating intention of control over the world.

Pragmatism has thus finally clarified that "science leads 'nature' back, as much as it is possible, to a prototype of formal 'mechanism' – not because nature in itself would be only a mechanism, but because it can be controlled and guided, even practically, by a living entity disposed to control, only in as much as it is a mechanism or it is widely analogous to it."[37]

Therefore Scheler can willingly grant the pragmatistic concept of knowledge a "relative rightness,"[38] since in his opinion the pragmatistic concept of knowledge neither claims ontological value for the mechanistic interpretation of nature (it even explicitly declares its motivating desire to control), nor excludes the metaphysical dimension, which it somewhat arrives at, even if inadvertently and partially, that is, from the side of "life."[39]

The pragmatistic reduction of knowledge of ideas to practical-work knowledge, manifests a validity of its own, albeit limited to the vital side of cognition, and Scheler even feels that he must formulate a "relatively pragmatistic" interpretation "of all the formal-mechanistic theories of nature"[40] enabling him to welcome them into the whole field of human knowledge: to Scheler, they constitute "the first foundation for the possible attainment of the ends of *homo sapiens* in man – and in particular for the attainment of these ends by the greatest possible number of men." In fact, in every formal mechanistic inquiry, only practical intelligence regarding the possible domination and control of the world is active, by means of vital movement: this intelligence wants nothing other than to be the supreme guide of human industry and technology and thus one disdains it only if one measures it with the false criteria of a purely theoretical cognition or of a knowing participation in nature, which do not appertain to it;[41] but neither do the latter constitute the totality of knowledge's virtuality.

On the other hand, it should be noted that also according to the just-described point of view regarding the vital relativity, or relativity to biological essence of mechanical-formal knowledge, pragmatism works in synergy with the phenomenological reflections of Scheler, according to which the ideal world, conquered by eidetic reduction, remains nonetheless contiguous with the real work, or that of life, in the same way in which the essential structure of phenomena becomes manifest only after the spirit has conquered reality's resistance and has raised itself above the incessant flux of life.

b) *The Eidetic Knowledge of Phenomenology*

Face to face with the world of life which pragmatic knowledge grasps in its aspect of controllable mechanism, there is also, according to Scheler, "a vast reign of thoughts, meanings or propositions, or their consequences, which by virtue of their constitutive *nature*, can 'change' nothing of the world which can be observed and experienced." In fact, these "assume and understand the *constant* forms of existence and in addition the *ideas* and the original phenomena themselves, in which *all* that is observable of the world is incorporated as in the *essential structure* of every 'possible' world."[42]

Scheler feels that Edmund Husserl's *Logical Researches*, in seeking such an ideal world, already proposed in and of itself an important confutation of the pragmatistic theory of thought. In fact, they examined the typological variety of cognition, which cannot be identified exclusively with positive science, since "knowledge" and collocation in the system of the forms of cognition also appertain to philosophy, and more specifically to metaphysics.[43]

For his part, Scheler holds that knowledge in its most general meaning represents the end of every cognition, that is, that which every cognition pursues and contributes to build: knowledge consequently cannot be defined in purely gnoseologic terms, because one does not want to lose its specificity by reducing it to the process that produces it. In order to grasp the idea of knowledge as a relation of being, the use of an ontological language is necessary. In fact, in knowledge, an entity takes part (= *teilhat*) in the being-so (*So-sein*) of another entity without modifying in any way the causal, spatial or temporal position of the being-so of the entity known or of the entity that knows. Such cognitive participation comes about by means of a specific agent which the ancients called "*mens*" and the moderns call "*spirit*": it constitutes an *X* in the sense that it can only be identified as the totality (*Inbegriff*) of the acts which make the being-so of an entity the "*ens intentionale*" for another entity. Vice versa, being-there (*Da-sein*) or "*ens reale*" is always beyond and outside the acquisition of knowledge.

Nevertheless, even though the acquisition of any type of knowledge does not involve the manipulative of reality, establishing exclusively that relation of being which is the cognitive participation in being-so, this acquisition does imply, on one hand, that in cognition, the being-so that is learned remains identical both *extra mentem*, that is *in re*, and *in mente*, that is as *ens intentionale* or object. And on the other hand, this acquisition implies that the learning processes in such can lead to knowledge because they are rooted in a taking-part *sui generis*, which only love can be, capable of transcending itself and its own being.

So also in the case of cognition and knowledge, as in the case of the previously examined relationship of theory and praxis, Scheler's inquiry leads to the opening of an apparently closed circle, as that of the modern conscience which, while it attempts to take possession of objects, in reality closes in on itself ever more tightly like a screw. Scheler observes acutely that looking closely one sees that knowledge of knowledge or *con-scientia* already presupposes the possession of an ecstatic knowledge, since consciousness by its very reflexive nature can exhibit givenness only by reflecting on something other than itself. Therefore the same possible knowledge must find grounding in a real tendency, in the entity-that-knows, in order to come out of itself and open itself to participate in another, even, in the words of Scheler, "to explode the limits of its own being for love."[44]

All the acts of our spirit, the intuitive acts of the having-imagines as much as the cogitative acts which carry meaning, contribute to the configuration of this knowledge of the being-so, or this possession of the ideal being. In the moment in which meaning coincides completely with the totality of partial intuitions, mediated by the modal functions of the senses, in addition to memories and expectations that lived experience of coincidence takes place in, there "the thing in its 'own' being-so appears evident in the spirit in an ever more adequate manner." Knowledge is made up of these essential, immediate and immanent to the conscience lived-experiences-of-coincidence (*Deckungserlebnisse*). For this reason knowledge is distinguished from all the operations of thought, observation and cognition which lead to it.[45]

The world of knowledge to which Scheler has arrived by following his phenomenological itinerary seems to correspond to the eidetic world already opened by Husserl's *epoché*. However, Scheler is not content to stop with what he has achieved, but also poses the teleological question with regard to the knowledge of essences: to what end does one seek knowledge? The pragmatists hold that one pursues knowledge for love of the useful or of practical domination, while their opponents affirm "*la science pour la science*." Instead, the nature of the knowledge illuminated by Scheler's preceding inquiry permits an ontological definition of the purpose of knowledge: it must consist in a "becoming" (*Werden*) or better still a "becoming-other" (*Anderswerden*) or in an expansion and an increase of the spiritual being present, by means of the cognitive participation in the being-so of another being.

In particular, Scheler identifies the three highest purposes of becoming to which knowledge must and can be oriented: 1) becoming as a development of the person, which is served by formative knowledge (*Bildungswissen*); 2) becoming of the world and of its essential and existential grounding, which benefits from knowledge of salvation (*Erlösungswissen*); 3) becoming of the world, in the sense of its transformation according to man's objectives, made up of the knowledge of positive science which is the knowledge of control or of production (*Herrschafts-* or *Leistungswissen*).

A precise ontological hierarchy orders the three supreme purposes of becoming, which knowledge serves, leading from the knowledge of control directed to the practical transformation of the world and the production of its possible means, to formative knowledge which pursues the unfolding of a microcosm of being and of being-so of the spiritual person in us, furthering participation in the essential totality of the world, according to our irrepeatable individuality. Finally, formative knowledge is oriented to knowledge of salvation, in which our personal soul seeks participation in the Supreme Being which, as it is the Foundation of things, in Scheler's words "'knowing' itself and the world, in us and for us, even arrives at the point of becoming timeless – to a certain manner of self-unification, to liberation from a 'tension' from an 'original contradiction' placed in it."[46]

Thus Scheler does not grant knowledge a purpose unto itself, unlike those who assert "*science pour la science*"; according to Scheler, they end up promoting a knowledge relegated to the purely contemplative sphere, objectively without meaning and purpose: they unknowingly fall into the service of the practical-technical transformation of the world.[47] Vice versa, in a correct hierarchy, the knowledge of control finds its teleological reference exclusively in "formation" (*Bildung*), or in the "free expression of all the spiritual forces of the person and in the increase of the spirit (subjective and objective)," which, however, in order to fulfill itself needs "the functionalization of the knowledge of essences in people and by means of people." Such formation of the spiritual essence of the human being then, in its fulfilling itself, enlarges to the knowledge of salvation, as much metaphysical as it is religious, revealing the "being of the macrocosm in the being of the microcosm of the human conscience" and thus clearly separating itself from the unilateral humanistic closure of oriental formative knowledge, for example, hetero-Chinese.[48]

Then for Scheler, eidetic, philosophical, formative knowledge occupies a middle position between that directed to control of nature and that directed to ultimate realities. In fact, it conforms not only to the true-false parameter, which is valid for every knowledge formulated in judgements, but also to the a priori or essential criterion, activator of the energies of the spiritual personality, and to the absolute reality criterion, which opens it to the knowledge of salvation that is definitive metaphysics. So for Scheler not even the knowledge of the essences can be considered sufficient unto itself. It must get a rigorous genetic and evolutionary distinction from knowledge of domination which is essentially related to life, but thus also philosophy, if it wants to arrive at the finality which applies to it and to unfold in microcosm the spiritual energies that refer themselves to it, needs to "make itself functional" in people and by means of people. In other words, philosophy, established in the ideal dimension of the spirit and directed to the "existence, essence and value of the absolutely real in all things," can well serve as orientation center and provide the map of the essential ontological structures along which to aim spiritual activities directed at the formation of the universe. But, like every other knowledge, philosophy only concerns the being-so of things and is incapable by itself of intervening in the dimension of existence. It needs the intervention of an agent or a complete personal entity that knows how to "unify in one single style, superimposing on top of each other ideal dynamic schemata for intuition, thought, comprehension, evaluation, manipulation of the world and of any accidental object which may appear therein, making them functional."[49]

In this way eidetic knowledge manifests itself not only as not being an end unto itself, but also as incapable of self-realization: like every other knowledge, it is at the service of the becoming-other of being and its achievement depends on being-that-knows. In particular, Scheler notes "a fourth relationship-of-knowing" regarding nature, in addition to that established by formal mechanical science, by philosophy and by cognition in its close sense. It is a fundamental "relationship-of-possession" (*Habensverhältnis*) that appertains to man and to every living being and is directed to "*natura naturans*." In it takes place a "purely dynamic and emotional comprehension of nature as a field of expression of that total living that creates images, that is creator of images based on materializing dynamic factors and on the eternal unique 'impetus' of which these factors are partial functions, no less than the total life itself." The way of this last dynamic foundation of nature is accessible only "to the human heart, to warmth and to the most elevated case of uni-pathy and of uni-action, with its 'forces' qua forces (not with its conceptual determinations). Neither the intellect, nor sensitivity, neither comprehension of ideas nor the vision of original phenomena can discern it."[50]

c) *Cognition and Work*

Scheler confirms the rigorous belonging of cognition and knowledge to the sphere of ideal being, drawing as much from inquiry on the gnoseology of pragmatism, as from the teleological prospects identified for eidetic knowledge as well. In fact, the pragmatism of cognition only apparently seems to identify itself completely with the factual flow: in reality it succeeds exclusively in taking on the point of view of biological essence, which remains spiritual, but is the most radical and consonant to pragmatism's intention of manipulating the world. However, not even pragmatistic cognition can proceed beyond the point of view of biological essence, since abolition of that last ideal reference would cause cognition, consciousness and control of nature to fail in a single blow. In fact, positive science had the value of "*Herrschaftswissen*" to date only because of a theoretical point of view, accurately selected to be effectively imposed on reality. That is, science has formed itself taking on the vantage point of a psycho-physical organism which, faced with the impossibility of modifying nature directly and practically, or spontaneously, has concentrated attention on that which in nature, vice versa, could be put in motion by a living being in general, thus meeting with work, in as much as it is a transforming activity which the living being is provided with. For this reason, nature has been represented as the prototype of a formal mechanism, since, if it moves by itself following constant and mathematically expressible laws, it will be able to be directed and guided to achieve ends which are more suitable to human development, following precisely the same modality.

Conversely, philosophic knowledge and the cognitive proceedings which appertain to it have been taken out of the isolation of the immobile ideal world and returned, as elements of "*Bildungswissen*" and "*Erlösungswissen*" to the teleological dynamic directed toward the full explication of the forces that pervade and constitute reality, that is of the centers-of-force, the vital-centers and the personal-centers.[51]

Therefore, none of the three knowledges is inactive: rather, all collaborate to formulate for being, in which they cognitively participate, ideal transformation projects, images for its becoming. And yet these knowledges, even so practically configured, also continue to maintain themselves in the ideal dimension of the spirit, be it finite or infinite, regarding its ideas, original phenomena, values and being-so: that is, they can make accessible for us neither existence (*Dasein*) nor contingent being-so (*zufällige Sosein*). The latter, in fact, belonging to the real dimension, and thus according to Scheler, coming from the dynamic principle of impulse, blind to the spirit, announce themselves in the living-experience-of-resistance to our will rather than to our intellect.[52]

In this way the two worlds of essence and life already revealed by Scheler's radical phenomenological reduction are confirmed in their phenomenological existence, but remain separate and distinct due to the evident ontological dishomogeneity that characterizes them. The metaphysical root of theory and praxis' modern constitution in a unitarian plexus sought by Scheler seems however to escape inquiry, which nevertheless proceeds, appealing to the phenomenon of so-called "functionalization" (*Funktionalisierung*).[53]

Beginning in *Formalismus in der Ethik und die materiale Wertethik* and continuing in the essay *Die Idole der Selbsterdkenntnis*, distinguishing the phenomenological point of view from that of the psychology of Brentano and Stumpf, Scheler affirmed that acts have their being in their commission by persons (*vollzogen werden*), while functions, attesting themselves on the psychic level, perform themselves (*vollziehen sich*), run (*ablaufen*).[54] Then in *Vom Ewigen im Menschen, Probleme der Religion* he focused on the relationship linking acts and functions, or the intentional level and the psychic level of the living experiences of consciousness in the human person, naming that relationship the functionalization activity. In fact, functionalization translates the essential contents reached by the intentional processes of the conscience into forms of thought (*Denkform*) to which the psychic functions can then conform themselves; the latter, taking on forms consonant with eidetic perspectives acquired over time, promote a true growth in the content of spirit, without spirit having to renounce being the depository of laws of choice and apriori functions.[55]

The antique category (*Kategorie*) of wisdom (*Weisheit*) carries out the functionalization of the cognition of essence to the law of use of the intellect, oriented to the contingent world of facts, or in other words, puts theory into practice. Only wisdom can unite cognition of the entity (*Erkenntnis des Seienden*), consciousness of value and systematic availability of the will (*sistematysche Bereitschaft des Wollens*) in order to comply with the demands of duty (*Sollensforderungen*) which only proceed from the synthesis of cognition-of-being and consciousness of value.[56] In fact, the wise person, unlike the pragmatist or the idealist respectively, sees and embraces the essences of the spirit as much as the resistance of the world of life and for this reason can attempt to harmonize them in itself in a way that is not rash.[57]

Thus the work of the wise person is able to prepare in the spirit the transformation of known essences into forms of thought, for which, as Scheler writes: "that which first was an object, becomes a form of thought on objects; that which was an object of love, becomes a form of love in which an unlimited number of objects can be loved."[58] In this the wise person does not apply (*Anwendung*) concepts, rules or laws on facts, but synthesizing non-reductively the various spiritual powers of the human being, enables the ideal realization of a possession (*Haben*) and an immediate seeing (*Schauen*) of things according to a determinate and new form and correspondence of sense; what takes place, in fact, is more a conformation (*Anmessung*) than an application[59] and prefigures on the level of the spirit, that is ontologically, the interpenetration that can happen between the two forces of the spirit and of life, in the realization of the metaphysical unity that reconnects the infinite real and ideal determinations to the world's one ground.[60]

However, Scheler acutely leads us to the remark that all this process of the spirit to give the "final cause" form to essences that can already be contemplated in themselves, would make no sense if work activity had not emerged from history in its ontological-metaphysical importance as an eminent "psycho-corporeal action of transformation of a material datum" (*Umformung eines Materials*): in fact, as Scheler had stated beginning with *Arbeit und Ethik*, since work does not autonomously possess purposes, its very own existence and activity need a link with the spirit which is the deposit of eidetic cognition and incessant producer of purposes.[61]

In this way, Scheler gives metaphysical amplitude to the observations on action in *Formalismus* in which he described the *iter* by which intentional acts translate into corporeal movements. There are seven passages, starting with the current presence of the object of action and leading to the actual accomplishment of the action. Once the content to realize has been identified, various levels of willing conform to it, from the most profound disposition (*Gesinnung*), through intention (*Absicht*), deliberation (*Überlegung*), proposition (*Vorsatz*), to decision (*Entschluss*). These are followed by the group of activities oriented to the body (*Tunwollen*) which leads to the movement of the members and is accompanied by states of sensations (*von Empdindungen*) and feelings (*von Gefühlen*). Here, with new correlates of sensations and emotions, "*Ausführung*" takes over; that is, the lived realization of the content itself which as a result (*Erfolg*) represents the objective taking place, or the fulfilling (*Erfüllung*) of the realization.[62]

The qualifying juncture of the just described psycho-physical process of passage from theory to practice, from cognition to action, understood generically as *Handlung*, is evidently situated at the fourth level, that of *Tunwollen*. In fact, here the acts of will change their own orientation and from the content of ideas and values, assumed in the decision of realization, turn to doing, taking on the form of "*Wollen des Tuns*"[63] and attributing to the will a double intentionality, since at this point the will is will to do what previously it wanted to decide.

A further sphere of human activity opens up if, with the lived realization of the content that one wanted to do (*Ausführung*), one concludes the process of acting as praxis, passing to consider those that Scheler calls "*Kausalfolgen*." It is no longer a depository of ethic values, nor does it belong to "my" acting, or the praxis of a person ethically described. It is rather "the result of my speculation and my calculation" (*Erfolg meiner Spekulation und Berechnung*) and therefore a "consequence of this doing" (*Folge dieses Tuns*), rather than of *Tunwollen*.[64] On the level of causal consequences it seems thus to delineate, from the metaphysical point of view, a kind of free zone, where the spirit's determinant action has not yet reached the flow of life. Therefore a need is felt here for a device, congruent both with the spirit and with life, qualified to resolve the contradiction between the two attributes of the one ground of the world. The importance of the specific industry of that particular human being-active that is work stands out: only through work, at least according to Scheler, can the interpenetration (*Durchdringung*) of the spirit and life complete itself in space and time, or, in other words, only through work can the manifold human and cosmic entities become one with the only divine identity and achieve their metaphysical destiny.[65]

University of Macerata

Translated by Sheila Beatty from Camerino

NOTES

1 The interest in the theme of work, stimulated by his first teacher, Rudolf Eucken, manifests itself at the very beginning of Scheler's speculations: already in 1899 Scheler publishes *Arbeit und Ethik* (now in: *Frühe Schriften, Gesammelte Werke* 1, Francke *Verlag*, Bern und München 1971, pp. 161–195), setting forth an acute and original phenomenology of work activity, and inspired by Eucken's work, *Der Kampf um einen geistigen Lebensinhalt* (Leipzig, 1886). The results of this early work remain substantially unchanged, even through the variations his philosophical thought will undergo over time. The work is defined as "arational activity" (*an sich arationalen Tätigkeit*), qualified by his fundamental boundlessness (*Unbegrenztheit der Tätigkeit*) (*Arbeit und Ethik, op. cit.*, p. 170). Later, in the 1920 *Arbeit und Weltanschauung* conference (now in *Schriften zur Soziologie und Weltanschauungslehre*, GW 6, 1963, pp. 273–289), the relationships between the conception of work and the vision of the world are delineated with greater historical sense, differentiating the structure of theory and practice, especially of antiquity and above all the medieval period, from the modern compeers of cognition and work. However, these analyses were prepared by the use of the notion of *Arbeitswelt* in the *Habilitationsschrift* of 1900, *Die transzendentale und die psychologische Methode* (now in GW 1, pp. 197–335). The concept of the world of work was introduced into the realm of philosophy by Eucken, in *Prolegomena zur Forschungen über die Einheit*

des Geistesleben in Bewusstsein und That der Menschheit (Leipzig 1885, p. 45). The world of work, referring to the activity of the spirit, included all practical, cultural and institutional realizations and thus opened to a closer comparison between the dimension of theory and that of praxis, understood in a broad sense, that is, also including productivity and practical-effective realization. For more on the Schelerian concept of work in his early writings see our *Il lavoro dell'uomo nelle* "Frühe Schriften" *di Max Scheler*, in *Annali della Facoltà di Lettere e Filosofia*, University of Macerata, 1988, pp. 193–220. Scheler's reflections on work represent something new in phenomenology, writes R. Sweeney in "Cognition and Work" – *Erkenntnis und Arbeit, Analecta Husserliana*, Vol. **XIV**/1983, p. 97.

[2] M. Scheler, *Erkenntnis und Arbeit*, GW 8, 1960, p. 193.

[3] Aristotle codifies the distinction between contemplative and practical life in his *Nicomachean Ethics* (EN, I, 5, 1095b 14–1096a 9), establishing as well the primacy of the first over the second (EN, X, 7, 1177a 12–1178a 9). Then, in his *Metaphysics* (I, 2, 982b 25–26, 983a 6–7), he also identifies the principal modulations of human life which pass from simple experience through art and praxis to contemplation. In the Middle Ages Saint Thomas Aquinas, in his *Summa Theologica* IIa IIae, p. 182, a. 1, adducing Saint Gregory in his *Moral*, I, VI, judges Rachael, or the contemplative life, to be higher than Leah with the bleary eyes. Scheler, for his part, in *Vom Ewigen im Menschen. Probleme der Religion*, criticizes the confusion between theory and praxis provoked by the Kantians, Fichteans etc. He remarks: "*da sie ja schon durch Erkennen leisten zu können, ja zu müssen glauben, was Sache ausschliesslich des Wollens und Handelns ist. Ihr erkenntnistheoretischer Voluntarismus hebt echtes Wollen auf*" (GW 5, note 1, p. 196).

[4] Cfr.: Scheler, *Erkenntnis und Arbeit, op. cit.*, p. 194.

[5] These statements demonstrate the differences separating Scheler from other contemporary thinkers in regard to the problem of the beginning of the modern historical dynamic. It is well known, for example, that Max Weber saw capitalism's starting point in the convergence of a series of innovative acts that "a young member of one of the entrepreneurial families" suddenly set out to accomplish, carrying into "the ancient, comfortable and calm existence," "the new spirit . . . of 'modern capitalism'" and its craze for "rationalization" (M. Weber, *Die protestantische Ethik und das Geist des Kapitalismus*, in *Gesammelte Aufsätze zur Religionssoziologie*, Mohr, Tübingen 1978, pp. 52–53). Such an important event is motivated by an "irrational element" (*ibid.*, p. 62), according to Weber, who exposes his analysis to the paradox that the most pressured and fundamentally pervasive rationality, which for him is that of capitalism, could originate in subjective actions dominated by irrationality. Scheler judges the results of Weber's study on the beginnings of capitalism dissatisfactory and explains them by the fact that Weber absolutized the methods of the social sciences and consequently excluded openings and deepened inquiries afforded by a philosophic approach (Scheler, *Max Webers Ausschaltung der Philosophie (zur Psychologie und Soziologie der nominalistischen Denkart)*, GW 8, *Zusätze aus den nachgelassenen Manuskripten*, pp. 430–438. See also our *Meraviglia e disincanto nel pensiero di Max Scheler*, in *Interpretazione e meraviglia, Atti del XIV Colloquio sulla Interpretazione*, Macerata 1994, pp. 53–64). In fact, the revolutionary actions of that young entrepreneur are only the means by which the advent of the new spirit of capitalism manifests itself in history. In the same manner as the voluntarism

of Duns Scoto (Scheler, *Beiträge zur Feststellung der Beziehungen zwischen den logischen und ethischen Prinzipien*, GW 1, pp. 26–27; ID., *Probleme einer Soziologie des Wissens*, GW 8, p. 117. See also our: *Duns Scoto-Scheler. Volontarismo e interpretazione del moderno*, in: *Via Scoti, Atti del Congresso Scotistico Internazionale*, Edizioni Antonianum, Roma 1996, pp. 1127–1144) and as the singular methodology that led to Liebig's discoveries, the behavior of the young entrepreneur represents, according to Scheler, the link of an explicative chain that can only find its ultimate reason in metaphysics. Moreover, Scheler also describes in *Vorbilder und Führer* (GW 10, pp. 255–344) the descent of the sociological classification of models from a rigorous axiological hierarchy which he matured in the ethical-religious personalism of *Der Formalismus in der Ethik und die materiale Wertethik* (GW, 2, pp. 573–596) and of *Vom Ewigen im Menschen* (GW 5). In the latter work in particular Scheler manifests the intention to dedicate a second volume (which, however, was never published) to the exploration of the "meaning of the personalistic model (*Vorbild*) in all its manifestations, for the religious and moral existence of humanity" (p. 9).

6 Scheler, *Erkenntnis und Arbeit*, *op. cit.*, pp. 194–195.

7 *Ibid.*, pp. 200–202.

8 Scheler demonstrates a singular ability to reveal the phenomenon of "missing." As early as *Wesen und Formen der Sympathie*, using the "consciousness of missing" (termed respectively *Mangelbewusstsein* and *Nichterfüllungsbewusstsein* for *Strebensakte*; *Leerbewusstsein* or *Nichtdaseinsbewusstsein* for emotional acts), he succeeds where Husserl failed in making a phenomenologically correct and rigorous affirmation of the existence of other "I's," or "of something that exists as sphere of the You – and of which only no example is known" (GW 7, pp. 229–230). It is not haphazardly that H. G. Gadamer attributes to Scheler a very deep and almost prodigious capacity for intuitive penetration (H. G. Gadamer, *Die phänomenologische Bewegung*, now in *Gesammelte Werke*, vol. 3, *Neuere Philosophie I. Hegel-Husserl-Heidegger*, J. C. B. Mohr (Paul Siebeck), Tübingen 1987, pp. 105–146). Regarding the "final" character vindicated by the philosophy of Hegel and of Marx cfr.: Scheler, *Arbeit und Ethik*, *op. cit.*, p. 165.

9 Scheler, *Erkenntnis und Arbeit*, *op. cit.*, p. 195.

10 The original tendency to gather signs of furtherness even in motifs such as theory and praxis which are well grounded in philosophical totalizing systems, comes to Scheler first from his teacher R. Eucken, spokesman for thinkers in post-idealistic spiritualism who pose the problem of the spirit and its power. Scheler's attention to extraspiritual powers with metaphysical valence comes from Nietzsche, Dilthey and Bergson. On the other hand, his interest in a pluralistic philosophical representation, similar in some features to that of W. James, finds fertile ground in his readings of Augustinian Christian authors (in addition to Augustine, this includes Duns Scotus, Malebranche and Pascal). Thus Scheler has a very clear consciousness of the philosophical distance between himself and Hegel and Marx. In fact, his own concept of historical development as *Kulturwachstum*, in which the forms of the spirit, as they become available, seek total involvement in new structures, is distinguished in *Probleme einer Soziologie des Wissens* (GW 8, pp. 36–37) from Troeltsch and Mannheims' *dialektische Wachstum* inspired by Hegel and his limited Europeanist horizon. In addition, shortly thereafter Scheler, expressing himself in the metaphysical terms, clearly declares his opposition to the monistic vision of history, which

he feels equalizes the philosophy of Hegel with that of Marx (p. 50). He accedes that both had notable impact on the masses, respectively in the Prussia of Minister Altenstein and the Soviet Russia (p. 72, note I), but at the same time he cannot keep himself from objecting to their same foundational prejudice, that is the conviction of "*Selbstmacht der Idee*," which renders vain every need to reopen the closed circularity of theory and praxis.

[11] Scheler, *Manuskripte zur Lehre vom Grunde aller Dinge, Drang und Geist*, in: GW 11, *Schriften aus dem Nachlass II, Erkenntnislehre und Metaphysik*, pp. 185–200. See also the chart on p. 222, *Die Gottesidee*, in which the two attributes known to us in the one foundation of the world are *Geist* and *Leben*. In addition: *Philosophische Weltanschauung*, GW 9, p. 81.

[12] Scheler, *Wesen als Werde-Entwurf und Wesenserkenntnis als Konstruktion*, in: GW 11, *op. cit.*, pp. 120–121; p. 91, n. 13. Cfr. the programmatic goal of a metaphysics that realizes a contemplated eidetic totality, expressed in *Vom Ewigen im Menschen. Probleme der Religion*, *op. cit.*, pp. 110–111.

[13] H. Marcuse as well, in his essay on the philosophical roots of the economic concept of work, developed, in 1932, an analogous intuition, even though he began with different methodological premises and used Heideggerian elements. Cfr.: H. Marcuse, *Die philosophischen Grundlagen des wirtschaftswissenschaftlichen Arbeitsbegriffs*, now in: *Kultur und Gesellschaft*, Frankfurt 1965.

[14] As G. Morra expresses in: *M. Scheler. Una introduzione* (Roma 1987, pp. 15–16). "The speculative exigencies expressed by Scheler in his first writings found full realization after he read Husserl's *Logische Untersuchungen*; however, at the same time the anticipation of phenomenological themes done in his first period allowed the Scheler of his second period to define right from the start a concept of phenomenology which was different from that of Husserl, even contrary to that of Husserl beginning in 1913 when his *Ideen I* convinced Scheler of Husserl's inability to rise above the limits of transcendental subjectivism." Scheler refers explicitly to Meinong and von Ehrenfehls in *Arbeit und Ethik*, *op. cit.*, when he attributes to them the merit of having begun the philosophical exploration of the field of value (p. 167).

[15] Cfr.: Scheler, *Der Formalismus in der Ethik*, *op. cit.*, *Vorwort*, the third edition, p. 17, which describes the metaphysical evolution as a passage to an ethical path toward the absolute.

[16] Scheler, *Vom Ewigen in Menschen*, *op. cit.*, p. 196.

[17] Scheler, *Phänomenologie und Erkenntnistheorie*, GW 10, *Schriften aus dem Nachlass I*, p. 477.

[18] Scheler, *Die Stellung des Menschen im Kosmos*, GW 9, *Späte Schriften*, p. 42.

[19] Scheler, *Idealismus – Realismus*, GW 9, pp. 204–208.

[20] *Ibid.*, p. 207. Cfr. as well: Scheler, *Vom Wesen der Philosophie und der moralischen Bedingung des philosophischen Erkennen*, in *Vom Ewigen im Menschen*, *op. cit.*, pp. 83–92.

[21] Scheler, *Erkenntnis und Arbeit*, *op. cit.*, p. 360: "*Aber der Geist allein und seine Idee, Urphänomene, Werte usw., sei er unendlich oder endlich, macht uns weder das Dasein, noch das zufällige Sosein irgendeines Gegestandes verständlich. . . . Die Realität ist in ihrer subjektiven Gegebenheit eine Erfahrung des ungeistigen, triebhaften Prinzip in uns: eine Erfahrung des einheitliche, wie immer sich spezialisierenden Lebensdranges*

in uns." A few pages later, Scheler continues: "*Realsein ist nicht Gegenstandsein, d. h. das identische Soseinskorrelat einer intellektiven Akte – es ist vielmehr Widerstandsein gegen die urquellende Spontanität, die in Wollen, Aufmerken jeder Art ein und dieselbe ist*" (p. 363).

22 Scheler, *Philosophische Weltanschauung*, *op. cit*, note 1 of p. 102.

23 Scheler, *Arbeit und Ethik*, *op. cit.*, p. 171, in which Scheler states that in the "modern vision of the world" a singular phenomenon takes place, by which "the ideal of positivistic logic could reasonably call itself a pure 'science of work' because of its strenuous attempt to expunge from itself every distance from the field of facts."

24 Scheler, *Erkenntnis und Arbeit*, *op. cit.*, p. 193. In addition, *Zusätze aus den nachgelassenen Manuskripten*, GW 8, p. 448.

25 Scheler, *Erkenntnis und Arbeit*, *op. cit.*, p. 212.

26 *Ibidem*. In note 1 Scheler refers to the W. Jerusalem's German translation of James' work, entitled *Der Pragmatismus* (Leipzig 1908). Jerusalem also was close to Pragmatism. Although critical essays on pragmatistic authors were already available, for example the work of E. Boutroux on *William James* (German translation by B. Jordan, Leipzig 1912), or the work of J. Goldstein *Wandlungen in der Philosophie der Gegenwart* (Leipzig 1911), Scheler holds that an in-depth historical survey of the movement is still needed.

27 *Ibid.*, p. 213.

28 *Ibid.*, p. 214.

29 *Ibid.*, p. 215.

30 *Ibid.*, p. 213.

31 *Ibid.*, p. 200.

32 *Ibid.*, p. 216. The motif of the natural and historical world as a plastic mass to form by work is also present in the *Arbeit und Weltanschauung* conference, now in GW 6, p. 280.

33 For the disruptive function of the voluntarism of Duns Scotus, see: Scheler, *Beiträge zur Feststellung der Beziehungen zwischen den logischen und ethischen Prinzipien*, GW 1, pp. 26–27. In addition, as early as *Arbeit und Ethik*, *op. cit.*, in 1899, Scheler notes how positivism already organized its own knowledge as "science of works," which "lacks the consideration of systems of objective purposes, that is, in this case: final fundamental methods and principles of science." In this positivistic science, vice versa, "there is a dominant opinion that fundamental methods and principles are a purely secondary result of scientific work, rather than that premise only from which fruitful work is possible" (p. 171).

34 Scheler, *Erkenntnis und Arbeit*, *op. cit.*, p. 239.

35 *Ibid.*, p. 240.

36 *Ibid.*, p. 241.

37 Thus writes Comte, in a passage quoted by Scheler: "The method does not admit to being studied separately from the research in which it is used, doing so means dealing with a dead study, unable to render fertile the intellect that takes flight. All one can say really, when one identifies the method abstractly, is reduced to some generic ideas so vague that they could never have any influence on the intellectual regime." Lipps arrives at the same idea – Scheler continues – when in defining philosophy, he states that "every particular science will necessarily indicate final generalizations, logically justified by its own research, in the same way in which the sides and edges of the same pyramid converge

in a vertex, without the need to artificially posit a vertex" (Scheler, *Arbeit and Ethik, op. cit.*, p. 171). In addition: Scheler, *Erkenntnis und Arbeit, op. cit.*, p. 241.
[38] *Ibid.*, pp. 239–242.
[39] *Ibid.*, pp. 244–245.
[40] *Ibid.*, p. 242.
[41] *Ibid.*, p. 245.
[42] *Ibid.*, p. 217.
[43] *Ibid.*, pp. 200–201.
[44] *Ibid.*, pp. 203–204.
[45] *Ibid.*, p. 204.
[46] *Ibid.*, pp. 204–206.
[47] *Ibid.*, p. 207.
[48] *Ibid.*, p. 206. It seems that here Scheler has resumed a path that reinserts in his metaphysical framework the relationship between microcosm and macrocosm outlined in the first drafts of *Formalismus* and then deleted from the third edition in 1926. Cfr.: Scheler, *Formalismus, op. cit.*, pp. 395–396.
[49] Scheler, *Erkenntnis und Arbeit*, op. cit., p. 209.
[50] *Ibid.*, p. 274.
[51] *Ibid.*, p. 359.
[52] *Ibid.*, p. 360.
[53] *Ibid.*, pp. 197, 198, 206, 209, 232, 301, 354, 355, 360.
[54] Scheler, *Der Formalismus, op. cit.*, p. 387, *Die Idole der Selbsterkenntnis*, GW 3, p. 235.
[55] Scheler, *Vom Ewigen im Menschen, op. cit.*, p. 198.
[56] Scheler, *Max Webers Ausschaltung der Philosophie, op. cit.*, GW 8, p. 431.
[57] Scheler, *Pragmatist, Idealist – und der Weise*, in *Zusätze zu "Erkenntnis und Arbeit,"* GW 8, p. 465.
[58] Scheler, *Vom Ewigen im Menschen, op. cit.*, p. 198.
[59] Scheler, *Die Formen des Wissens und die Bildung*, GW 9, p. 108.
[60] Scheler, *Erkenntnis und Arbeit, op. cit.*, pp. 359–361.
[61] Scheler, *Arbeit und Ethik, op. cit.*, p. 168; idem, *Zusätze, op. cit.*, p. 448. In addition, p. 452: "Denn nur sofern es geistige Akte gibt, die jene Ursprünglichkeit und Selbständigkeit besitzen und in denen wir am Sein und Leben der Dinge unmittelbar Anteil gewinnen können, hat es ein Sinn auch eine eigene geistige Kultur als Selbstwert zu erstreben und die Arbeitszivilisation als ein zur Befreiung dieses ursprünglichen Geistes in uns dienendes Mittel aufzufassen. Gibt es dies nicht, so wird auch gelten müssen, dass die praktischen Förderung dieser Zivilisation das Alpha und Omega unseres Daseins als Endziel darstellt."
[62] Scheler, *Der Formalismus in der Ethik, op. cit.*, p. 137.
[63] *Ibid.*, p. 139.
[64] *Ibid.*, pp. 137–138.
[65] One of the last volumes of the published GW (number 11, *Schriften aus dem Nachlass II*) reports the *Manuskripte zur Erkenntnis – und Methodenlehre der Metaphysik als positive Erkenntnis (Auseinandersetzung mit Gegnern)*. Here on p. 90 we can read a "practical" definition of metaphysics: "Metaphysik, das heisst: Sich in die ewige und zeithafte Produktion der Natur und der Geschichte selbst hineiversetzen und alles Gewordene nach-

und miterzeugen." The "practical" definition of "speculation" on the same page is also illuminating: "Die Spekulation ist es, die das (Metaphysik) vor allem nach ihrer ideenhaften Seite will, die dionysische 'Technik' nach jener der Realität (Begriff der Spekulation: Nacherzeugung der Wesenheit)." Regarding *Durchdringung* of spirit and life as the metaphysical task of man and his work, see: Scheler, *Die Stellung des Menschen im Kosmos, op. cit.*, p. 55. See also: *Die Formen des Wissen und die Bildung, op. cit.*, pp. 101–103. In *Erkenntnis und Arbeit* as well Scheler underlines the dynamic character of philosophical-metaphysical cognition and the importance of work in relation to it: "Erst in Verlauf der Arbeit an der Welt lernt der Mensch die zufällige obkektive Bilderwelt und ihre Gesetz kennen. . . . In diesem Sinne ist die Arbeit – und nicht die Contemplatio – in der Tat die wesentlichste Wurzel alle positiven Wissenschaft. . . . Aber – der Mensch besitz auch noch eine ganz andere Erkenntnismöglichkeit . . . es ist die philosophische Erkenntnishaltung. . . . Erst in der grössten Spannung zwischen beiden Haltungen und erst durch Überwindung dieser Spannung in der Einheit der Person aber wird die eigentliche pilosophische Erkenntnis geboren" (p. 362).

FERNANDO R. BORNAETXEA

INTERSUBJECTIVITY

From Phenomenology to Ethnomethodology

1. INTRODUCTION

One of the most widespread debates regarding ethnomethodology is about its phenomenological character. Let's make it clear that phenomenology is a philosophy directed to the study of essences, in contrast with the philosophies of life and existence which emerge from it. Not one of the post-Husserlian thinkers who regarded themselves as phenomenologists, such as Heidegger or Sartre himself, are fully committed to that radical philosophy, and as Husserl himself puts it "they have slipped back into anthropologism or into transcendental psychologism."

Ever since it came to be, Husserl's phenomenology has been a source of inspiration for investigators in the social sciences. That is why Max Weber's or Georg Simmel's sociologies present "intimate basic connections" (Wolff in Bottomore, 1978, p. 572) with phenomenology apart from having both been elaborated in Germany within the same historical framework. It could be said that American research workers such as James, Cooley, Mead or Thomas have adopted part of the phenomenological method (J. Douglas, 1970, pp. 3–44). All of them emphasize the importance of consciousness and subjective significance in the interpretation of social action – though having a different approach on the "natural attitude" and "intersubjectivity."

The importance of the "natural attitude" and "common sense" in sociology necessarily refers to the interpretative tradition and, particularly, to the phenomenology and works of Alfred Schutz. The latter having applied Husserl's method to social studies put that new and original research field at the disposal of specialists in the social sciences. The ethnomethodology springing from Garfinkel has undertaken deeper research in the analysis of interaction, further developing the study of the productions of common sense.

In this paper we are particularly interested in Schutz's as well as the ethnomethodologist stand with regard to intersubjectivity. Schutz believed that this theme had not been resolved by Husserl's philosophy, though

A-T. Tymieniecka (ed.), Analecta Husserliana, Vol. LIV, 167–177.

he used it as a starting point for this own analyses. The solving of this problem largely determined his conception of the foundation and the methodology of the social sciences. The point was to set up a conception of the subject and its relationship with the other which would allow one to know the social phenomena and would therefore legitimate the project for a comprehensive sociology.

2. THE PHENOMENOLOGY OF NATURAL ATTITUDE

This objective was set with Schutz's first analyses, which demonstrate how the intellectual climate of the time required a refocusing of intersubjectivity. In his book *Der Sinnhafte Aufbau der Socialen Welt. Eine Einleitung in die Verstehende Soziologie* (The Construction of the Social World Provided with Sense. Introduction to the Sociology of Understanding) dated 1932, Schutz tackled the problem of intersubjectivity, avoiding the dangers of empathic comprehension and answering the critiques coming from the logical empiricism that limited the apprehension of the other to what is externally observable (Chapter 1, sec. 3: "The Pre-given Character of the Self and the Other and the Postulate of the Comprehension of Subjective Meaning"). Schutz saw in the treatment of intersubjectivity two conceptions of comprehension opposed to each other and equally unsatisfactory. In the first case, the assumed accessibility of the other brings us to the conclusion that through Dilthey's empathy or Max Scheler's "direct intuition" it is possible to grasp right away the "subjective meaning" that the other gives to its own actions. In the second case, what emerges from the experience is nothing but the body of the other and its noticeable transformations.

Between those two extreme attitudes, Schutz proposed to elaborate a theory of "subjective meaning" that goes beyond the boundaries of purely subjective comprehension or purely objective knowledge. He meant to systematize the analysis of the "life-world" which takes place in the "natural attitude" – an analysis that Husserl had included in his philosophical research – and to solve the problem of intersubjectivity within that "phenomenology of the natural attitude." For that purpose, Schutz's work was based on post-Husserlian phenomenology, the greatest effort in which, as can be seen in Merleau-Ponty or Sartre, was that of trying to avoid the direction taken by transcendental philosophy and that of not ending philosophical thought with the constitution of a subject of the theoretical contemplation but with a subject that is in relationship with the pre-given world.

3. TRANSCENDENTAL SOLIPSISM

In his constitutive phenomenology Husserl ran into the question of solipsism. This had hardly been tackled in *Ideen* I, had later been taken up in the Fifth Cartesian Meditation; the problem of intersubjectivity covered up that of "transcendental solipsism."

The founder of phenomenology describes its nucleus as being the method of suspension (epoche), which designates the suspension of the natural thesis of the existence of the world, this being a fundamental task for the constitution of a pure knowing self. This method of suspension is thoroughly critical in that it grasps objects in a mode absolutely intrinsic to their perceived status and therefore without further unexamined or unexaminable presuppositions. Later Husserl would turn away from the critical programme of a presuppositionless phenomenology to a phenomenology of objects and relations given in a "co- and pre-knowledge," the horizons of which unfold through our interpretative being-in-the-world. From that perspective, the life-world is not a simple correlate of consciousness, nor is it thematically present to us; we are present to it as already meaningful.

Husserl characterized the "natural attitude" as being the belief in an objective, as well as space-time and intersubjective world, and he gave as a subsequent task its systematic and complete description: "But neither do we propose at the present time to go on with the work of taking the pure description up to a systematic and complete characterization, or to exhaustive limits of what is found in the natural attitude (nor by any means do we ask to do it in all the attitudes that need to be harmoniously interwoven with the latter). Such a work – scientific as it is – can and must be fixed as a goal, and this is a highly important work, though it has hardly been noticed so far" (Husserl, 1913). Husserl tried to discern how transcendental consciousness takes form, through the suspension of the belief in the existence of the external world. After all, the transcendental self can constitute itself only when the reduction gets rid of all the relative meanings of other subjects. Husserl's attempt clearly shows an egological point of view, the constitution of the other can only take place if there has been a previous constitution of the transcendental self. Thus, intersubjectivity derives from the self as a fundamental sphere. Within that sphere, the subject discovers contents which are linked with his body and with himself as a unity of soul and body, as a psychophysical unit. It can then constitute itself analogically to the other in relation to the subject.

In his attempt to constitute intersubjectivity based on the transcendental reduction, Husserl ran into insurmountable difficulties. The reduction will open the way to a pure self, to a private sphere devoid of complete significance of the other. Transcendentally, the others are nothing but mere transcendental subjectivities. Likewise the other is susceptible to applying the transcendental reduction to himself, and therefore, to constituting himself as a private sphere disconnected from the other. According to Schutz, solipsism is here unavoidable. What Husserl describes is nothing but "monads without windows" which are not related to each other, and such a cosmos of "monads" would in no way be intersubjective.

4. SCHUTZ AND SARTRE

Schutz did agree with certain analyses of post-Husserlian phenomenology (Schutz, 1962, p. 175ff), while sticking to his own viewpoint. As far as Sartre is concerned, he does agree, for example, with *Being and Nothingness*, for its famous analyses of the other and the "regard d'autrui" following from Sartre's finding that Husserl had not solved the problem of intersubjectivity by way of a transcendental solution. He also admits with Sartre that Husserl did not tackle the problem with a concrete individual subject, but only with a knowing and conscious one, one therefore susceptible to becoming a pure conscious self. Like Sartre, he rules out the question of the existence of the other, for he considers it as given to the world (pre-ontological existence in Sartre). At the same time, he is opposed to Sartre when Sartre considers the objectification of the other. If Sartre is the thinker who gets closest to a philosophy of the other as an other, that is to say, not as an analogically constituted other in relation to the subject, it's at the cost of an objectification (the subject is perceived as nothing more than an object by the other, and reciprocally, the other is viewed as a mere object), which prevents approaching the other as subject or "concrete subjectivity in the freedom of his concrete action." Moreover, Sartre is forced to postulate the interchangeability of the subject and the other, insofar as he rejects any analogical constitution. That does not solve the problem of intersubjectivity, and for Schutz this is a mere replacement of Husserl's "transcendental solipsism" by a "practical solipsism" where "free mutual interaction has no place."

5. THE RECIPROCITY OF PERSPECTIVES

Though they failed to establish a transcendental intersubjectivity, Schutz thought that Husserl's analyses were legitimate at the intra-mundane subject's level. His interpretation of the Fifth Cartesian Meditation made him consider that this work "does not demonstrate how the other constitutes itself as transcendental subjectivity, but only how it constitutes itself as a mundane psychophysical unity." Schutz's work is precisely an attempt to "state explicitly a philosophy of the mundane reality" or – in more formal language – a phenomenology of the natural attitude.

Schutz set forth the natural attitude as being based in a series of assumptions that are "taken for-granted"; they are tacit, they precede and constitute the very characteristics which we attribute to the life-world. The fundamental assumptions are "the sense of reality" and "intersubjectivity." We imagine a world where social and natural objects are stable, in spite of the changes occurring in the physical perspectives from which they are being observed, and in spite of the changes in the forms and in the various behavioural expressions in the case of animated objects. We can have access to that world thanks to a "list of available knowledge," the origin of which is fundamentally social. Moreover, Schutz put forward that intersubjective understanding can be achieved through an active process in which the participants assume the "general thesis of reciprocity of perspectives" (1962, pp. 11–13). According to this thesis they treat their experiences as "being identical to all the practical effects," in spite of the different perspectives, biographies and motivations that mean the actors have not undergone the same experiences of the world.

For Schutz, the other constitutes itself in its corporeity and subjectivity, only with reference to the self as the "zero point." The self would be its origin and point of reference. Furthermore, Schutz sees the time dimension as the core of the phenomenon of intersubjectivity. The "simultaneity" of two internal times, two co-presences, growing old together, establishes an objective time in which actions are shared. To that simultaneous experience one must add the acceptance of the fact that we share motivations and symbolizations of the common world (reciprocity of perspective) through which we become directly aware of having a common experience that, even if it is transient and fragmentary, is fundamental for social life.

The following three explanations will give us a better understanding of this viewpoint.

In the first place, Schutz considers that Husserl's analyses are not fully conclusive, insofar as the constitution of the other is based, above all, on the "perceived body." Now, as long as we stick to an analogical constitution of the other's body, we are adopting a form of the empirical position, that is to say, of behaviorism, which confines itself to the actions and the external body movements of the other. On the contrary, Schutz intends to accept the other as a subject, that is to say, as the origin of an internal attribution of meanings.

In the second place, the Schutzian theory is not a theory of the other as a "double" of the self. At this stage Schutz takes up again a fundamental moment in Husserl's phenomenological description: from the outset, the other is perceived in the form of the "there" as opposed to the "here" proper to the perceiver. It's an "Ego" which coexists in the world of the "there," and is therefore an "alter ego." The other cannot be perceived other than in relation to the self of which I am the ego, but it cannot be conceived as a reduplication of the self, and in just that lie all of Husserl's methodological problems in trying to make sure of the other's comprehension.

In the third place, intersubjectivity is to be taken in the strict sense of a relationship between individual subjects. That position allows us to get away from the transcendental subject as well as from an intersubjectivity that is supposed to be in the individual subject represented as "ourselves." Schutz (1962, pp. 151ff) had found in Max Scheler a philosophy aiming at a transcendental constitution of the other. According to Schutz, Scheler's only achievement in his investigation of transcendental intersubjectivity was to postulate a trend of general consciousness prior to the constitution of the self, a sphere of the *nous* pregiven to the sphere of the self, without fixing methodological foundations for reaching such a conclusion. Coming back to Husserl, Schutz found again an egological principle of analysis which is, once transfered to the level of the intramundane subject, at the origin of the study of the interindividual relationships taking place inside the daily life-world.

6. THE GENERALIZED OTHER

The approach of the interactionists such as Mead or Cooley is something different, though their notion of the "generalized other" is related to Schutz's idea of intersubjectivity. The interactionist relation between the self and the other maintains that the nature and the content of the other have influence over the self and the self influences the other reciprocally. The self achieves this by taking on the other's identity and attitude as well as by playing his role and by allowing him to do the same. However, the "generalized other" has no relation as such with a "particular individual person"; there is no interaction, the self "takes" the attitude coming from the "generalized other," interprets it, understands it and then incorporates it into its own universes of signification. In that sense, it looks like the community's attitude, which manifests itself directly or indirectly as an element of social control and as the abstract formulation of society's "ethos."

> The organized society or the social group that gives its unity of self to what is individual can be called the "generalized other." The attitude of that generalized other is the whole community's attitude (Mead, 1932, p. 154).

According to Schutz, the more anonymous the typifying construction is, the more remote it is from the singularity of the like implicated individual and the fewer are the aspects of his personality and rules of conduct that fit into the typification as significant elements in relation to the purpose for which the type has been built:

> When he built the Other as a partial self, as someone playing roles or fulfilling characteristic functions, the corollary is the process of self-typification that takes place whenever I interact with him. The whole of my personality does not get involved in such a relationship, but a few layers of it only. When I typify the Other's attitude, I am typifying my own attitude, which interacts with its own to the point of becoming a passenger, a consumer, a participant, a reader, etc. This self-typification is what William James and George H. Mead see as a basic distinction between the 'self' and the 'Other' with regard to the 'Social Self' (Schutz, 1962, p. 48).

Schutz's concept of the "alter-ego" has been criticized from the interactionalist point of view as presenting a self similar to that of the windowless monads of Husserl, but one having unidirectional mirrors (Perinbanayagam, 1975). Symbolic Interactionism is not a Parsonian sociology but on this point (intersubjectivity) interactionists make the same analysis as Talcott Parsons. For G. H. Mead the important point is the negotiation of norms or definitions in the interaction, in the social world in which we live. In contrast, ethnomethodology is characterized by its taking these norms and definitions as the actors' resources for establishing the situation.

7. ETHNOMETHODOLOGY

Ethnomethodology asserts, as does Schutz, that intersubjectivity is an assumption which has not been studied yet but is essential to the natural attitude. It tends to withdraw from the latter insofar as it reduces the temporal subjectivity so characteristic of Schutzian phenomenology. For Garfinkel the situated social interaction is the privileged place where intersubjectivity is being produced and reproduced. The central idea of intersubjectivity is that for any given reality, it can only be considered as an individual or collective construction of the "members" who elaborate it.

From this point of view intersubjectivity appears as another operation rather than an intersection of sets. In order to see how to reach that operation, it is essential to analyze the way in which the typification takes place, even the way "members" are perceived and their characteristics are being given so that they may appear normal or stable. This work is carried out through the strategies of introduction of the "members" and the work of interpretation of the language in each social context.

Ethnomethodologists contemplate studying how intersubjectivity is constituted by the actors in their situated practices, that is to say, describing the processes by which each "member" gives meaning to his own action and the other's action so that they arrive at mutual comprehension or incomprehension. The notion of member is the key point in the ethnomethodological approach to intersubjectivity. "Member" does not refer to an ego, nor to a self, but to the good command of natural language. That good command does not refer to its syntactic or semantic correctness but to its communicative competence. The "member" is exclusively related to the activity in process for which historical or identity questions are set by a selective operation, a typification produced at the time when the interaction takes place.

As far as its appreciation of this aspect of the daily life-world is concerned, ethnomethodology owes more to Schutz than to Mead. The interactions believe in a system of intersubjective and trans-situational meanings, whereas ethnomethodologists consider the meaning of the signs (language included) as the product of interactive practices. In fact, as Schutz and Garfinkel have pointed out appropriation of the "true meaning" seems to be a prospect rather than a possibility. The relationship between signifier and signified cannot be neatly defined and when a precise signifier is used in a concrete situation with a specific cultural reference, the ethnomethodological description of the interaction should show it in the first instance.

Nevertheless, the radicality of the ethnomethodological approach drives it out of phenomenology. Garfinkel is only interested in the members' knowledge in the "making" of the situated action, whereas for Schutz the social distribution of knowledge, the socialization of the knowledge of common sense or the stock of available knowledge are elements which form part of his "constitutive phenomenology of the natural attitude." For ethnomethodologists, however, action has priority over knowledge.

Ethnomethodology, by the fact that it takes language to be the fundamental element in "making," that it is more interested in methods than in persons and, finally, that it expels the human subject from the centre of social life, takes us closer to the perspective of structuralism.

8. ETHNOMETHODOLOGY AND STRUCTURALISM

One of the subjects debated in France in the 1960s within the framework of structuralism was that of phenomenology. Structuralist thinkers rejected phenomenologists' subjective consciousness as a starting point for their analysis. For them, language operates according to its own set of rules and is independent of human creativity. The subject is a mere product of the linguistic and symbolic impersonal order through which communication is made possible. However, the idea that linguistics provides general methods having possibly wider application has lost supporters. The "linguistic turn," so characteristic of modern social philosophy and theory, is interested not in that application but in the intersection between the language and the constitution of social praxis. Garfinkel, like Wittgenstein and others, has emphasized the fact that knowing a language, being a "member," knowing a way of life, makes you apply certain methodological strategies in connexion with the "indexical" qualities of the contexts in which social practices are carried out.

The expression "the subject's being put off centre," though coming from post-structuralist literature, derives from many sources: from psychoanalysis, where important parts of the self are unconscious; from Heidegger, who claims the primacy of Being over consciousness, or from Nietzsche himself, who was critical of Cartesianism and of any philosophy that takes consciousness as the basis of knowledge. For those authors, as well as for Saussure, Lévi-Strauss, Foucault, Barthes or Lacan within the framework of structuralism, there is no "thinking self," a nucleus of stable selfhood accessible to oneself. The self is immersed in a system of anonymous meanings. The "discourse of the Other" in Lacan's terms, enables the use of "self."

The "subject's unbalance" has to be taken into account by anyone who has some interest in modern social philosophy. However, in ethnomethodology there are at least two elements of improvement on the structuralistic elaboration of that perspective: the emphasis on action, hence the notion of "member" that, above all, is agent, and the notion of "indexicality," which shows that the signified is created and remains as such through the use of methodical resources. As already mentioned, the signified is not constructed through the interplay of the significants, but in the intersection of the production of signifiers and worldly objects and events, organized by the individual who acts. That capacity of organization is not usually expressed verbally but is remarkably elaborated and shows up when human action is approached in terms of practical consciousness, yet remains invisible when operating in contrasting terms between the conscious and the unconscious. Besides, "informal conversation," instead of "parole" with all its formal connotations, is the space-time par excellence of the language-praxis interaction. Garfinkel has, better than anyone, demonstrated that informal conversation is governed by the indexicality of the context and the methodic resources used by the members to create a "meaningful" social world. The signified does not lie in the syntactic or semantic rules but is built up contextually thanks to a good command of the rules of the occurrences which take place in a specific social situation.

CONCLUSIONS

In short, ethnomethodology has assimilated the important contributions structuralism has made to modern social theory. It considers the "linguistic turn" as a turn towards the investigation of the intersection of language and praxis. Besides, it adopts the subject's being put "off centre" when it does not accept the hypothesis that subjectivity is the immediate deriving basis of experience. Because it keeps in view the capacity to be the agent of each "member" in an interaction and his sense of belonging to the situation in progress, it maintains (ambiguous) relations with phenomenology.

UPV/EHU
San Sebastián
Spain

REFERENCES

Bottomore, T. and Nisbet, R., *Historia de análisis sociológico* (1978) (Buenos Aires: Amorrotu, 1988).
Douglas, J., *Understanding Everyday Life* (Chicago: Aldine, 1970).
Garfinkel, H., *Studies in Ethnomethodology* (Englewood Cliffs, N.J.: Prentice Hall 1967).
Giddens, A. *et al.*, *La teoría social hoy* (Madrid: Alianza, 1990).
Husserl, E., *Ideas* (1913) (Mexico City: F.C.E., 1949).
Husserl, E., *Meditaciones Cartesianas* (1931) (Madrid: Paulinas, 1979).
Husserl, E., *La crisis de las ciencias europeas y la Fenomenología Trascendental* (1934–37) (Barcelona: Crítica, 1991).
Mead, G. H., *Espíritu, persona y sociedad* (1934) (Buenos Aires: Paidós, 1964).
O'Neill, J., "From Phenomenology to Ethnomethodology," *Current Perspectives in Social Theory* **1** (1980): 7–20.
Perinbanayagam, R. S., "The Significance of Others in the Thought of A. Schutz, G. H. Mead and C. H. Cooley," *The Sociological Quarterly* **XVI** (1975): 500–521.
Rogers, M. F., *Sociology, Ethnomethodology and Experience* (Melbourne: Cambridge Univ. Press, 1983).
Schutz, A., *Fenomenología del mundo social* (1932) (Buenos Aires: Paidós, 1966).
Schutz, A., *El problema de la realidad social* (1962) (Buenos Aires: Amorrortu, 1974).

CHRIS NAGEL

INTERSUBJECTIVITY AND THE INTERNET

No technological innovation is as much anticipated as the universal spread of communication mediated by computers. In every way, computers have changed the way we live in the world, especially in the developed world; these changes have come with astounding speed and without much thought. But we must ask ourselves about this new medium, and about how it conditions our interrelations. How do we encounter one another through computer communications? What impact does the computerization of communication have on community?

The answer to these questions lies in the evolving status of the field of technology surrounding the central idea of computer-mediated communication (CMC). Both the fields I bring into intersection here – phenomenology and CMC – are too broad to allow me to be complete. The delimitation of my topic comes as a result of fit. The phenomenologies of Schutz and Merleau-Ponty seem best equipped to yield a fruitful study of the experience of others on the Internet – the largest global network supporting CMC. As for the Internet, the most vivid examples of interpersonal communication are to be found in Usenet and Internet Relay Chat (IRC). I will define Usenet and IRC in giving my descriptions below.

This phenomenology of intersubjectivity over the Internet will focus on three issues: (1) the status of corporeality or of body-subjects on the net; (2) the status of others or of intersubjectivity on the net, including the status of identity; (3) the possibility of the formation of Internet "communities." Taking cues from both Schutz and Merleau-Ponty, authentic intersubjectivity should require bodily simultaneity, or co-presence in one another's situation. The experience of bodies and of time in CMC are therefore of key importance to phenomenological understanding.

There is some urgency to my research, given the tremendous push by major media, telecommunications and computer corporations to sell the idea of Internet community. A careful and deliberate attempt to come to terms with the Internet is important to both the Internet itself and the hype surrounding it. (A mass-mailing of CMC software by a major corporation asks "Do you want to be better informed, more con-

A-T. Tymieniecka (ed.), Analecta Husserliana, Vol. LIV, 179–197.

nected, more productive, entertained? Just run this program.") Not unlike other technological devices, especially communication devices, the Internet is sold as a cure-all. The tremendous response to this campaign, and the ever-increasing number of people "on-line," shows us that citizens of highly siliconized societies believe the hype, and also feel that in their everyday lives they lack the capabilities the Internet grants them. Typical is the explanation offered by Howard Rheingold in *The Virtual Community.*[1] Rheingold claims that the Internet gives back to suburbanites a venue to meet informally; he cites Ray Oldenburg's book *The Great Good Place* to the effect that suburban Americans suffer from having only work and home places. The sense of being part of a community is produced by CMC. This paper will in part evaluate this claim in relation to phenomenological understanding of intersubjective community. In concluding, I will bring my phenomenological interpretation to bear upon naive analyses of Internet communication.

THE TERMINAL-TEXT BODY

To reach the Internet, you must be in front of a computer terminal connected by cables to a mainframe. Leaving aside the possibility of technical difficulty,[2] and assuming the terminal is powered up, what you see is in most cases a blank screen, and below the screen, a plastic keyboard connected to the terminal by a cable. Presuming now further that you know how to operate the equipment, that you have assented to the culturally evanescent desire to engage in CMC and acquired the specific knowledge of how this particular terminal functions, you begin typing the appropriate commands and passwords. When you reach access, a message scrolls up onto the screen giving technical information about the status of the network, the mainframe, your terminal, and your account (for example, the number of users, the last time you used your account, etc.). Once again, you must know what to do with this information, and what command to enter next, to effectively use the machine.

The technical knowledge described in part above is presumed for every action you undertake on the Internet. Your terminal gives a minimal report back regarding the effectiveness of your action, but nothing more. The terminal distinguishes proper and improper commands, but these commands have no contextual relevance to the medium itself: they are abstract. Furthermore, the situation on the Internet depends entirely on understanding these abstract commands. It is a situation with a paucity of signification. Only totally explicit and perfectly typed commands are recognized.[3]

To use the terminal means to be able to read the terminal screen and type on the keyboard. The body one occupies in the Internet situation is a terminal-text body. One's body-subjective being in the world is manifested in the Internet only in these ways, which are modulations of seeing and of typing. The eyes and the hands are the immediate organs through which a body-subject is open to the situation of the Internet. That situation refers back to eyes and hands. The physiognomy of the Internet situation is extremely limited in this way; however, within these limitations, given the proper technical knowledge, there are virtually limitless variations on what can be typed and read, and how text may be presented. Usenet and IRC – my two examples of Internet intersubjectivity[4] – are particular forms of the presentation of text on the computer screen.[5] Usenet is "a way of managing multiple public conversations about specific topics, conversations that are not located or controlled in a central site but spread throughout the system," according to Howard Rheingold.[6] One subscribes to specific Usenet "newsgroups" to receive the articles posted to that group. Internet Relay Chat uses the various machines connected by telecommunications lines to transfer or relay text, which anyone may receive by being linked to the relay channel at the time the text is processed. It has been compared to citizens' band (CB) radio because like CB, public messages are available as soon as technically possible to everyone who is connected to the same channel. The goal of IRC is an approximation of real-time face-to-face public conversation across distances.[7]

In both cases, another aspect of the terminal-text body is a delay or lag between the time of an action and the time that the action takes effect. My own actions and the actions of other users are subject to a time lag. The terminal side of my body is linked to other computers through networks which are not visible, and so the lag time between my action and the response of the network – of someone else's response to my typed statement, for instance – is an invisible temporal dimension of my body-subjectivity. Even a breakdown of the network is not immediately evident, so in an important sense I have no evidence of my body's efficacy in the Internet situation. My body is in partial absence, a dimension of bodily time exists which is unrecoverable to my understanding.

The limitations of the terminal-text body provide me with a short range of possible actions.[8] I can type and see text. That I can type text does give me a certain degree of expressive potential, but the modulations of my expressivity are radically determined by the medium. The style, the emblem of the relationship to being of a terminal-text body-subject can only be expressed in explicit symbols. The difficulty of this limitation, and to some degree the anonymity of it, are obviated by what have come to be called "parenthetical smilies": symbolic faces at 90° angles to the horizontal line of the text: –). Parenthetical smilies come in many forms, some of which have distinct contextual meanings; appended to sarcastic or ironic comments one finds a parenthetical smiley who is winking; –). On the IRC channels I observed, approximately one in eight lines of text included a smiley. In addition to smilies are abbreviations of words and phrases that have become symbols in themselves. "In My Honest Opinion," a phrase used to denote that the utterance to follow is meant frankly and earnestly, is typed as the acronym, "IMHO." Smilies and abbreviated expressions occur frequently, especially in Internet Relay Chat. IRC users can also type a command which has the terminal represent a physical action of theirs (as text). The others' terminals will report to them that the user has done something, or in other words, other users can literally read a user's action. All of these appropriations of the terminal-text body occur in the following example:

(krelos2) Which Japan thing? I assume you mean the tariff deal?
(Dava) krelos: yep
(krelos2) I feel if they want our markets open to them they should open theirs to us.
(Dava) krelos: me too
> Re: Japan. I think we're trying to find someone else to blame
*** Action: johnjuan thinks we need to return to the gold standard
(Dava) krelos: i think it's time the US stood up for itself
(Dava) whoop: partly but not really IMHO
(Dava) :–)
(krelos2) Well, I agree that is what happened in the 1970's with the auto industry, but that delusion did not last long. And now US products are better for it . . . but open markets should be both ways.

The terminal-text body lends itself to an interpretation along the lines of Merleau-Ponty's phenomenology of the body-subject. The being-in-the-world of the body-subject is, according to Merleau-Ponty, existence and transcendence, the continual act of appropriating one's situation and transforming it. The possibility of this transcendence is founded on the anonymous structure of the lived body, which is lived not as a set of determined conditions, but as variable modes of living. The situation of life is ambiguous, lacking complete determination, and hence calls for our transcending appropriation. In short, the situation of being-in-the-world is an ambiguous and personal life founded on the anonymous structure of the body underlying it. In transcending, the body-subject alters the situation, modulates itself. One such modulation is speech, which is a development of the body-subjective capacity to verbalize. This ambiguous sound-making capacity is modulated to produce the sounds of a particular language, and this language's sounds become sedimented as a mother tongue. The anonymous structure of the lived body founds the ambiguous lived situation, which founds in turn the transcendence which can be sedimented, and becomes a further level of anonymity as the habit body.

The terminal-text body produces a certain modulation of the vector of transcendence described by Merleau-Ponty as proper to the body-subject. In the first place, the terminal-text body's situation is marked by the complete explicitness of its physiognomy, along with a varying, unpresented temporal gap. The ambiguity of the Internet situation is limited to the unpresentable, contingent and indeterminable time lag; everything else must be given and present entirely explicitly, because computers are incapable of processing ambiguity. From this situation of determined capacity, Internet users (terminal-text body-subjects) create ambiguity and produce a context of greater significative depth, by using text (alphanumeric symbols) in non-textual ways. A parenthetical smiley, rather than substituting for a non-verbal cue,[9] is an effort to bring the meaningful context of explicit utterances back into play. The flexibility of Internet users and their disregard for the conventions built into the design of network software,[10] are further efforts to embody meaning. Instead of transcending from the less explicit, ambiguous anonymity of the meaningful milieu towards the more explicit, denotative anonymity of universal language, Internet corporeality exhibits a transcendence from the anonymity of explicit structure to the personal level of ambiguity. The terminal-text body is an alphanumeric presentation on screen, which either is or is not manifested, depending on (a) whether it is properly typed, and (b) whether the computer network is operative. These absolute, all-or-nothing conditions are entirely unlike the lived ambiguity of being-in-the-world. But we cannot outrun our physical natures, only transcend them and our situation. Corporeality is the condition for entry into the Internet, where it is limited to the capacity to transmit and receive alphanumeric symbols. In the use of parenthetical smilies, abbreviations, and the reconfiguring and misuse of the network software that typifies the history of the net, users have continually attempted to create ambiguity. Body-subjects can live neither solely in the anonymous generality underlying potentiality, nor in the anonymous universality of sedimented structure; as Merleau-Ponty says, "In both cases we are imprisoned in the categories of the objective world, in which there is no middle term between presence and absence."[11] In the real world, life is already ambiguous; the Internet has been made ambiguous by the body-subjects who have made it a medium of expression.

INTERSUBJECTIVITY

Internet Relay Chat is a system for holding conferences between users who are all on the same channel. On IRC, a dialogue appears on your terminal screen between the participants who have joined that channel. The system reports changes of the topic, of the participants, and statements typed by other participants onto their terminals. You can also exchange private messages to other users that the rest of the participants cannot read. IRC users attempt to approach the system like a face-to-face conversation. They "speak" frankly and openly about the topic, and their statements are reported on one another's terminals in a linear order, roughly the same as the order of face-to-face conversation. Rheingold offers this description of IRC exchanges:

> Quick wit is necessary because rapidity of response becomes important in this written medium in the same way it is important in face-to-face conversation. IRC is a dynamic form of communication Somewhere in the world, a human being has typed those words on a keyboard, no more than a couple of seconds ago; if you know the right words to say in response, you can leap into the conversation and make that person and others around the world laugh out loud, grow angry, feel lustful.
>
> [. . .] You can treat IRC as a spectator sport, never venturing into the flow. Or you can show the IRC tribe how fast you are with a well-worded rejoinder by keeping up with other participants in a rapid interweaving of cleverly linked comments. . . .[12]

Rheingold also notes the "[i]nitial absence and subsequent reconstruction of social context"[13] on IRC:

IRC is what you get when you strip away everything that normally allows people to understand the unspoken shared assumptions that surround and support their communications, and thus render invisible most of the web of socially mediated definitions that tells us what words and behaviors are supposed to mean in our societies.[14]

The presentation of others through IRC is a presentation of text on a terminal screen. I read their statements and respond by typing my own. The presentation of others is always textual; the text has a certain ambiguity of situation developed through the transcendence of the digital format of computerized data transfer.

According to Alfred Schutz, authentic intersubjective relations are possible only in the case of simultaneity – the sharing of time and space.[15] Shared time and space entails the presence of common objects, which are within the reach of all of us in our intersubjective situation.[16] Schutz' discussion of simultaneity comes to define it as the melding of three dimensions of time: the private event in my time, the private event in the other's time, and the public event we both perceive. In this melding, common objects serve as the basis of public time. Schutz' example is of a bird seen in common, thus within the 'reach' of sight; the private side of this public event and object is the difference in our meaningful approach to the bird (as Schutz puts it, our private events of wanting to shoot it or simply to enjoy it). On IRC, the terminal screen reports back all the public messages exchanged by the participants in the order they are processed, all private messages directed at you, and announces changes to the channel (other users joining or leaving the channel, changes in the "topic" of the channel by particular users). The public events of the main body of the dialogue and the announced changes to the channel constitute the situation of the conversation on IRC. The common object, within reach of everyone, is the text itself. IRC seems thus far to fit Schutz' definition of authentic intersubjective experience.

Unlike when speaking to one another, we do not see each other. Unlike face-to-face or radio or telephone transmissions, the levels of mediation by computers in IRC always entail a lag of several seconds or up to several minutes. The dialogue appears only at first to be in perfect linear order. More than cursory experience of IRC shows that one of the major obstacles to communication on IRC is lag. IRC users strive to respond quickly to what appears on their terminal screens, but this is a vain attempt to recover the time lost to the mediation. In addition to the problem of lag itself is the problem that there is no feedback to tell a user that there is lag, or how great the lag is. A user reported to me that he responded to the dialogue on his terminal, and the dialogue following his response fit his statement, so he responded again, and continued in this vein for some time. Eventually, the dialogue on the screen made less and less sense, and he realized there was a significant lag between what he was reading and his own responses, despite their apparent linear fit. It turned out that the lag was over ten minutes, and none of the users with whom he was communicating were still online! The disruption of the dimension of time, the absorption of time by the machines which mediate all acts and perceptions online, has two consequences for the possibility of intersubjectivity. In the first place, lag breaks apart the three temporal dimensions of common objects. The object – text – is public, but the private events of each of the users reading the text are not temporally unified in the public event. There is no common object in Schutz' sense. In the second place, lag precludes co-presence with others, since we cannot distinguish the simultaneous (and hence authentically intersubjective) from the contemporary, the present from the past, and therefore all interactions with others are left without the full co-presence of intersubjectivity. Others are never presented with evidence as other selves, as alter egos. In the first place, they may not be "selves" at all, but software bots; in the second place, even real others are never co-present, and may be presented with great enough lag that although they appear to the terminal-text body as co-present, they are already gone. There is no evidence of intersubjectivity on the Internet. There may always be no real others.

According to the received view of the Internet, the question of the possibility of real others is not legitimate. Of course there are no real others, so the objection runs, since on the net there are only virtual others, a virtual community. The use of the term "virtual" in describing relations with others over the net further emphasizes the interpretation I've offered here. The *Oxford English Dictionary* defines *virtual* as "that is so in essence or effect, although not formally or actually; admitting of being called by the name so far as the effect or result is concerned."[17] Others which are not intersubjectively real others formally or actually, are for phenomenology – as for real life – not others. The question is, what are unreal others, and in what sense are they others?

Schutz suggests that inauthentic others are still made part of a community by a contemporary recovery of them. He considers this a kind of intersubjective fiction. Those others we constitute fictionally are susceptible, to a greater degree, to our subjective imaginative flourishes, and there is less they can do to present their independence of our constitution. As we read fictional descriptions of characters in novels, we can be aware that those characters are imagined by the author; however, there is a power of fiction and of characters in fiction to elicit responses from us, to capture our imaginations. A fictional empathy can even be produced, which makes us cheer the protagonist and despise the antagonist. The sense of a character may be so well absorbed by a reader that the reader can imagine the character's reaction to other situations, can compare the character to real people, can wish to meet the character. All of this is quite like the way we know a real person, except for the crucial fact that fictional others are ideally constituted rather than perceived.

The limitations of perception and of the presentation of others on the Internet leads me to suggest that others on the net are fictionally constituted in consciousness, rather than perceived. Our fictionalization of them amounts to an imaginative "fleshing out" of more or less entire persons on the basis of the perceived terminal-text body. (Perhaps it is the way these fictional others fail to meet our expectations when we meet them that accounts for the disappointment and awkwardness described as common to such meetings.[18] It is certainly more intelligible than the explanation offered by some that Internet users become so dependent on the terminal-text body that real life communication is difficult for them. This explanation implies that for Internet users the basic mode of interaction has become the terminal-text body rather than the real body. Even in the extreme cases of individuals who are "addicted" to the Internet, the real body must be central, since it is the seat of those modulations – seeing and typing – which are the foundation of the terminal-text body.)

Others on the Internet are not virtual others, they are fictions. As fictions, they have an important effect on the lives of Internet users. They are a particularly stimulating set of fictional types because they come and go in ways that are not fully predictable (these others are, after all, usually real people). There is, however, in the final analysis, no evidence of their reality, of their being alter egos. The Internet is a solipsistic situation.

Users on IRC take up nicknames; the names they create for themselves appear on the screen with their dialogue, like the printed text of a play. Nicknames are an option to using one's institutional account number as a name, and serve users as personas. The terminal-text identity of a user on IRC can be discovered (unless the user knows how to prevent discovery) by typing a command which has the system report back the user's institutional account number. But in general, this account number says little about the user. Furthermore, the user may be allowing someone else to use the account, may be deliberately hiding his or her identity or playing an identity role, or may even be a computer-generated persona or "bot." Although bots are generally banned from IRC, and using someone else's account draws suspicion, IRC users seem to revel in role-playing, especially on channels devoted to deviant sexual practices. Over more pedestrian channels, one's identity is rarely questioned.

Identity is problematic over the Internet in a way that it does not appear to be in ordinary life. When I speak to you face to face, I perceive you, I have an immediate experience of your presence as an alter ego. There are no qualities in my perception of you which are not analogues of my own qualities; the abnormalities you may present remain analogizable.[19] Should your presence suddenly alter in a drastic way, that would be beyond my analogizing capacity. There must be a threshold of analogizability in our experience of others in the Husserlian view of intersubjectivity. If your identity undergoes rapid transmutation, it does not seem fitting to refer to it as an identity.

On the net, every user is the same – a string of alphanumeric characters appearing on a terminal. Variations of style are impossible to determine from this alone. The keys to identity must be in the explicit text – one's style is foreshortened to variations in spelling, grammar and vocabulary. Yet even these variations have little or no expressive value. Especially at the high speeds with which users type at their terminals, linguistic style is flattened. It is difficult enough just to maintain the patience and the audience to present a complete thought or an argument, as my experience on the IRC philosophy channel has shown. The niceties of syntax and punctuation are in general completely abandoned by users, and so a simplified, sparsely-punctuated, highly abbreviated idiom has developed, with highly stylized manners. The textual body inhabited by Internet users (as dictated by the technology) reduces the expressivity of the body to the capacity to write lines of text; the norms of Internet use further erode the potential to express oneself.

My encounter with others on IRC as a persona shared with two friends shows us the puzzling status of identity on the Internet. I and two other people combined as a single persona one evening, and each of us contributed to the dialogue reported as belonging to our namesake. Presentationally, there was nothing about us that could appear as unanalogizable. To other users, we appeared as a single user, and there was no way to make a direct determination that there were three of us operating this puppet. The manifestation of our persona as lines of text maintained a terminal-text unity as the foundation of the fictional individual our audience constituted. There was one typist, so the style of our terminal-text presented no abnormal shifts or alterations. Only because our three roles within the user persona were completely at odds, and only after we took to making very obscure references from our varied technical specialties (law, computer programming, and philosophy), did we ever face suspicion about our genuineness.

COMMUNITY

Lack of real identity appears to be a problem for Internet users; however, the net's lack of co-present corporeality attracts users. Physically impaired and remotely located people can acquire a quasi-community online. Groups threatened with persecution in our society can communicate – albeit with fictional others – in safety. The question this raises for my analysis is whether real others are needed for community; or, to put it another way, given that there are no virtual others in the experience of the Internet, can there be virtual community?

One of the correspondents in my research is a young gay man. The problems of anti-homosexual prejudice and violence, and his distaste for stereotypical gay meeting places – bars and clubs – led him to the Internet homosexual community. For homosexuals in the United States, information, literature, and open community are difficult to find, especially outside of major cities. The transcendence of the spatial body towards the terminal-text body of the Internet opens the possibility of a community with less fear of persecution. He has made a few Internet acquaintances with whom he has made regular CMC contact for some time, but he complains that when he meets these friends face to face, their ease of relating online fades. In general, he says, acquaintances online are not very dependable or solid, and have some sense of reality missing about them. Despite the irreality of other people (as discussed above in the section on intersubjectivity), the "community" he encounters online is important.

Schutz claims that communities extend beyond the simultaneous co-existence of intersubjectivity; "contemporaries" can be brought into the community ritualistically. Such contemporaries can be members of previous or succeeding generations. In any case, they are not intersubjectively real others. Contemporary environmental ethicists sometimes extend community to include future generations and even animals, though future generations are not real others, and it is doubtful that animals could be so construed. There is no reason to claim, on the basis Schutz gives us, that Internet community is not a real community. What is peculiar about Internet communities compared to those in everyday life is that they are based on no authentically intersubjective experiences of other members of the community.

In *The Virtual Community*, Howard Rheingold discusses the community spirit to be found in computer bulletin boards and Usenet groups. Usenet is a network of messages sent and received among various segments called newsgroups. The Usenet situation is similar to subscribing to a magazine: indeed, you subscribe to newsgroups on your terminal and the articles posted to the group are saved under that group's heading. You do not interact directly with other people in Usenet, but rather send messages back and forth that are publicly available. You select which newsgroups to read, which articles and follow-up responses to read, and whether and what to write and publish in the group. Usenet is comparable to conventional print media. The option of publishing and distributing your own text is available through vanity presses. Usenet is a universal, computerized vanity press. One joins the Usenet community strictly as a matter of choice and remains active by contributing to a thread of articles or posting a new article. If Usenet constitutes a set of communities, it is marked by the second peculiar feature that its communities are established only voluntarily.

Usenet users are quick to punish and correct abnormal behavior, a phenomenon known as flaming. As an investigation of the norms of Usenet, I subscribed to a group devoted to flaming, and sent an article to the group in which I made a very unsophisticated flame. I was immediately flamed in response and declared a "newbie" (an uncouth new user). If this had not been the reaction I was expecting, it would have had a powerful impact on my Usenet behavior.

"Why are people so aggressive on the Internet?" asked the subject line of a Usenet article. Most of the responses were flames, some of which described the respondent violently attacking the inquirer. The community for which the Internet receives praise has its distinct advantages, as I have already stated. One of the advantages is vouchsafed by the anonymity of the computer mediation, the lack of stylistic expression possible through a terminal and the textual representation of others. The anonymity also offers the potential for open abusiveness without consequence. Anyone from anywhere can join a channel or newsgroup and get help and encouragement for the problems of growing up gay, for example; by the same magic, anyone from anywhere can invade that channel or newsgroup and antagonize those communicating.

FREEDOM

It would not be out of place for me to discuss freedom in the present context. Internet users do describe a feeling of freedom – to state whatever crosses their minds, to be witty, to create a masque – as a lure to the Internet, but I am not speculating about their motivations. The fact is that this feeling exists, astonishingly, for those who spend many hours transcending in no more free a way than reading text and typing. The freedom of the terminal-text body is mythologized by the metaphors of "cyberspace," of "surfing" the net, and so forth. Nothing that one does on the net comes near to the myth; nevertheless, there is an important sense of freedom among users.

Freedom, in this case, means freedom in the transcendent sense given the term by Merleau-Ponty. To be free is not a political or physical condition, but an essential feature of being-in-the-world. Every act is transcendent, hence free. Merleau-Ponty's concept of freedom extends to both acts which open the body-subject to the world, and acts which bring the body-subject into recession. The anorexic girl is no less free than Merleau-Ponty's never-identified "normal" subject. Freedom can mean the transcendence of the situation in such a way that the body-subject is opened further; but an important form of freedom is enacted in recession from the world. The Internet provides an example of each.

Rheingold refers many times to the fact that the Internet has grown up organically into the "anarchy" that users congratulate themselves for maintaining today.[20] Internet is the result of many generations of changes to the basic technology of CMC "home-brewed" by users themselves. Early CMC networks were developed for purposes of governmental bureacratic control, but the lack of a central location of the computer network – the fact that the network was supported by the full network of computers, which would be able to reroute data if a machine in the network broke down – made the network open to misuse. Early network communication almost immediately overflowed the bounds of the machines' capacities with unofficial transfers of information. Eventually, those engaging in unofficial communication developed software to handle the load. This story has been repeated at every stage of the Internet's development. In each case, the urge to communicate outruns the structure of the network and the software, and a group of software hackers puts together a new system. The new systems are immediately misused again, taken up for communicative purposes that are not well suited to the networks.

This great effort has produced a network whose nominal aim is to bring people in touch with one another; the result is a network which systematically precludes the possibility of authentic intersubjectivity. Yet the net has not failed. My correspondent told me that on the net he feels "freer to approach people. People are on the net because they want to simply communicate." As a result, "conversations are somewhat wittier than in real life. . . . And people are nicer; there's more of a spirit of friendliness (when they aren't ripping each other's throats out inexplicably)."[21] The irreality of others on the net yields a freedom of relation. Others are constituted fictionally, which places them under the command of a mental construction. As Clifford Stoll puts it, "[e]lectronic communication is an instantaneous and illusory contact that creates a false sense of intimacy without the emotional investment that leads to close friendships."[22] They cannot be faced as real, so they cannot break through my effort to recede from conversation, they cannot pose a threat. It is a persona of unfixed and unreal identity to whom one communicates. This inauthentic relation underlies the easy community and harsh aggression online. The other is an unreal other, there is no real harm.

The transcendence of the remarkably strict limitations of the terminal-text body has been the condition for the development of the Internet. That same set of limitations is the condition of the transcendent relation to others found in Internet intersubjectivity. CMC is a situation of the body-subject as a terminal-text body. The terminal-text body is of course still transcendent, but the transcendence of the terminal-text body cannot extend beyond the limitations of the situation of CMC. In short, transcendence cannot mean the supernatural leap from one's situation. Body-subjects cannot exist everywhere or nowhere; they cannot exist in cyberspace. Freedom is not the arbitrary absolute power of a pure subject – no such power, no such subject exists. The Internet surfer and his virtual waves belong to myth. The Internet belongs to the real world, as a mode of inauthentic intersubjectivity of real transcendent body-subjects, who go online to exercise the subjective freedom afforded a body-subject in inauthentic recess from others. Freedom from others online produces the peculiar intensity of these relationships – an intensity without reality. The limitation of the terminal-text body prevents authentic intersubjective openness to others; this situation can be transcended, but not overcome. We cannot co-exist in a mode mediated by computer communication; we cannot co-exist in cyberspace.

NOTES FOR A PHENOMENOLOGICAL CRITIQUE OF NAIVE ANALYSES OF THE INTERNET

One of the most highly-reputed studies of CMC is Starr Roxanne Hiltz and Murray Turoff's book *The Network Nation*, originally published in 1978, updated and republished in 1993. Their analysis relies heavily on statistics and a view of communication ascribed to Erving Goffman. According to their conceptual model, the content of communication is what is explicitly expressed; all that is not part of this central and discrete meaning is relegated to the status of "cues." Thus, from their perspective, CMC is identical to face-to-face communication insofar as it retains "content"; the only difficulty for interpretation is the lack of "cues."

Not only does this analysis seem shortsighted in its conception of communication, it also appears false on the basis of the present interpretation. I subscribed to the Usenet group alt.flame to observe; after a week, I posted an article in order to see how the other participants would respond. There were four responses within 24 hours, one of which confidently (and accurately) declared me a "newbie." I asked an Internet veteran about this, and he explained to me that it is actually quite easy to spot a new user, simply on the basis of the way they behave on the net. My post contained no "cues," nor did anyone else's; but this is not an obstacle to anyone's determining my status as an Internet user. Yet Hiltz and Turoff (and no doubt many other net enthusiasts) would tell me that the disoperative cues are those connected to every kind of invidious stereotype. Over the Internet, they claim, one is not identifiable by appearance or rank. Your social class, your race, your age, your gender, are irrelevant because they are unknown.

> There are some persons who . . . are not skilled at interaction rituals, who can nevertheless seem particularly suited to communication in this new medium. One . . . is the person who is constantly interrupting others, talking over them, interjecting comments that are "off the subject," or otherwise aggressively deviate from the norms of polite face-to-face conversation. What might seem to others to be obnoxious behavior in the face-to-face condition is apt to seem scintillating or original in computerized conferencing since no one is actually interrupted by such out-of-order contributions.[23]

Hiltz and Turoff recount (as does William Rheingold) many examples of computerized communication improving conference discussion, by permitting individuals to reverse their usual status in everyday life and take on a persona more powerful, more thoughtful, more well respected, etc., online. In short, those who find it difficult to succeed in ordinary life seem drawn particularly to the computer's possibilities of transcendence. Perhaps, but a new stereotype, and new biases and exclusions, are immediately evident. The net is not a place of free exchange of ideas with no discrimination; discrimination is simply shifted to what does appear: CMC sophistication, obeying the norms of the net, vocabulary choice that reflects dialect (there are newsgroups whose participants are homogeneously American, British or Australian, for example). Status-based discrimination is not very different over the Net than in face-to-face relations. In both modes, we can make efforts to hide our status liabilities.

Moreover, if non-verbal cues are derivative upon communication proper, then two extant facts become unintelligible: (1) an effort is made to transcend the explicitness of alphanumeric code, towards ambiguity; (2) a "netiquette" exists, and violators are punished – sometimes only because they have not followed a newsgroup's idiomatic form.

The case of my posting to alt.flame shows that there is no distinction to be made between the content and the ancillary cues of communication. Content and cues are just more and less articulated gestures of the same movement of communicating. My deliberately unsophisticated article did not contain any explicit information that could only be interpreted to mean that I was an inexperienced user, but the explicit content failed to recognize the idiom of the newsgroup.

The naive analyses and publicity hype of the Internet are based on a fundamental misunderstanding of the nature of community and communication. The reduction of the perception of others to the perception of the terminal-text body conditions interrelation by forcing expression and perception through alphanumeric encoding. But the lived situation of CMC is one already made ambiguous, and this is already meaningful at a level that is not explicit, although it does, of course, depend on what is presented. Terminal-text manifestation is always gestural, even if only to a degree that is still inauthentically intersubjective.

Furthermore, the hype of the creation of online communities is espoused in ignorance of the fact that online communities are founded upon the real community which has produced them and produced the conditions which make them attractive. The intersubjective community of those living in the siliconized world forms the cultural milieu which found the possibility of CMC, as idea, as technology, and as an aim for praxis.

Duquesne University

NOTES

[1] Rheingold, Howard, *The Virtual Community* (New York: HarperPerennial, 1993). Hereafter VC.

[2] The day I wrote this, I had trouble making hotel reservations because of computer malfunctions at two different hotels. In addition, I failed in my attempt to post an article to a Usenet group because of problems with my machine. Like the reservation clerks, I reacted sharply to my equipment failure. Replacing the old stand-by excuse "my dog ate my homework," I have heard numerous students tell me that "the computer ate my paper." Other people have suffered grief when their computer "died," or become enraged as the machine treacherously turned against them. On the other hand, computers and software are sold on the merits of being "user-friendly." Two current advertising campaigns emphasize the "friendliness" of the machines. In one, a young woman sips her morning coffee while her computerized assistant, who has a pleasant if mechanical voice, and is given a puppy-shaped icon on her computer screen, reports on all that "he" has done for her so far this morning, ending with, "I'm still working on those playoff tickets." "Good boy," she responds. The other campaign sells a helpful guide called "Bob" who appears on the screen to present explanations of steps to achieve a particular aim with certain software packages.

All of these examples point to our mythological everyday comprehension of computers. They are not machines, in our naive conception, but entities imbued with intelligence, with animation, with emotional responses. Computers are not taken up in the everyday world as machines at all, but as creatures with lives of their own. The magical superstitious relation to computers is taking longer to wear out than the initial reaction to other technological appliances (telephones, for example), perhaps because the complexity of the computer is something intangible; but the production of computer mythos cannot serve to enhance our understanding. The selling of "friendly" machines has been extended, along with the myth, in the field of CMC. Now computer and telecommunications companies sell friendly machines and friends to go with them. (Rheingold notes how the WELL waives fees for "hosts" whose job is to encourage users to join, and remain active on, the WELL. The WELL is "in the business of selling the customers to each other . . ." *The Virtual Community*, p. 43.)

[3] Some commands may be abbreviated, but this is not a recognition of implicit or tacit meaning on the part of the terminal or computer. Only abbreviations which are as explicit as full commands are accepted.

[4] In addition to Usenet and IRC are the Information Gopher and the World Wide Web. The Gopher is comparable to an online directory of information of whatever type – anything that can be and has been uploaded to a computer system somewhere. Gopher projects include the creation of an online library, the construction of an online Library of Congress card catalog, and online phone books and e-mail directories. The World Wide Web is a system for allowing Internet users to reach a prepared presentation – a "page" – giving information about the person, corporation or institution that has put the page together. The Web is increasingly popular since it has multimedia capability – that is, a person with a multimedia terminal can not only read text, but see images and hear sounds digitized on a computer and assembled on the page. Rheingold, in describing the Web and multimedia machines, calls this technology "a solution looking for problems to solve" (*The Virtual Community*, p. 194). In the two years since Rheingold's book, entertainment corporations have begun to sell products for multimedia with the author's signature file, inviting you to "check out my web page!" Neither the Gopher nor the Web are directly relevant to my discussion of intersubjective experience.

[5] As has been recently scandalized in the American popular press, there are digitized pictures encoded and sent over Usenet groups, but this is clearly not a form of intersubjective experience in the sense taken up here, viz., of encountering other persons or communities of persons.

[6] VC, p. 69.

[7] VC, p. 176.

[8] See Stoll, Clifford, *Silicon Snake Oil* (New York: Doubleday, 1995), pp. 8, 18.f, for two of his many comparisons of activities possible with a computer and those possible without.

[9] Such is the interpretation which follows from Rheingold's and Hiltz and Turoff's analyses. See the final section of this paper, "Notes for a Phenomenological Critique of Naive Analyses of the Internet."

[10] See VC, Chapter Two, "Daily Life in Cyberspace: How the Computerized Counterculture Built a New Kind of Place," pp. 38–64.

[11] Merleau-Ponty, Maurice, *Phenomenology of Perception*, trans. Colin Smith (London: Routledge, 1962), p. 80.

[12] VC, p. 177.

[13] *Ibid.*

[14] *Ibid.*, p. 178.

[15] Schutz, Alfred, *The Problem of Social Reality (Collected Papers, Volume I)*, edited by Maurice Natanson (The Hague: Nijhoff, 1962), p. 315.

[16] *Ibid.*, p. 316.

[17] *The Compact Edition of the Oxford English Dictionary* (London: Oxford University Press, 1971), p. 3639.

[18] My correspondent mentions this. Accounts of the same phenomenon can be found in Stoll and Rheingold.

[19] Husserl, Edmund, *Cartesian Meditations*, trans. D. Cairns (The Hague: Nijhoff, 1960), pp. 50ff.
[20] See VC, Chapter Three, "Visionaries and Convergences: The Accidental History of the Net," pp. 65–109.
[21] Personal interview. My own experience is the same. Accounts published by Rheingold and Stoll are similar, and even the more controlled experimental research of Hiltz and Turoff shows the same open friendliness and open aggression.
[22] Stoll, p. 24. See also p. 58.
[23] *Network Nation*, p. 102f.

PART FOUR

THE SENSE OF CITIZENSHIP, POLIS, RIGHTS IN THE MORAL INTERPLAY BETWEEN THE SELF AND OTHERS

MERCEDES GALÁN-JUARÉZ

HUMAN RIGHTS AND PHENOMENOLOGY

There are at least two good reasons for connecting phenomenological research and the problem of human rights. The first is of a general nature: phenomenology has been *important to legal philosophy* from the very beginning of its development. The second is, that modernity has widely accepted the rights language, that is the language of rights and duties, as *appropriately expressing the essence of man.* Both reasons have their own difficulties and philosophical as well as legal-theoretical consequences. However, both are fascinating themes for research to phenomenologists.

PHENOMENOLOGY

Let us begin with the *first* observation. Classical texts in phenomenology, from the second and third volumes of Edmund Husserl's *Ideen* and in later years his *Krisis* to the works of Sartre, Merleau-Ponty, Schutz and Gurwitsch are considered important features of legal discourse. An outstanding phenomenological approach to jurisprudence has developed in Germany, with Reinach, Gerhard Husserl, F. Kaufmann, J. & W. Schapp, and T. Litt as its founding fathers. One has also to mention France with Kojeve's interpretations of Hegel, with Poulantzas, Kalinowski and Amselek. South America has the work of Sergio Cossio and his egological philosophy as well as the oeuvre of Miguel Reale. All these authors in phenomenology deserve the full attention of legal philosophers today. Modern sociology and social theory, symbolic interactionism and cultural anthropology have taken advantage of the above-mentioned body of knowledge.

Most important appear to be thoughts parallel to those developed in Kelsen's *Pure Theory of Law*. Kelsen shows phenomenology at work! His analyses were directed to the phenomenon of normativity. In this context Kelsen needs to specify the juridical. One could conclude with B. Waldenfels (*Phänomenologie in Frankreich*, 1983, Ch. VII) that the *Pure Theory of Law* prepared the way for a more elaborate *phenome-*

A-T. Tymieniecka (ed.), Analecta Husserliana, Vol. LIV, 201–205.

nology of law. This is first of all because the *eidos* of law in general was his central issue, and furthermore because he determined the eidos of jurisprudence to be *normativity*. The legal world, a world of legal facts, is implicitly a world of norms, norm-propositions, and evaluations. Facts and values are, phenomenologically speaking, on an equal semantic level. Facts are by no means objective data; they are rather *expressions* and *propositions* meeting the basic requirements of legal discourse as a *normative* discourse. Facts and norms correspond. This Kelsenian intuition and *insight* as well as the *research on this specific correspondence* is truly phenomenological.

ORDER

A basic issue of a phenomenology of law is the problem of *order*. Human Rights are in this regard an important concept to discuss. They challenge the legalistic conceptualization of order in so far as vertical structures are given less importance, and horizontal concerns – a problem of law and legal theory in its own right – are brought to the fore.

The traditional hierarchic understanding of order is indeed based upon the *metaphor of the vertical* in law and legal thinking. Research into the many dimensions of that metaphor is an exquisite phenomenological theme. Hierarchy was for centuries the original scheme for law and even a cornerstone for attempts to define law *sensu strictu*. Is it necessary to say that the human creative condition, the theme of this congress, is a prime theme for my exploration of the relationship between phenomenology and the concept of Human Rights? Even the self-individualization of the individual in the occidental hemisphere is shaped by the concept of Human Rights and, more generally, by rights language and its implicit social forms. The phenomenological problem is that *a basic image*, a basic *scheme* appears: the image of the necessity of the vertical. This image does not stand on its own: it is in itself *a dimension of the more general image of man that dominates legal discourse* and contemporary society. At this point, phenomenological interest and the power of phenomenological analyses could become extremely helpful. Is it the final word to say that horizontal concern and coordination can only be understood as humanity, solidarity, philanthropy or *caritas*, but not as relationships that find articulation in the language of the law, that is basically: in rights language? Does the law's verticality conceal the virtues of social life's horizontal relationships?

A fundamental insight in this context is that some social structures are based upon *necessity* while others are based on *freedom*. The first archetype is the classical idea of the state, the second, the international legal order. Is the legal order of the European Union or the legal realm of Human Rights an example of the latter kind? If one is prepared to accept this idea and to do research into the structures of the example, one will discover that in structures which accept freedom as a basis for legal relationships the vertical is not absent.

This outline of legal theory and the philosophy of law leads to a *second* consideration with regard to the connection between phenomenology and the maintenance of the concept of Human Rights. *The practice of Human Rights and theory of law both accept the presupposition that the essence or an essence of man can be expressed in what we call today rights language.* That is a phenomenological motif, but is it correct to acclaim this position? In the case of a phenomenology of law, there is no discussion of the content of that essence – this is left to philosophers and ethicists. There is, however, the problem of *expressivity* in its own right. We repeat: Is rights language compatible with the phenomenality that is envisaged in Human Rights?

Two questions are involved. The first is the so-called universal character of Human Rights. The problem is debated on an almost universal scale. Characterizations of 'rights language' are therefore phenomenologically interesting. The diversity of relevant discourses appears to be the substance of clarification. Phenomenology thus focuses on the relationship between two concepts. The universality of the concept of Human Rights is a *subjective activity* and an *a-subjective event*. Human Rights, thanks to their objectivation, can be claimed by individuals and they are constructed by institutions which are determined according to the same patterns of thought. This individualistic creativity can, however, only occur within the courses of a-subjective events. Subjects are cultural constructs, and the fundamental processes of individualization in occidental culture show their importance. How are the two related? 'Rights language' suggests how to solve this problem: rights are subjective and a-subjective activities at the same time. The riddle is in the coincidence of these activities.

It is phenomenologically important to consider that human activity never results in *things (res)* but in *history and historicity*. In as much as things are created, one could say, things witness human history. Our world bears the marks of human creativity in all regards. Law identifies history as a major feature of human existence. Laws and legal rules are like *things* in so far as they reify behaviour and history; rights – especially human rights – are conceived as things in legal practice; they are philosophically spoken *images of an interpreted future*.

RIGHTS, FACTS, REFERENCE

This is phenomenologically important. Human rights are legal facts, but also images of man that have become expressed in legal language. They imagine a legally determined and interpreted future situation of concrete individuals. This point of view involves two very important dimensions.

(a) The first is, that the bearer of Human Rights claims a value and a dignity *beyond* traditional norm-oriented legal concepts. However, *rights must be claimed within the existing laws as an institution*. Courts and rules, judgements and cases unfold within the boundaries of that institution.

(b) In bearing Human Rights and accepting the 'rights language' as an instrument for defending and realizing human dignity, individuals who claim human rights surpass their traditional references to norm-propositions. *The true reference of human rights activities is reference to an image – not to a norm*. This implies a change in the basic pattern of legal thought, a change which challenges the dominant legal paradigm. That change can be adequately analysed through a phenomenological approach.

The image of man in Human Rights parallels the general legal image of man. This is not so by nature, but this adaptation results from the requirements of the rights language itself.

The connection between phenomenology of law and the concept of Human Rights shows how legal thinking and jurisprudence need certainty, stability, coherence, and therefore tend to resist paradigmatic changes. This might be one of the many reasons why the human 'rights language' never openly refers to the image of man, determined by law and legal thought patterns but only refers to such an image *as if* it involves a norm. In transforming the image into a norm, often a norm-proposition, the very concept of Human Rights succeeds in becoming accepted as a truly legal thought pattern. Phenomenological analyses could underline the specificity of this referential structure and determine the social consequences both in understanding and in governing social life through legal practice.

Universidad Complutense
Madrid

GRACIANO GONZÁLEZ R. ARNAIZ

DERECHOS HUMANOS Y CONDICIÓN HUMANA

Notas sobre "Maneras de Pensar" la Fundamentación de los Derechos del Hombre

Somos cinco amigos; cierta vez salimos uno detrás del otro de una casa, primero vino uno y se puso junto a la entrada; luego vino, o mejor dicho, se deslizó tan ligeramente como se desliza una bola de mercurio, el segundo, y se puso no lejos del primero; luego el tercero, luego el cuarto, luego el quinto. Finalmente todos estábamos de pie, en una línea. La gente se fijó en nosotros y señalándonos decía: los cinco acaban de salir de esa casa. Desde entonces vivimos juntos, y tendríamos una vida pacífica si un sexto no viniera siempre a entrometerse. No nos hace nada, pero nos molesta, lo que ya es bastante; ¿por qué se introduce por la fuerza allí donde no se le quiere? No le conocemos y no queremos aceptarle con nosotros. Nosotros cinco, en verdad, tampoco nos conocemos ahora, pero lo que es posible y admitido entre nosotros cinco es imposible e inadmisibble en ese sexto. Además somos cinso y no queremos ser seis. Por otra parte, qué sentido puede tener esta convivencia permanente, si entre nosotros cinco tampoco tiene sentido, pero nosotros ya estamos juntos y seguimos estándolo, pero no queremos una nueva unión, precisamente en razón de nuestras experiencias. Pero, ¿cómo enseñar todo esto al sexto, puesto que largas explicaciones implicarían ya una aceptación en nuestro círculo? Es preferible no explicar nada y no aceptarle. Por mucho que frunza los labios, le alejamos empujándolo con el codo, pero por más que lo hagamos, vuelve siempre otra vez.[1]

Salvo que se tome la determinación de dar por zanjada la cuestión del "fundamento" en la reflexión sobre los derechos del hombre – desde posturas realistas[2] o postivistas, al estilo de N. Bobbio[3] – el contínuo "revival" de la polémica hace aconsejable, cuando menos, la "vuelta del pensamiento" sobre la misma, en el bienentendido de que lo que nos jugamos en ella es determinante para la manera que tenemos de entendernos como individuos y como ciudadanos.

Es cierto que la carga de la prueba cae de nuestro lado, pero no es menos cierto que una "continuada" lectura de los derechos del hombre exige un afinamiento jurídico constante al que no es ajeno la búsqueda de "otras" razones a cuya luz interpretar las "situaciones humanas" que niegan esos derechos, les legitiman o les prolongan.

Entiendo por ello, que en este empeño han de coincidir, al menos, una perspectiva jurídica, una perspectiva política y una perspectiva ética o filosófica. El orden aquí no es consecuencia del "puesto" que ocupan en el ordenamiento de las ciencias, ni de la importancia o supeditación de una perspectiva a otra. Sin embargo, es preciso dejar claro ya desde

A-T. Tymieniecka (ed.), Analecta Husserliana, Vol. LIV, 207–218.

el principio, que una actividad jurídica o política, que son siempre prácticas por antonomasia, deben ser – en el sentido de que merece la pena que sean – también filosóficas en la medida en la que participan de la urgencia de tener que "dar razones" de lo que hacemos o decimos.[4]

Por ello, la relexión sobre el fundamento, remite siempre de una u otra manera – bien es verdad, que a su manera – a la perspectiva ética "encargada" a partir de este momento de otorgar aquellas razones "primeras" en las que poder apoyar, denegar, legitimar . . . unas determinadas "lecturas" de los derechos del hombre, como en el caso del relato de Kafka, y en las que, a mayores, está implícita una determinada concepción del hombre.

Las diversas teorías éticas compiten en noble lid por dar las "mejores" razones, pero son razones "segundas" – que no secundarias –; es decir, a la ética se le requiere o bien para que aporte el "fundamento absoluto"[5] a algo que aparece como dado o cuando menos, se le exige que se ponga de acuerdo en las "razones relevantes" con las que solventar los problemas que un ejercicio de tales derechos lleva consigo.[6] Tareas ambas de imposible cumplimiento como es obvio en el primer caso, mientras que el segundo requerimiento olvida el hecho del pluralismo cognoscitivo.[7]

A la vista de estos planteamientos la pregunta, por lo que se refiere al plano característicamente filosófico del tema, tendría una doble perspectiva: la primera consistiría en abordar la "manera de pensar" la cuestión del fundamento desde la conciencia o el ser por parte de la filosofía. Y es que pensada en radicalidad, esta modalidad terminaría negando la posibilidad del disfrute de tales derechos a nadie que no fuera él mismo. El **déficit** de **calidad humana**, es decir, moral, de una realización así concebida arruinaría cualquier intento de fundación de los derechos del hombre.

Por ello, la segunda perspectiva aborda el cambio de rumbo en la investigación filosófica en la medida en la que reconce que la "**situación humana**," que en la modernidad aparece ligada al disfrute de los derechos del hombre, hay que verla por relación con el "**acontecimiento antropológico**" (8) que translucen dichos derechos y que es anterior, en el sentido de previo, a la reflexión. Dicho "acontecimiento" consiste en decir que el Yo se levanta ya como **ser moral, es decir**, como un ser que tiene que **responder** de y por los demás. Esto explicaría que la pregunta por los "derechos del hombre" se convierta en la "cuestión primera" por cuanto en ella se juega la **calidad humana** o no de una determinada realización. Así pues, en ambas lo que está en juego es la posibilidad de **pensar** lo moral como lo determinante para una "situación" que de verddad pueda denominarse humana.[8]

1. UNA MANERA DE PENSAR

A tenor del texto de Kafka que preside nuestra reflexión y la ilumina, entiendo que no es ocioso preguntarse por el **sentido** de la inteligibilidad, en la medida en la se reconoce que "ponerse a pensar," i.e., el hencho de tener que dar explicaciones, "implicaría ya una aceptación en nuestro círculo" del sexto, que no tiene el rango de amigo y a quien no se le quiere reconocer derecho alguno.

La cuestión aquí no es que no se pueda apelar la "consenso" teniendo como base la aceptación de un diseño procedimental, tan caro al derecho por otra parte. El problema es tan primario como decir que "no le conocemos y no queremos aceptarle"; i.e., no queremos "entrar en razón."

En este sentido habría que plantearse cómo una voluntad libre o autónoma que reivindica el derechoe del hombre, puede imponerse a otra voluntad libre, sin que tal imposición suponga el *efecto* de una violencia sufrida por esta voluntad. La cuestión se solventairia si la **decisión** de esa voluntad libre se conformara a una **máxima de acción** capaz de ser universalizada sin contradecirse. En la tesis kantiana que avala este planteamiento, lo que pone de relieve la capacidad de universalización de las máximas es la **razón** que habita una voluntad libre capaz de hacerse *respetar* por las otras voluntades que son libres gracias a su racionalidad.

El problema es si esta *voluntad libre*, traducida por Kant como **razón práctica**, termina por "doblegar" esa parte incoercible de la espontaneidad de la que hace gala la voluntad entendida como libertad. El texto de Kafka es un desmentido rotundo a esta posibilidad y manifiesta la "situación" de riesgo en la que se encuentra todo intento de fundamentación que se quiera hacer desde aquí.

La situación es tan comprometida que o apelamos a una situación **anterior – en el sentido de previa** – a la posibilidad concedida a un **YO** de "ponerse a pensar" o hemos de declararla como situación originaria – donde se da el sentido de una inteligibilidad sin sentido, sin salida – cuya "originalidad" consiste el negar la condición humana – sus derechos – a alguien.

¿Qué clase de **calidad humana** podríamos adscribir a esta manera de pensar, que no es más que reflejo de un modo de ser? La práctica unanimidad en definir la "situación humana," al menos en occidente, por su relación cons los derechos del hombre, ¿avalaría una reflexión en la que un "tercero" – si el **YO** lo quiere – podría o tendría que quedar excluído?

El texto de Kafka, ciriosamente descrito bajo el rótulo de **comunidad**, describe a la perfección esa "situación humana" en la que, para Levinas,[9] se inscribe la reflexión filosófica en tanto que pensamiento sobre el Ser, es decir, en tanto que referido a un "territorio" – la casa –, como lugar desde el que es preciso dar cuenta de todo.

La "territorialidad" de la conciencia, en la que la filosofía moderna encuentra **su** casa, explica el sentido de unas leyes por las que se rige y que son, en un sentido literal del término, leyes de carácter **oikonómico**: de abastecimento y de sentido.

En un contexto así descrito, la comunidad de los cinco ocupa un lugar investido de poder – en tanto que posibilidad de afirmarse en él – gracias al cual poder tener un *sitio*. Aquí, "tener sitio," es poseer un derecho merced al cual la comunidad se constituye como tal.

Lo realmente curioso y llamativo de esta "situación" es que es ese derecho de la comunidad, de los cinco amigos, – que habitan un lugar – el que deja afuera al sexto, al que se le reconoce, no obstante, como poseedor de significación human, puesto que vuelve siempre otra vez."

La consecuencia no puede ser más penosa para quienes han empeñado su vida en una lucha por los Derechos del hombre y a la vez, paradójica, para quienes se han empeñado en la tarea de seguir pensando esos mismos derechos. Si la situación "real" es la descrita en el texto de Kafka, tendremos que concluir que "a través de la expresión del derecho de la comunidad el hombre no tiene derecho".[10] Es más, si el derecho es *formalidad* – como muy bien ha recordado Kelsen – entonces tal formalidad no puede ser derecho del hombre, puesto que con la humanidad manifestada por el sexto individuo, se desgarraría el tejido de esas formas a causa de la terquedad con la que este individuo irrumpe una y otra vez en el "lugar."

La violencia inscrita en la negación de los derechos del otro – de aquel **que no es** por relación con ninguno de los otros cinco – mo es, en esta perpectiva sino el fruto de una manera de pensar – manera de ser – en la que todo debe pasar el fielato de la conciencia. Saber del otro, es ya "captarle," estar capacitado para dar cuenta de él y a la larga, temor del otro – de lo Otro – que exorcizamos "divinizando" el territorio, lo natural, que a partir de ahora representa lo Sagrado (Heidegger).

Esta "resacralización" de lo natural, en donde campan por sus respetos las fuerzas obscuras y sin nombre – anónimas –, explica contundentemente la "vaciedad" de un concepto como el de humanismo y declara expedito el camino para que "se pueda hacer de todo con el hombre."[11] Incluso invirtiendo "lo visto" – captación – por "lo escuchado" – la voz del Serl –, a nadie se le oculta esta violencia primera del discurso que al decir Yo tiene que referir todo a sí mismo para no perder la cordura. i Curiosa "resacralización" en donde el "otro" ha de pasar a la condición de **víctima**, para que el **Yo** pueda seguir siendo tal;

De manera que estamos en disposición de decir que ni las leyes "económicas" de la comunidad garantizan la realización de los derechos del hombre, ni el modo de pensar la identificación es capaz de superar "la repugnancia a lo desconocido del psiquismo del 'otro' ";[12] esa especie de *negación o de aversión universal* a lo-que-no-sea-yo en la que se inscribe el sostenido – no sé si tendríamos que decir buscado – fracaso comprensivo y a fuer de comprensivo, fracaso político, de los derechos del hombre.

2. LA MORALIDAD DEL ENCUENTRO

¿Dónde situar, entonces, el **sentido** de la inteligibilidad? ¿Dónde se da la significación del **sentido** que significa la razón?

La pregunta es pertinente en tanto en cuanto expresa el primer balbuceo de la filosofía, empeñada siempre en **dar razones** – en la pregunta del "por qué" que clausura el círculo mudo del "porque sí . . ." cerrando así el paso a todo tipo de visceralismo,[13] que no de apasionamiento –; pero a su vez, entiendo que la pregunta es, fenomenológicamente al menos, relevante si por fenomenología entendemos "la manera de abordar una noción haciendo valer la concrección de una situación en la que originalmente adquiere sentido."[14]

La unanimidad en definir la "situación humana" por relación con los derechos humanos, como anteriormente hemos dicho, avalaría una pregunta como la anterior que es previa, – en el sentido de originaria y no en el de relevancia, oportunidad o eficacia, sin que ello suponga descrédito alguno para dichas preocupaciones –; y que es "anterior," entendemos, a planteamientos metafísicos u ontológicos,[15] y también a los más específicos de filosofía del derecho[16] o de filosofía política.[17]

En todos ellos, la pregunta por el fundamento se resume en el problema de la **justificación** – del deber ser – al que terminan por declarar como una "cuestión abierta." Y es que reducir la **justificación** de los derechos del hombre "a una necesidad que se obtiene por inducción, a partir de la extensión del interés que suscita el derecho del hombre y del consentimiento general que tal expansión supone"[18] es intentar justificar algo por lo que ya hay; es decir, apelar de nuevo a una manera de pensar, en la que todo adquiere **sentido** a partir del Yo que "capta" – tradúzcase este Yo como Ser, Estado, Dios . . . –.

Pues bien, de resultar correcto el análisis que hemos llevado a cabo, la posibilidad de descubrir el **sentido** de la razón o de la inteligibilidad, ha de tener que ver con algo previo al movimiento de la razón que acude a sí misma para der cuenta de sí e identificarse, y con algo previo a la "positivación" en exigencias – derechos – de aquello a lo que apela para poder decir **Yo**, llámese esto libertad, autonomía, respeto. . . .

Dicho lo cual, parece obvio decir que el **sentido** de la inteligibilidad mantiene estrechas relaciones, si no es que se agota, con la manera de abordar y comprender esa "situación humana" que, de manera eminente, es puesta de relieve en los derechos del hobre.

De manera que tratar de mostrar el **sentido** de la inteligibilidad y pensar los derechos del hombre son dos momentos "correlativos" en la medida en la que estos últimos "exteriorizan" esa situación humana que vamos a denominar **socialidad primera**, en la que reposa ya un "Yo" que se levanta – y no puede no levantarse – más que como *ser moral*; es decir, como un ser que **tiene que responder** a/de "otro" que no es él.

Bueno será recordar, para no sacar conclusiones demasiado apresuradas,[19] que esta lectura de la "situación humana" se ha llevado a cabo por medio de una conciencia que parte del Renacimiento y Que culmina en la modernidad. Conciencia que es consciente de que "no inventa" tales derechos sino que más bien plantea que tales derechos del hombre o el respeto de los mismos *no proceden del rigor o de la gracia de Dios*,[20] i.e., que tales derechos no son conferidos.

De la misma manera hay que situar en el "haber" de la modernidad el **a priori** de la conciencia de un derecho original. Para decirlo en otros términos, la modernidad se ha apercibido de que la reivindicación incondicional de la libertad humana y de los derechos que ella implica – a pesar del peso de la tradición, de las determinaciones físicas y políticas e incluso de la violencia – **requiere** "la posibilidad de asegurar el disfrute efectivo de estos derechos";[21] lo que, desde luego, no está garantizado de entrada y lo que propicia, por ello mismo, la necesidad de traducir en cada momento la urgencia y modalidad de compresión de los diversos derechos[22] y también la lucha política en su favor que ellos mismos amparan.

Es este doble aspecto del **a priori**, que podríamos denominar aspecto reactivo y efectivo, de tales derechos, el que otorga a la mentalidad de hoy la posibilidad de referirse a ellos como derechos más legítimos que cualquier legislación y más justos que cualquier justificación; al punto de que se convierten en **medida de todo derecho y, sin lugar a dudas, de su dimensión ética**. En este sentido, tal **a priori** daría cuenta del significado del carácter de **independencia** y de **prioridad** que tienen tales derechos. **Independencia** en la medida en la que dichos derechos se reclaman "independientes" de la fuerza con la que la naturaleza adorna a los hombres; "independientes" de las determinaciones sociales y políticas. . . Y **prioridad** en la medida en la que "miden" el grado de "relatización humana" o no, es decir, sirven de referente moral.

¿Cuál es, entonces, el problema? El problema es que la modernidad que ha atisbado como nadie la importancia y la urgencia de estos derechos, no ha dado en el clavo de su comprensión. Y no lo ha hecho porque su descubrimiento central, que es el Yo o la Razón, se le ha ido de las manos. Es más, para apuntalar el Yo, ha tenido que ir construyendo a golpes de autonomía, libertad y objetividad una subjetividad que tuviera en sí misma su **propio sentido de ser**. El resultado ha sido que esta primera alternativa – que era en gran medida ya moral – se le ha vuelto contra sí misma hasta llegar a reducir "lo humano" a mero producto de una Razón ya independizada. Y lo más curioso es que este apuntalamiento del Yo ha adquirido la forma de una frontal oposición a todo lo que no fuera sí mismo, como si para ser Yo necesitara quitarsse de enmedio a los demás, expulsarles de su "territorio."

En este contexto, lo que nos planteamos es si puede reconducirse esta situación. Y lo que sostenemos es que la "situación humana" que ponen de relieve los derechos humanos avala y permite una manera de entender la subjetividad, no sobre las bases de la actividad, de la potencia y de la violencia, sino sobre la base de la moralidad. Decir que el Yo es un **yo ya moral**, es reconocer que cuando **se pone a pensar** ha vivido ya una serie de acontecimientos – a esto la sociología lo denomina proceso de socialización – de los que se levanta, en tanto que ser que piensa, no gracias a la pregunta ¿quién o qué soy yo?, sino más bien gracias a la pregunta más entrañable, y que es por eso primera, de ¿está justificada mi existencia? La "carga moral" de esta pregunta es la que nos permite decir que el **sentido** primigenio, previo a la reflexión – que es un movimiento segundo – viene otorgado **por la moralidad que preside este encuentro gracias al cual la subjetividad sale del anonimato merced a la escucha de la voz del otro que a la que no tiene más remedio que responder**. Por eso decimos que la pregunta primera, que inaugura la reflexión filosófica, no es "quién soy yo?" sino otra más originaria que se plantea con radicalidad ante el "otro" al interrogarse: "está justificada mi existencia?"

En un contexto así descrito, la filosofía remite a una **socialidad primaria** por relación con la que la **conciencia** es ya tener consciencia de "tener que responder" a la llamada de alguien que es anterior a mí y que se me manifiesta como **rostro al que tengo que contestar convirtiéndome, de esta manera, en responsable de él.** La virtualidad de poder traducir esta **socialidad primera**, a la que remite el **sentido** de la razón, en los términos de una "situación humana" *positivada* en los derechos del hombre es lo que convierte el discurso filosófico – irremediablemente segundo – en un discurso sobre la *justicia* entendida como "tener que justificarse ante el otro."

Testimoniar la referencia del **sentido** a algo previo al Yo, es reconocer su an-arquía y su "altura" gracias al "espacio moral" en el que el ptro jamás puede ser sacrificado – ser víctima – en la medida en la que la moralidad del mismo lo impide por más que no por ello lo imposibilite. Por eso, aunque el **sentido** sea éste, en absoluto *garantiza* la moralidad de unos comportamientos ni la *eficacia* política de los mismos.

Por eso hay que "volver a pensar" una y otra vez los derechos humanos por más que traicionemos en "lo dicho" **la moralidad** en la que se sostienen. No pensar aquí, sería una traición a la posibilidad de plantearse siempre el "por qué" que exterioriza ya esa dimensión de moralidad que alberga toda actividad del pensamiento. Pero pensar sin reparar en el **sentido** que antecede a la propia pregunta es condenar la **moralidad** a tener que ser "deducida" por relación con el ser, el interés, la convención. . . o la "razón de estado." Que esto es posible, lo vemos todos los días, pero no por ello deja de ser deseable – lo que es ya una opción moral – pensar **de otra manera** el sentido de lo humano del que son un testimonio los derechos humanos.

En ello estamos, pero sólo será posible cuando nos apercibamos – ¿sentimiento intelectual de Kant? – de que el sentido de los derechos humanos pasa por **la moralidad del encuentro** en la que el Yo se descubre como ser moral, es decir, como una subjetividad "precedida" por los otros. Dar cuenta" de que los otros me preceden, de que "pasan por delante de mí," se convierte así en la primera tarea de la filosofía por relación con la cual los derechos del hombre serán del hombre, cuando sean, de verdad, derechos del otro hombre; en nuestro caso y para terminar aludiendo al relato de Kafka, cuando sean derechos del sexto hombre – del otro –.

Universidad Complutense
Madrid

NOTES

[1] F. Kafka, *Beim Bau der chinesischen Mauer* (trad. cas.: *La Muralla China. Cuentos, relatos y otros escritos*. Alianza, Madrid 1990 (7ª reimp.), 129–130.

[2] Una exposición de esta postura puede verse en: A. Ollero Tassara, "Un realismo a medias: el empirismo escandinavo" en: *Revista de Ciencias Sociales* 25 (1984) citado en: A. Ollero, "Consenso y disenso en la fundamentación de los derechos humanos" en: G. Peces-Barba (ed.), *El fundamento de los dderechos humanos*. Debate. Mardrid 1989, 245 en nota 13.

[3] La figura de N. Bobbio es paradigmática en este contexto. Bien es cierto que también en él confluyen una serie de dinámicas derivadas de la fundamentación, pues si bien es cierto que se parte del reconocimiento de la "ilusión" de encontrar un fundamento absoluto como alternativa a otros (cfr. N. Bobbio, "Sobre el fundamento de los derechos del hombre" en: *El problema de la guerra y las vías de la paz*. Gedisa, Barcelona 1982, 117–128); la alternativea es invertir esa tarea de justificatión en esfuerzo por su **protección**, lo cual conllevaría acudir a una situación **de hecho** (Declaración universal de 1940), a la que no cabe prolongar, sólo acatar, y que en términos de filosofía se traduciría como irracionalismo ético basado en la afirmación de que los valores no se fundamentan se asumen. Por otro lado, la necesidad de trascender críticamente unos valores positivados por la dominación – en contra de las tesis de Kelsen – parece presuponer una suerte de **iusnaturalismo ateórico**. Todo esto da cuenta de los numerosos problemas que plantean las tesis de Bobbio.

[4] W. D. Hudson, *Modern Moral Philosophy*. MacMillan Press, London 1970 (trad. cas.: *La filosofía moral contemproánea*. Alianza, Madrid 1974) en donde dice: "La suposición de que lo fundamental de tal discurso es suministrar a agentes libres *razones* para la accción es algo esencial para el discurso moral tal y como lo conocemos (De no ser así), en mi oponión, los cambios serían radicales. Y uno de los más radicales sería que el discurso moral tal como lo conocemos y las formas de vida que constituye se convertirían, en el mejor de los casos, en pálidos recuerdos. Me atrevo a afirmar que su muerte haría que cualquier cosa que viniera luego no fuese reconocible como lo que entendemos por vida humana. El fin de la moralidad sería el fin del hombre" (o. c., 340).

[5] Los "límites" de todo intento de fundamentación son siempre los mismos. Desde el

punto de vista lógico o deductivo, la insuperabilidad de lo que H. Albert acuñó con el término de **el trilema de Münchhausen** (cfr., H. Albert, *Traktat über kritische Vernunft.* J. C. B. Mohr, Tubingen 1968). Desde el punto de vista de la **eficacia** y **resolución** del tema, su declaración de sinsentido o su abandono. Y es que el carácter "pedestre" de tener que ir de razón en razón – de denuncia en denuncia (cfr., E. Bloch, *Derecho natural y dignidad humana.* Aguilar, Madrid 1980, investigando la carga ideológica de los derechos como triunfo del burgués sobre el ciudadano y reivindicando un cambio de perspectiva en el socialismo que había rechazado el derecho natural poniéndole en relación con la utopía social (p. XI)) – para muchos, es algo inservible i.e., no se puede perder más tiempo en ello, urgidos por "lo real."

6 Es llamativo que las "razones relevantes" que debían aportar los diversos sistemas filosóficos en lo tocante a interpretaciones y justificaciones de la proclamación de la ONU, no se solventara, como muy bien describió J. Maritain, atendiendo a discusiones entre los diversos sistemas sino más bien atendiendo a un acercamiento pragmático, en donde era más importante "ser capaces" de plantear y proclamar unos derechos "sentidos" como humanos que esperar a una "aclaración imposible" en el terreno de la discusión filosófica (cfr., J. Maritain, *Los derechos del hombre. Estudios y comentarios en torno a la nueva declaración universal.* México-Buenos Aires 1949, p. 17).

7 A. Ollero, *a. c.*, 249.

8 Me refiero extactamente a la cuesión que deja planteada el profesor Fdez-Galiano en el colectivo citado recogido por G. Peces-Barba (ed.), *El fundamento de los derechos humanos* y que reza como sigue: "Entiendo, pues, que los derechos fundamentales se hallan en la dotación jurídica de todos los hombres con anterioridad a toda norma, por lo que no tienen en ésta su origen o fundamento, que habrá de encontrarse, por tanto, en una instancia suprapositiva" (a. c., 168–169 con no disimulados ribetes de iusnaturalismo) y a la aseveración de la que parte el profesor A. Ollero cuuando dice: "A mi modo de ver, existe un fundamento, **intuitivamente captable**, que permite privilegiar determinadas exigencias éticas hasta hacerlas merecedoras de relevancia jurídica" (*a. c.*, 239, negritas mías).

9 Las tesis que se sostienen en este trabajo tienen su referencia en la obra levinasiana, a la que me remito para poder entender el alcance de lo que aquí se propone. Por lo que respecta al tema concreto de su reflexión sobre el tema de los derechos del hombre, cabe señalar: E. Levinas, *Hors Sujet.* Fata Morgana (s. L.), 1987, 173–187 y "Droits de l'homme et Bonne Volonté" en: *Entre nous. Essais sur le penser-à-l'autre.* Grasset, Paris 1991, 231–235.

10 A. David, "Levinas en France" en: J. Greisch and J. Rolland (edits.), *Emmanuel Levinas. L'éthique comme philosophie première.* Les Éditions du Cerf, Paris 1993, 303.

11 E. Levinas, *De Dieu qui vient à l'idée.* Vrin, Paris 1982, 84, en un contexto de crítica del pensamiento heideggeriano.

12 E. Levinas, *L'au-delà du Verset.* Édit. de Minuit, Paris 1982, 223.

13 A. Cortina, "Una tería de los derechos humanos" en: *Etica sin moral.* Tecnos, Madrid 1992, 239–253.

14 E. Levinas, *Entre nous. Essais sur le penser-à-l'autre.* Grasset, Paris 1991, 258.

15 Por la relevancia en sus planteamientos destacaríamos dos estudios en esta perspectiva: J.-M. Palacios, "El problema de la fundamentación metafísica de los derechos

humanos" en: *Revista de Filosofía* 6 (1983) 257–273 en donde se refiere a la contribución de J. Maritain a un fundamentación de los derechos humanos sin descuidar apuntes sugerentes; J. Seifert, "The fourfold root of Human Dignity and of Human Rights" (texto de la conferencia pronunciada en los cursos de verano de la Univ. Complutense en julio de 1993 y que se resumen de un estudio más amplio denominado: *Los fundamentos filosóficos de los derechos humanos* (trad. al cas. por M. Crespo) o la obra de R. Spaeman, *Das Natürlische und das Vernünftige*. Piper Gmbh & Co. GK verlag, Munchen (trad. cas.: *Lo natural y lo racional. Ensayos de antropología*. Rialp, Madrid 1989).

[16] Tradicionalmente la filosofía jurídica se ha referido a dos maneras sobresalientes en las que se dirimen la punga sobre la fundamentación que es otra manera de llamar al **sentido** de la inteligibilidad que hemos dicho. Tales referentes serían: el Iusnaturalismo y su contrario, el iuspositivismo jurídico a las que se añadira el intento de fundación ética de los derechos basada en la idea de dignidad humana (cfr., E. Fernández, *Teoría de la justicia y derechos humanos*. Debate, Madrid 1984, cap. III y su consideración como derechos morales); aunque otros prefieren otra clasificación; así por ej.: A. E. Pérez Luño, *Derechos humanos, Estado de derecho y constitución*. Madrid 1984, en el cap. III habla de posturas objetivistas, subjetivistas e intersubjetivistas para intentar dar cuenta del fundamento; otra es la de G. Peces-Barba Martínez, *Curso de derechos fundamentales. Teoría general*. v. I, Madrid 1991, pp. 35–51.

[17] Nos referiremos, a título ilustrativo, a dos reflexiones, puesto que la perspectiva de la filosofía política está "complicada" con los anteriores saberes. La primera reflexión es la de A. Cortina llevada a cabo desde la ética discursiva y cuyo objetivo es "superar" el trilema en el que había caído la reflexión jurídica, en alusión a la división tripartita de E. Fernández aludida en la nota anterior (Cfr., A. Cortina, *Etica sin moral*. o.c.); la segunda perspectiva es la de L. Ferry y A. Renaut, *Philosophie politique*. vol. III "**Des droits de l'homme à l'idée républicaine**." P.U.F., Paris 1988 (2ª edic.) para quienes la alternativa es "volver a la lectura republicana de los derechos del hombre" (p. 166ss), en la medida en la que tal idea apaciguaría la tensión entre **derechos-créditos** (libertades reales) y **derechos-libertades** (libertades formales) sustituyendo tal tensión por la de **derechos-participaciones** (derechos políticos) de participación efectiva (Existe traducción cas.: *Filosofía Política. III De los derechos del hombre a la idea republicana*. F. C. E., México 1991). En este contexto, cabe citar también los tres artículos de J. Habermas recogidos en: J. Habermas, *Escritos sobre moralidad y eticidad*. Paidós, Barcelona 1991, en particular el último articulo.

[18] E. Levinas, "Droit de l'homme. . ." a. c., 233.

[19] La especial virulencia del debate modernidad-postmodernidad, prolongado en la polémica entre neo-consevadores y críticos (y sus diversas versiones) hace muy difícil mantenernos en una postura de equilibrio. A pesar de ello, mantenenos lo dicho: – que es en la modernidad donde resplandece la conciencia de ese derecho original y que es en la modernidad donde se toma conciencia de que-tal desechs ha de pourse ca práctica. De veavera que es "contra" la modernidad como planteamos una manera de pensar distinta que como YO, a la luz de los frutos "espúreos" que un determinado concepto de RAZON ha generado, del que da cuenta el "fracaso" en el disfrute de esos derechos.

[20] Cfr., E. Levinas, *Entre nous*. o. c., 178. Esto es así, hasta tal punto que invierte el sentido del discurso "teológico" puesto que de entender los derechos como **deducidos** de las *verdades sobre Dios*, se pasa a comprenderlos como "revelación" de "una original

venida de Dios a la idea de hombre" (E. Levinas, o. c., 176) que sólo es posible partiendo de "la evidencia de los derechos del hombre" (*Id.*, 178).

[21] E. Levinas, *Hors Sujet*. Fata Morgana (s. l.) 1987, 179.

[22] En este contexto cabe prolongar, casi se debe decir que es ineludible llevar a cabo, la reflexión sobre el significado (contenido) y la urgencia que van adquirienddo los derechos del hombre por relación con el contexto socio-político. Es más, a nuestro entender es uno de los brazos en los que se ha de apoyar las ciencias jurídicas, aunque solo fuera por que las declaraciones en las que se enuncian los derechos humanos son categorías históricas. La discusión, a este respecto, se centra en la comprensión de las "**generaciones**" de los derechos. Así en términos generales, hablamos de derechos humanos de la **primera** generación, más atentos a proteger la esfera privada de los individuos de la injerencia de los poderes públicos; con posterioridad hablamos a derechos sociales, culturales y económicos más acordes con la visión de una sociedad del bienestar, para terminar con la discusión acerca de si hay lugar para hablar ya de **tercera** generación de derechos y en qué términos. La disparidad respecto de este punto es enorme (Cfr., A. E. Pérez Luño, "Los derechos humanos de la tercera generación" en: *El Basilisco* 5 (1990) 3–10 con una bibliografía básica sobre el tema y también I. Ara Pinilla, "Los derechos humanos de la tercera generación en la dinámica de la legitimidad democrática" en: G. Peces-Barba (ed.), *El fundamento de los derechos humanos*. Ed. Debate, Madrid 1989, 57–65).

OLIVER W. HOLMES

PRINCIPLES OF NATURE, HUMAN ASSOCIATION AND THE POLITICS OF EQUALITY

Systematic Thought in L'Encyclopédie

> Human knowledge and human power meet in one; for where the cause is not known the effect cannot be produced. Nature to be commanded must be obeyed; and that which in contemplation is as the cause is in operation as the rule.
>
> Francis Bacon*

The general function of the *Encyclopédie*, revealed in the subtitle as the *Systematic or Reasoned Dictionary of Sciences, Arts and Crafts*,[1] was the device by which Diderot, d'Alembert and other *philosophes* organized their experiences in the light of an acceptable systematic spirit. The editors' devices for organizing a variety of intellectual innovations in uniform molds made their methods and principles attractive to those groups in eighteenth-century society who would control change by embarking upon it from the security of a new intellectual order. This new approach to human knowledge, at once critical, comprehensive, ordered and useful, encountered criticism and opposition in France and abroad. After the publication of the second volume, the Attorney-General of France, Omer Joly de Fleury, charged in the Decree of the King's Council of State of 7 February, 1752, that "in these two volumes," the *Encyclopédie*'s intellectual purpose was "to insert several maxims tending to destroy Royal Authority, to institute the spirit of independence and revolt, and, in obscure and ambiguous words, to erect the foundations of error, of the corruption of morals, of irreligion and unbelief."[2] De Fleury's sentiments may have stemmed from d'Alembert's characterization of the revolutionary changes that occurred in "natural philosophy," "from the principles of the secular sciences to the foundations of religious revelation." This revolutionary espousal of human knowledge, coupled with revolutionary activities of the French Revolution, prompted the editors of the *Encyclopedia Britannica*, in 1801, to accuse "justly" the *Encyclopédie* of "having disseminated far and wide the seeds of anarchy and atheism."[3]

Though it is true that Diderot and d'Alembert promoted indepen-

A-T. Tymieniecka (ed.), Analecta Husserliana, Vol. LIV, 219–258.

dence of thought and viewed their collective enterprise as revolutionary, and Diderot challenged the arbitrary exercise of "royal authority," the charges of "irreligion" and unbelief, of "corruption of morals," of "anarchy and atheism," were exaggerated, as a closer reading of the various articles in the *Encyclopédie* will testify. A closer reading of these articles will present evidence that points in a different direction, and will reveal instances where Diderot, d'Alembert and others perceived in the principles of nature the rational basis by which to connect individual wills to the common utility of society. Indeed, they viewed nature as having emerged from a definite "authorship" or definite principles of organization, or both.

The scope and range of the articles in the *Encyclopédie* represent an active reflection of its historical moment. In exhibiting the complex unity of the elements of nature, human nature and the manner in which they become manifested in human association, and the political ways of life made possible through social relations, the Encyclopedists provided at once an articulation of the phenomena of nature and human nature in the attainment of their potential and a compelling defense of the dignity of human nature and social life. The notion of a nature which operates in accordance with laws refers both to biological and physical phenomena and to human behavior. The laws of nature, then, and the idea of a common humanity based upon the rational and natural or biological unity of all humankind, become a tendency that appears consistently as the main characteristic of the *Encyclopédie*'s organizing principles. This study aims to assess these connections in exploring whether there is a perceptible and unitary principle in reality for Diderot, d'Alembert and other key contributors, and in identifying conceptions of nature as the basis for rules of human action. For if a coherent system in nature exists, they argued, an individual's reason should discover the system and transform it into moral conduct. Thus, descriptive nature and the laws of nature provide prescriptive rules for human action. The idea of a certain correlation between laws of nature, rationality, and morality had had an enormous appeal to eighteenth-century philosophers and a great influence in the political principles espoused in the *Encyclopédie*. Some indication of the important political implications of this influential idea of analogy between the rational, the natural and the moral will be the cumulative effect of the essay.

In various theoretical forms, the idea of nature attained a great deal of prominence in eighteenth-century thought. Diderot, d'Alembert, de

Jaucourt, Boucher d'Argis, and other contributors to the *Encyclopédie*, inspired by the Baconian and Cartesian visions of science and equipped intellectually with Newton's "new scientific method," set upon the task of discovering a natural order that comprised both the individual and general facets of experience. The successful synthesis of mathematical analysis and experimentation, and the identification of the natural with the rational upon the entire realm of thought, accounted for their contribution to the idea of nature. Newton encouraged all scientists who were willing to listen to observe nature before applying mathematical reason. Some of the contributors, for example, d'Alembert, remained under the influence of mathematical physics. Others, like Diderot, became more affected by the distinctive manner in which "life" operates. However much the contributors differed in their method of analysis, they were in fundamental agreement about adopting a method and the important role the concept of nature performed in their respective analyses. Nature became, for them, both a scientific ideal and a normative force in human associations.

The closely observed natural phenomena, the eternal order of nature, are visible manifestations of the physical world. These natural phenomena become tangible images which are interpreted in symbolic terms. Such symbolic interpretation of the physical universe read Providential intention in a universe that has been understood to be governed by the order of natural laws. This mental attitude, derived from biblical and classical writings, revealed a certain credence in signs. The *philosophes* who wrote for the *Encyclopédie* demonstrated this credence in that they pointed to a fundamental relationship with origins as a vital energy. For Diderot and d'Alembert, origins were a spontaneous force instead of an ideal principle. The interest in origins explains, in part, their emphasis on the scientific method as the process by which to apprehend nature. Hence, Bacon, Descartes, Newton and Buffon, among others, became the important models with whom they identified.

Francis Bacon's contribution to scientific thought in the seventeenth and eighteenth centuries lay in popularizing the notion of experimental technique. In his general approach to science, Bacon criticized what he considered "narrow empiricism" and "dogmatic" theories, espousing a combination of experiment and theory. Bacon's "new" philosophy compared the radical empiricists with ants, characterized as individuals who accumulated a vast array of facts after having burrowed away at a narrow course of inquiry. The philosophers were equated with spiders who, through the exercise of logic, spun cobwebs from their own bodies. The "authentic" scientists, in contrast, were likened to bees who extracted substance from flowers and thereby transformed it into honey. "The men of experiment," he averred, "are like the ant; they only collect and use; the reasoners resemble spiders, who make cobwebs out of their own substance. But the bee takes a middle course; it gathers its material from the flowers of the garden and of the field, but transforms and digests it by a power of its own."[4] Where the mechanists applied a single mechanistic method to natural phenomena, Bacon advocated observations over a wide range of phenomena. Bacon's scientific approach draws upon a human analogy, instead of a mechanical one, to describe a natural phenomenon which carries his assumptions within the organic tradition.

Through the application of the scientific method, Bacon strove to combine the theoretical interpretation and the technical control of nature. Descartes was familiar with Bacon's stress on scientific method, and supported these objectives for his own work. However, he thought that Bacon initiated the procedure with the empirical facts of the natural world instead of the general principles which provided a basis for deductive inquiry. Although Descartes's views on the mathematical method developed with the physical sciences, he also acknowledged that there were certain fundamental ideas "given by intuition" which provided the surest starting point for deductions of a mathematical character. These ideas were those of extension, motion and God. The concept of God became fundamental to his natural philosophy, in that God, for him, made extension and put motion into the universe. However, this motion has been put into the universe only once, at the moment of creation, and therefore the amount of motion in the world was conceived as constant and eternal. Descartes arrived at the principle of the conservation of momentum through this sort of conception, and explained the notion of motion in terms of Divine Action:

> He created matter along with motion and rest in the beginning; and now, merely by his ordinary cooperation, he preserves just the quantity of motion and rest in the material world that he put there in the beginning.[5]

The concept of "design" and of a Divine Creator implied that once God had created the world, its physical properties remained untampered. The concept of a Divine Design also presupposed that God ruled the world through "laws of nature," which had been formulated at the outset, in Descartes, by employment of mechanical analogies to describe what God had created. He informs us that "the rules of mechanics . . . are the same with those of nature," and in discussing the proof of God's existence, he draws upon the analogy of a machine-maker. The concept of design also implied the "clear and distinct" idea of a Creator who has endowed all individuals with the equal capacity to discern truth from error. The general ordering powers of the mind are a natural ability possessed by each individual, the appropriate exercise of which depends upon an adequate procedure of analysis:

> Good sense is the best shared thing in the world: for each of us thinks he is so well endowed with it that even those who are the most difficult to please in all other respects are not in the habit of wanting more than they have. It is hardly to be expected that everyone is mistaken in this; rather it testifies that the capacity to judge correctly and to distinguish the true from the false, which is properly what one calls common sense or reason, is by nature equal in all men, and consequently that the diversity of our opinions does not come from some of us being more rational than others, but solely from the fact that we conduct our thoughts along different lines and from not examining the same things. For it is not sufficient to have good mental powers; but the principal matter is to apply them well.[6]

Method, therefore, provides the key to success in science; this very method, which is "a more powerful instrument of knowledge than any other that has been bequeathed to us by human agency, as being the source of all others."[7] Disillusioned with ancient authority and learning, Descartes joined Bacon in the quest for a new method. Through experience, mathematical analysis and experimentation, the new method strove to understand and to control nature. In the successful search for knowledge, the individual scientist attained power, a power achieved over nature rather than over individuals.

The "modern" theories of Bacon and Descartes hardly exhausted all the possibilities of natural phenomena, albeit the methods of observation and analysis they put forward encircled reality with the purpose of exhausting all possible knowledge of natural phenomena. This approach to knowledge, which placed emphasis on the role of experience, experimentation and mathematical analysis, became the intellectual purpose of Diderot and d'Alembert and, by extension, the *Encyclopédie*. Having been inspired by the Baconian and Cartesian visions of a science emerging from nature, Diderot and d'Alembert turned both to them and to others for intellectual guidance.

Between 1751 and 1772 Bacon's encyclopedic ideal was revived and, in several respects, surpassed in the seventeen volumes of text and eleven volumes of plates of Diderot and his "men of letters." The affinity that Diderot and d'Alembert felt for Bacon's classification of knowledge becomes apparent in the *Encyclopédie*'s "tree of knowledge," opposite the frontispiece in the first volume, which outlines the intention of the enterprise. Indeed, in the "*Prospectus*," Diderot acknowledged the "principal obligation" for their *encyclopedic* tree that was owed to Bacon. As a consequence, articles in the *Encyclopédie* covered crafts and trades as well as philosophy, literature, the arts and the sciences.

In his "*Discours préliminaire*," d'Alembert declared the twofold objective of the *Encyclopédie*:

The work whose first volume we present today has two objectives: as an *Encyclopedia*, it has to expose as much as possible the order and the connection of human knowledge; as a *Reasoned Dictionary of the Sciences, Arts and Trades*, it has to contain on each science and on each art, liberal or mechanical, the general principles that are the basis of each, and the most essential details that make up the body and substance of each. These two points of view, of *Encyclopedia* and of *Systematic Dictionary*, therefore, will shape the design and the division of our Preliminary Discourse (*Discours préliminaire*, I).

Thus, as an encyclopedia, the editors' purpose as to illustrate the interpenetration and coherence of all human knowledge; as a "reasoned dictionary," the editors' purpose was to present, systematically, the principles of all knowledge derived from theoretical sciences and practical arts. The extended purpose of the encyclopedic enterprise was to create a compendium of human knowledge that may remain for posterity. In his article on "*Encyclopédie*," published later, Diderot declared the objectives at which the enterprise purported to aim:

The aim of an Encyclopedia is to reassemble all knowledge scattered over the surface of the earth, to expose to men the general system with which we live, and to transmit it to men who will live after us; so that the works of past centuries have not been useless works in the centuries that will follow; that our descendants become more educated, that they become at the same time more happy, and that we will not die without having deserved well of humankind (*Encyclopédie*, 635a).

This strong sense of purpose and anticipation of their contribution to "Posterity" was apparent in the epistolary to d'Argenson where Diderot and d'Alembert claimed "it is to the enlightened nation of Men of Letters" and "to the free and unselfish Philosophers" that they and "several well known Savants and Artists" have been "willing" to devote their collective "talents" (*Encyclopédie*, vol. I: 7–8). The "society of men of letters and artists," which extended well beyond the *salons* of Mme. Geoffrin, Mme. du Deffand, and Mlle. Lespinasse was assembled to "change the common mode of thinking," and thus, for Diderot, characterized the revolutionary spirit of the *Encyclopédie* ("*Encyclopédie*," 642a, b).

All the variegated energies of the mind are concentrated in the collective creation of the *Encyclopédie*, and articulated through the collective voices of the talented "society of men of letters and artists." The expressive form of this collective enterprise, for Diderot and d'Alembert, was revolutionary. Thus, the encyclopedic enterprise accepted the implicit mission to continue the struggle along the revolutionary path paved by seventeenth-century philosophical and scientific truths. "Once the foundations of a revolution have been laid down," d'Alembert remarked in recognition of this revolutionary moment, "it is almost always the succeeding generation which completes that revolution" ("*Expérimentale*," 299b). D'Alembert's observation of the revolutionary implications of their mission points to yet another objective of the *Encyclopédie*: to document the historical moment and to provide a "reasoned history" of the recently founded "society" of learned scholars.

In his "*Discours préliminaire*," d'Alembert invited a comparison between the political world of revolutions and the political implications of revolutions in science. He proposed a novel approach to the issue in the context of Descartes' contributions to "modern" science. D'Alembert characterized the concept of revolution, concerning Descartes' contribution to "modern" science, as having taken the "daring" stance

> to show good minds how to shake off the yoke of scholasticism, of opinion, of authority, in a word, of prejudices and barbarism; and by this revolt whose rewards we are reaping today, Philosophy has received a service from him, perhaps more difficult to render than all those that Philosophy owes to his illustrious successors. We may regard him as a chief of conspirators, who had the courage to protest first against an arbitrary and despotic power, and who, in preparing a resounding revolution, laid the foundations of a more just and happier government which he had not been able to see established (*Discours préliminaire*, xxvi).

Another particular and distinctive character of the concept of revolution was elucidated by d'Alembert in the form of specifying the scientific sense of the word, hitherto the physical rotation of celestial bodies, as an intellectual process characterized by the notion of "successive generations." This latter sense of the word, "revolution," provided the conceptual framework within which to develop his "reasoned history" and, thereby, to resist "arbitrary and despotic" authorities in thought: An intellectual history that was dynamic, provided process and demonstrated different lawful principles.

D'Alembert's article, "*Expérimental*," understands the sense of revolution within the sciences in terms of the development of calculus by Newton and Leibnitz. d'Alembert and his contemporary, mathematician Alexis Claude Clairaut, along with Voltaire and Maupertuis, were some of the first to receive and expound Newton's theories in France. The mid-eighteenth century was a period marked by an efflorescence of intellectual activity in the mathematical sciences of astronomy, mechanics, electricity and magnetism. In 1747 Clairaut read a paper at the Paris Royal Academy of Sciences before the public Assembly, entitled "System of the World in the Principles of Universal Gravitation." Clairaut extolled Newton's methods of mathematical calculations and the latter's "famous book of Mathematical Principles of Natural Philosophy as having ushered in the epoch of a great revolution in Physics."[8] D'Alembert shared this sense of signaling a new era in the mathematical sciences as evidenced in his article, "*Expérimental*." He identified himself and his cohort of *philosophes* with this "century of philosophy," and acknowledged the development of a "true system of the world."[9]

D'Alembert describes natural philosophy, or "experimental philosophy," as the point of view "that makes use of the experimental way in order to discover the laws of Nature" ("*Expérimental*," 298a). The descent of Newton's natural philosophy, proceeding beyond the scientific genealogy of Bacon and Descartes, cleared the path of shadows that clouded the visions of preceding generations. After Newton, the "new methods of philosophizing" provided findings that recognized variety in physical reality. "Nature is so varied and so rich," d'Alembert observed, "that a simple collection of completed facts would advance incredibly our knowledge; and if it were possible to push forward this collection to the point that nothing of it was missing, perhaps it would be the only work to which a physicist ought to exercise self-restraint" ("*Expérimental*," 298b). For d'Alembert, "the principles and the method" of which these physicists make use will provide the appropriate means by which to discover the "good order" in "nature" ("*Expérimental*," 301a). A serious study of how the "new science of nature" developed will reveal "the genealogy and the filiation of our knowledge, the causes which must have given birth to it and the characteristics that distinguish them." This new approach entails "going back to the origin and generation of our ideas," and demonstrating how natural philosophy's analytic method operates on the implicit assumption that all of our knowledge has a firm basis in "the history of our thoughts" ("*Discours prélim-inaire*," p. I., "*Eléments des sciences*," 495b–496a).

Traditional metaphysics and its transcendental concern with locating the ultimate nature of reality will have to yield to this new natural philosophy which establishes that the ultimate basis of knowledge rests in the "certain simple, recognized facts" of everyday sense-experience. This grounding in the "observation of daily phenomena" provides the modern physicist with the "necessary guiding principles" for penetrating the "facts of nature" ("*Expérimental*," 300a, b, "*Eléments des sciences*," 493a). The emphasis by d'Alembert on the empirical and factual aspect of natural philosophy demonstrates his departure from traditional metaphysical speculations of reality. "A single phenomenon," he averred, "a single fact, well observed and well developed, teaches more than all the hazardous conjectures, which we shall never be inclined to verify, and which, without enlightening the Philosophers, exercise their imagination and their leisure" ("*Cosmique* (Qualité)," 292b).

After having initiated the intellectual inquiry from the point of view of single phenomena, "the philosopher's art consists in adding new links to the separate parts, in order to render them the least distance that is possible; but he must not flatter himself that he will not always remain empty in many places." For as "we perceive some continuous parts," we recognize immediately that "everything is connected in Nature," and ought to acknowledge that "in a great number of cases the continuity escapes us." To minimize potential misperceptions and, hence, "to form the links of which we speak, one must allow for two things: the observed facts which form the substance of the links, and the general laws of Nature which form the connection. I call *general laws*, those laws that appear circumspect in a great number of phenomena because I am very careful to say *in all [phenomena]*" ("*Cosmologie*," 294a). In other words, the intuitive dimension of philosophy requires the synthetic function of observing individual facts and establishing a coherence between them. Through the "necessary guiding principle" of the "good laws of Nature," d'Alembert's philosophic ideal seeks the ultimate unification of all knowledge. For, he suggested, "one cannot doubt that all the bodies of which this universe is made up, form a single system, of which the parts are interdependent and have interrelations that derive from the harmony of the whole" ("*Cosmiques* (Qualités)," 292b). In a methodological manner similar to Descartes, d'Alembert utilized mathematical analysis to explain the physical and "mechanical" sciences in terms of a chain of reasoning. "All the sciences," Descartes averred in his *Opuscules de 1619–1621*, "are interconnected as by a chain. No one of them can be completely grasped without the others following of themselves."[10] The mathematical ideal thus engenders the ideal of unifying knowledge.

Once the disparate "facts of nature" are observed, the philosopher articulates these experiences and, in the process, provides a voice for nature. Through this process, the philosopher's function consists in listening to nature's voice. Nature appears before and speaks to the observer *qua* thinker who, in turn, organizes these observations into a coherent system of thought designed specifically to record and to interpret all that has been experienced in nature. The systematic aspect of thought provides the basis for which philosophy "contains the fundamental principles of all human knowledge." Systematic thinking enables the theorist to categorize the "single recognized facts" of "daily observation." From this perspective, d'Alembert proposes, "the elements of all the Sciences would be confined to a single principle, of which the main consequences would be the *elements* of each particular science. The human mind, participant in that case of supreme intelligence, would see all knowledge as joined together under an indivisible point of view. Such are the laws of movement, which are the continuation of the impenetrability of the bodies, and the source of several of the effects that we observe in Nature" ("*Eléments des sciences*," 491a, b).

The detailed observation of the "facts of nature" requires a pursuit of the path of everyday experience to discern the "isolated and floating" truths inherent in reality. This perceptual pursuit of daily phenomena reflects how the study of nature becomes involved with penetrating the "impenetrability of bodies." To understand the body of empirical phenomena connected by the general laws governing motion, d'Alembert proposed proceeding beyond the limits of abstract mathematics to mechanics. Therefore, the elements of this science rest on philosophical principles which are categorized as "primary principles" and "secondary principles." Primary principles form ideas which demonstrate the proposition that "when one will consider Metaphysics in a more broad sense [of the term], as the universal science that contains the principles of all the other sciences [Physics, Geometry and Mechanics]; because if each science has and only has observation for its true principles, the Metaphysics of each science consists only in the general consequences that arise from observation, presented under the most extensive point of view that one can give them . . . I will be careful to define it [Metaphysics], as the sciences wish it, the *science of ideas.*" D'Alembert's philosophy defines metaphysics in this manner because he characterizes "philosophy" further as "either the science of facts or the science of chimera" ("*Eléments des sciences,*" 492b–493a).

Philosophic principles grounded in the "science of facts," simple as they may be, attain success in studying nature by distinguishing between the various propositions that are put forward:

> If the propositions that form the general unity of a science do not immediately follow each other, one would observe the places or the chain is broken, and the propositions that form the head of each part of the chain, are those which must enter within *the elements.* With regard to the same propositions that form a single portion continuing the chain, one will distinguish in it two kinds: those propositions that are only simple consequences, a simple translation in other terms of the preceding proposition, must be excluded from *the elements,* since they are evidently contained in it. Those positions that borrow from something, not only from the preceding proposition, but from another original proposition, would appear to have to be exclusive by the same reason, since they are implicitly and correctly contained in the propositions from which they originate ("*Eléments des sciences,*" 492a, b).

Since human reason cannot attain the wisdom of the "Author of Nature," human thinking can aspire to comprehend the manner in which "single facts" are linked by philosophic principles. For "the principal usefulness that we must derive from *Cosmology,*" d'Alembert explained, "is to raise us by the general laws of Nature, to the knowledge of the author, from which wisdom has established its laws. . . . Thus *Cosmology* is the science of the World or of the Universe considered in general, understanding that it is a being composed, and yet simple by the union and harmony of its parts; of all, that is governed by a supreme intelligence, and by which the springs are united, set in action, and modified by this intelligence" ("*Cosmologie,*" 294b). The theistic implications that the principles of Nature demonstrated God's ordinance become apparent here. D'Alembert's concept of a world created by a "supreme intelligence" emphasizes a universe that was governed entirely by "general laws of nature," and implies a mechanism set in motion which becomes intelligible to human reason. The concept of "authorship" thereby provided more of a "natural" than a Christian explanation of the world.

However, Diderot opined that the mathematical sciences had peaked in influence and in importance. Where d'Alembert remained under the influence of mathematics and physics, Diderot became more affected by the distinctive manner in which "life" and "natural history" operate. After d'Alembert abandoned the "society of men and letters," Diderot wrote to Voltaire explaining why he thought his former collaborator "did a foolish thing" in quitting the *Encyclopédie*:

> The reign of mathematics is no more. Tastes have changed. It is the taste of natural history and of letters that dominates. D'Alembert, at his age, will not throw himself into the study of natural history, and it is very difficult [for him to] create a work of literature that corresponds to the celebrity of his name. A few articles in the *Encyclopédie* would have sustained him with dignity during and after the edition. That is what he has not considered, what possibly no one will dare tell him and what he will hear from me, because I am made to tell the truth to my friends, and sometimes to the unconcerned, which is more honest than wise.[11]

The tone of the letter demonstrates Diderot's candor and sense of revenge and disappointment over d'Alembert's "desertion" from the *Encyclopédie.* It also conveys Diderot's interest in a scientific method that places emphasis on the biological aspects of everyday experience. The physics of movement, though an important principle of explanation, was insufficient in explaining the vital dimensions of life. Thus Diderot turned away from mathematical formulas to undertake the direct study of nature.

Diderot preferred to pursue the "middle course" of Bacon's bee, collecting material from a variety of phenomena, and analyzing the substance observed in the natural realm. The influence of Buffon directed Diderot's interests away from mathematics and its abstract mechanical principles towards the more tangible and practical chemistry of biological "living bodies." The turn to natural phenomena renders intelligible the connective links of "life" and "natural history" instead of the "elements" of "propositions" and the "*chain of judgments*" so characteristic of "abstract reasoning." From this perspective, the word "encyclopedia," for Diderot, "signifies *chain of knowledge*; it is composed of the Greek preposition . . . *of*, and of the substantives . . . *circle* and . . . *knowledge*" ("*Encyclopédie*," 635a, "*Raisonnement*," 776b).

In the Foreword to Volume II, the editors of the *Encyclopédie* announced that Buffon had agreed to contribute the article on "*Nature*" in an effort to illuminate "a vague enough term." By the time the volume was prepared, for whatever reason, Buffon failed to make the contribution. Despite the absence of Buffon's participation, Diderot read his *Natural History* upon which he relied to prepare the article on "Animal."

"The word animal," he explained, "represents a general idea, formed from particular ideas that one has made from a few particular animals" ("*Animal,*" 468b).

Buffon's discussion of the method of natural history emphasized the numerous proliferation of forms, and how each form graded imperceptibly into others. Living bodies constituted a system of physiological processes which produced a variety of external manifestations. The ability to discern and to comprehend these processes provides the challenge for natural history and its method. Hence, the natural historian pursues the task of discovering the laws regulating natural phenomena.

This tendency to generalize particular observations becomes apparent through the manner in which Buffon assigns a unique place in nature to humans. For Buffon, the natural, indeed, divine gift of "intellectual power," faculties of thought and speech, had set humans apart from the "lower" forms of life. Though humans possessed external physiological properties that resembled higher apes, Buffon established that they differed from apes internally by the very fact that humans developed speech and thought patterns which were evidenced in their cultural achievements. Therefore, he argued,

> There is an infinite distance between the faculties of man and those of the most perfect animal. . . .
>
> The immense distance that the kindness of the Creator has placed between man and beast. Man is a rational being, animal is a being without reason; and as there is hardly a middle course between the positive and the negative, as there are hardly intermediary beings between the rational being and the being without reason, it is evident that man is of a nature entirely different than that of an animal, that man only resembles an animal by external appearances, and that to judge him by this material resemblance is to be taken in by appearances, and to close voluntarily the eyes to the light that ought to make us distinguish appearance from reality.[12]

Diderot assigned a different place for humans in nature. Where Buffon identified intellectual power as the internal signs that separate humans from animals, Diderot viewed humans as forming a part of the graded "chain" of "living bodies." Though Diderot granted humans higher developmental status, they remained "always an animal." Moreover, he contended,

> if it is true, as one can hardly doubt, that the universe is one single and unique machine, where everything is linked, and where beings raise themselves above or lower themselves below each other, by imperceptible degrees, in such a manner that there is no gap in the chain . . . where from nuance to nuance one passes from white to black

without noticing it, it may be a true image of the progress of nature; it will be very difficult for us to determine the two limits between which animality, if it is permitted to express oneself in this manner, begins and ends ("*Animal*," 468a, b).

Diderot's article on "*Animal*" reflects two rather contradictory views of the organic world, both of which were inherited from classical antiquity and elaborated by Aristotle. One view presented the organic species as a hierarchy of creatures with comparatively enormous discontinuities between their ranks. The other view perceived the variety of animals and plants as several links in a chain of creatures, the gradations between the various stages being imperceptible and continuous. The "natural" system of classification, in contrast to the "artificial" system, was unable to reconcile the two conceptions and, therefore, attempted to incorporate the diverse organic species into a classificatory framework of natural families, in which a continuity of creatures was maintained, and through which several characteristics were studied to assess the affinity of the organisms within a natural family.

Buffon's popular *Natural History* opposed the artificial method which classified organic species into discontinuous orders:

The first truth that comes out of this serious examination of Nature, is a truth perhaps humiliating for man: it is that he [man] ought to line himself up in the class of animals, which he resembles through all that he has of physical properties; and even their instinct would appear to him possibly more sure than his reason, and their industry more admirable than his arts. Examining then successively and by order the different objects that make up the Universe, and sitting at the head of all created beings, he [man] will see with astonishment that one can descend, by almost insensible degrees, from the most perfect creature to the most formless matter, from the more organized animal to the most brute mineral; he [man] will recognize that these imperceptible nuances are the great work of Nature; he [man] will find, these nuances, not only in sizes and in forms, but in movements, in generations, in successions of all kinds. . . .

But Nature proceeds by unknown gradations, and consequently cannot totally lend itself to these divisions, since Nature passes from one species to another species, and often from one genus to another genus, by imperceptible nuances; in such a manner that [one] finds a great number of equally divided intermediate species and objects that one does not know where to place, and that necessarily disturb the plan of a general system.[13]

Buffon's discussion of the appropriate and inappropriate ways of studying natural history makes clear that this sort of order, provided by a "general system," necessarily becomes arbitrary and "artificial." Diderot carried this position another step in accepting the idea of the continuity of nature and of the insensible progression of vegetable and animal life. "The march of nature," he explained, moves by "degrees of nuances, and often by degrees of imperceptibility," and, as such, "no line exists" that can clearly "separate animal life from vegetable life" ("*Animal*," 469a). Diderot also proceeded to remove the barrier established by Buffon that separates the individual, who thinks, from animals, who do not think. "The spectacle of nature within individuals," Diderot claimed,

> the state of this faculty to think, to act, and to feel, resides in some men in a high degree, in a lesser degree in other men, continues weakening them in proportion that one may follow the chain of being on descending, and apparently understand one another within some very remote point of the chain: placed between the animal kingdom and the vegetable kingdom, a point from which we will approach more and more by observations, but which will never escape us; experiences will always be on this side of them [observations], and the systems will always proceed beyond them [observations]; experience walking step by step, and the spirit of system always skipping ("*Animal*," 470b).

The range of metaphysical problems inherent in affirming the "imperceptible" unity of inanimate and animate life was not broached here. Diderot mentions that the philosophical discussion of the "metaphysical" tendency of the "question" will be discussed in the article "Âme." The article on "*Animal*" emphasized the unity of life and the close correlation of experience, feelings and thoughts in relation to reality. In emphasizing the lines that connect inanimate and animate, animal and human existences, and thereby discerning a certain order in nature, Diderot was perceptive to recognize the epistemological tendency of the "metaphysical question" that may arise ("*Animal*," 470b). Therefore, the questions surrounding the notions of body and soul, and mind and body, had to be addressed in the long article, "*Âme*," as well as the problems it addressed, it had its own particular nuance. In French, the word signifies "soul," "feeling," and "mind." Because of these connotations, we may understand how Diderot, in the course of his inquiry into where in the body *âme* resided, covered a broad range of subjects concerning theology, psychology and philosophy.

The subject matter was conducive to direct scientific findings along a path leading to metaphysical questions. Since Descartes proposed it, the question of the "seat of the soul" has been of particular interest to "Physicians, Physiologists and Musicians," and metaphysicians, after the experiments of La Peyronie, have come to realize that the contents of thought rely solely on sensory "impressions made on the bodily organs." Indeed, for Diderot, the dependency was reciprocal. In his discussion of certain case histories drawn from "observations" found in the "Transactions of the Royal Academy," in which he correlated religious superstitions with physical disease, Diderot identified "two other very proper facts to demonstrate the effects of *the soul* on the body, and reciprocally the effects of the body on *the soul*" ("*Âme*," 343a).

As in the discussion of the "imperceptible lines" connecting inanimate and animate life, Diderot put forward the notion that soul and body, mind and body, are inextricably connected in a reciprocal relationship. "In the meantime," he proffered,

> let us consider how much their functions stretch: very little; an upset fiber; a drop of extravasate blood; a light inflammation; a fall; a contusion: and farewell to judgment, reason, and all this insight of which men are so vain: all this vanity depends on a fine thread well or poorly placed, sound or poor judgment.
>
> After having employed so much space to establish the spirituality and the immortality of *the soul*, two sentiments very capable of making man proud of his future condition; that it may be permitted for us to use some lines to humble him on his present condition through the contemplation of futile things from which the qualities he most values depend. He does it in vain, experience does not leave him and doubt on the connection of the functions of *the soul*, with the state and organization of the body; he has to agree that the inconsiderate impression of the Midwife's finger may be sufficient to make a fool of Corneille, when the bony box that encloses the brain and the cerebellum was soft as dough ("*Âme*," 342b).

Diderot's idea of the "imperceptible lines" that connect all living forms provides some insight into how he may have been drawn to write about classical philosophy, specifically those aspects of thought that addressed the issue of man in relation to nature. The Stoics represent a philosophical attitude of life based on several ideas expressed by Diderot and other collaborators in the *Encyclopédie*. As a philosophical system of carefully enunciated doctrines concerning the individual and the natural world, Stoicism and Eclecticism were two important cases in point. For the Stoics, the fundamental basis of human behavior rested on the postulate that the individual should "live according to Nature," and Nature was that Force, Providence, Reason, Fortune or Fate which determines that the "order of things" will be as they are. When confronted by adversity, hope for the individual was found in subordination to this pervasive and life-sustaining Power or Force. For Diderot and certain of the Encyclopedists, this "subordination" was not to the "authority" of tradition and political institutions, but to reason. Reason in this context connotes a vital principle rather than an abstract mathematical principle. The coherence of this principle, in turn, implies the unity of humankind and the brotherhood of humanity, and the projected equality of individuals. Through these ideas, inferences may be drawn concerning citizenship and political associations. The Stoic ideal, "to live according to Nature," also implied that Nature contained a code of laws which the philosopher discerns by a rational process. Therefore individual behavior ought to be directed toward "perfecting" reason. In possessing reason, the Stoic attained "autarky" or self-sufficiency by way of identifying with Nature or the Universe.

To stimulate critical thinking and the acquisition of this form of independent judgment, Diderot challenged his contemporaries to think for themselves. His article on the Chaldeans provided the context in which he warned against the dangers of constraining rational thought because, as he stated, "man is born to think for himself." Therefore, "one must be very little of a philosopher oneself, so as not to feel that the best privilege of our reason consists in believing nothing by the impulse of a blind and mechanical instinct, and that it is to dishonor reason, to place it in shackles as did the Chaldéens" ("*Chaldéens*," 21b). Diderot elucidated the sort of philosopher he had in mind in his article on eclecticism:

> The eclectic is a philosopher who, trampling underfoot prejudice, tradition, antiquity, universal assent, authority – in a word, everything that subjugates the minds of the crowd – dares to think for himself, to go back to the clearest general principles, to examine them, to discuss them, to admit nothing but on the evidence of his experience and of his reason, and of all the philosophies, that he has analyzed without respect and without partiality, develops a particular and domestic philosophy that belongs to him: I say *a particular and domestic philosophy*, because the ambition of the eclectic is less being the teacher of humankind, than its disciple; reforming others than reforming himself; knowing the truth, than teaching it.
>
> . . . The sectarian is a man who has embraced the doctrine of a philosopher; the eclectic, on the contrary, is a man who hardly recognizes a master: thus when one speaks of the Eclectics that it was a sect of philosophers, one assembles two contradictory ideas. . . . The Eclectics are among the philosophers who are the sovereigns on the face of the earth, the only ones who have remained in the state of nature where everything belongs to everyone ("*Eclectisme*," 270a, b).[14]

Thus, an "unprejudiced," rational analysis of the evidence, independent of traditional ideas about the sources of authority, turns Diderot and his kind of philosophy towards the study of nature. If we pursue his line of reasoning, the question becomes, what can we expect to learn from nature? Stoic philosophic principles, particularly the idea of "life according to nature," proved to be attractive to Diderot and the Encyclopedic enterprise. Individuals were enjoined to pursue nature and, thereby, identify themselves with the immanent reason or *logos* that pervades the universe. In discussing these general principles, Diderot noted that, according to Stoic philosophy, "wisdom is the science of human things and divine things; and philosophy, or the study of wisdom, is the application of the art that guides us to it. This art is one, it is pre-eminently art; the one being virtuous. . . . There are also three kinds of philosophies: natural, moral and rational; where one observes nature, one attends to morals or manners of people, [and] one improves understanding. Each of these practices necessarily influences the other." The principles of Stoic moral philosophy apply the art of understanding to life's purpose. "Man's purpose," the Stoics maintain,

> should be to shape his behavior to the laws of nature. Nature is only something else universal reason arranges; to form his behavior to that of nature, is to see himself as a part of the great whole, and to tend to its harmony. God is the main part of nature; man's soul is a particle of God; the law of nature or of God, is the general rule by which everything is coordinated, moved, and invigorated; to live according to nature, to imitate the divine, to follow the general order [of things], is the same thing under different expressions.
>
> . . . Sovereign bliss consists in listening [to nature]: then one chooses that which is suitable for general nature or for God, and one rejects that which contradicts its harmony and its law. . . . The nature of man's good fortune is the same as the nature of God's good fortune. It is virtue. . . . A thing is good in relation to the nature of a being: a reasonable created being can only be happy by the purposes [which are] analogous to reason.
>
> . . . Therefore, there is one immutable and eternal law in the universe, an order united by causes and effects, chained by so necessary a link, that all which has been, and will be, has not been otherwise; that is destiny ("*Stoïcisme*," 526b, 530a, b, 528a).

Diderot derived the Stoic ideal of the "wise man," the self-ruling and free individual, from these passages. The Stoics were among the first, during the Hellenistic period, to understand that the "wise man" is a citizen of the universe. Several of the contributors to the *Encyclopédie* reveal a similar understanding and purpose in their articles.

The idea of Nature's overriding purpose, of God and providence, to connect all living forms in the universe requires a rational mind that is engaged in everyday experiences; requires thinkers who can and will discern the disparate elements and several links that come together to form the larger "chain of being": Thinkers who will study human nature, as they study physical nature, in accordance with the principles of the mathematical and physical sciences as then understood. The method comprises what "science" and "reason" offer to explain the moral basis of human behavior. This orientation to nature underscored the human capacity to make choices, and to shape the fundamental basis of human society in the necessary form that will reflect the presupposed harmonious order in nature. Human faculty has been provided a central place in the process. Hence, Diderot, Rousseau, de Jaucourt, and Boucher d'Argis accentuated this aspect of Stoic thought in their respective articles on human nature and society.

In the article on "Reason," Diderot defined three principles of necessity: geometric, physical and moral. He characterizes science as being constituted by an experience that testifies to the immanent rationality of nature. The virtue of reason consists in the principle of continuity and uniformity it provides, and in being essentially a human faculty. Reason, Diderot writes, is the "natural faculty which God has provided men, in order to know the truth and light that it follows, and to any order of matter that it applies. . . ." We may also understand reason to mean

> the chain of truths which the human mind can naturally attain, without being assisted by the lights of faith. The truths of reason are of two kinds: One is what one calls *eternal truths*, which are absolutely necessary . . . the necessity is logical, metaphysical, or geometrical, that one would not know how to reverse without being led to absurdities. There are other sorts that one can call positive truths, because they are the laws that God took pleasure in giving to nature, or because they are all the laws that depend on nature. . . . Thus one can say that physical necessity is founded on moral necessity, that is to say on the choice of the sage, worthy of his wisdom, and that one as well as the other ought to be distinguished by geometrical necessity. This physical necessity is what makes the order of nature, and consists of the rules of movement and of some other general laws, that God has established upon creating this universe. The laws of nature are always subject to the dispensation of the legislator . . . ("*Raison*," 733a, b).

The "natural faculty" of reason becomes the instrument by which the reflexive thinker *cum* philosopher apprehends the perceptible lines of nature through observation, and formulates general principles based on this testimony. As reason is inherent in both the individual and the philosopher, for Diderot, the reflective man epitomizes such a philosopher. "The *philosopher*," he remarked, "forms his principles [based] on endless particular observations. People adopt a principle without thinking of the observations that produce it: [people] believe that the maxim exists, so to speak, by itself; but the philosopher takes the maxim from its source, examines its origin, knows its proper worth, and only makes suitable use of it" ("*Philosophe*," 509b).

"The philosophic spirit therefore," Diderot continued, "is a spirit of observation and of exactness, that relates everything to its true principles; but this is not the only spirit that the *philosopher* cultivates, [the philosopher] carries further his attention and his care" ("*Philosophe,*" 510a). Diderot's philosophical observations led him to formulate the following principle:

Man is hardly a monster who should live only in the depths of the sea, or in the far end of the forest: The sole necessities of life render him the dealings of other necessities; and in some state where he may be able to find his needs and well-being engage him to live in society. Thus reason requires of him that he knows, studies, and works to acquire sociable qualities. . . . Our philosopher who knows how to divide [his time] between the withdrawal from and the [social] intercourse with men, is full of humanity. . . .

The philosopher therefore is an honest man who acts on everything through reason, and who combines a mind of reflection and of exactness with morals and sociable qualities. Enter a sovereign out of a *philosopher* of such moral fibre, and you will have a perfect sovereign. From this idea it is easy to conclude how much the sage imperceptible of the stoics is far away from our philosopher: such a philosopher is man ("*Philosophe,*" 510a, b).

The sources of strength and wisdom are clearly derived from those natural qualities inherent in the faculty of reason, and in the human capacity to exhibit those social qualities that are reflective of moral principles. The true nature of the reflexive thinker, then, presupposes the divine gift of reason operating in individuals, with both its sovereignty over senses and emotions, and its sovereignty over social relations. An important conclusion derived from this assumption is the notion that "natural reason" provides the foundation of all human institutions and thereby becomes identical with moral principles. Articles by Diderot on "Political Authority," "Natural Law," and "Society," by Rousseau on "Moral and Political Economy," by Boucher d'Argis on "Law of Nature or Natural Law," and by de Jaucourt on "Natural Liberty" and "Natural Equality" advanced the ideal of nature's laws as the rational organizing principle of human society, and convey contrasting social philosophies surrounding Diderot's postulate: "men are made to live in *society*. . . . Such is indeed the nature and the constitution of man, that outside of society, he would not know how to preserve his life, how to develop and to improve his faculties and talents, nor how to obtain a true and secure happiness" ("*Société,*" 252a). Furthermore, these articles attempt to answer the abiding eighteenth-century question: do social needs and sociable qualities draw individuals together in society, or does self-interest motivate individuals to form communities? These issues, among others, underlay the various "chains" of "propositions, truths, judgments, knowledge, and living bodies." The collaborators of the *Encyclopédie* combined to construe a coherent systematic political philosophy that connected individuals together within the "moral bonds" of society. These issues also draw our attention to how the Encyclopedists continued to appeal to the Laws of Nature and to Nature's God as the founding principles of human association and, thereby, connected nature's overriding purpose to the moral purpose of human society. Boucher d'Argis, a specialist in the philosophy and history of law, offered his contribution to these issues in the article, "Law of Nature or Natural Law."

In his article, Boucher d'Argis stated that rationality subjects humans to the general rules of natural equity. "One understands more often by *natural law*," he explained,

> certain rules of justice and equity, that natural reason alone has established among all men, or to be more exact, that God has engraved in our hearts.
>
> Such are these fundamental precepts of law and of all justice, to live honestly, to offend no one, and to return to every one that which belongs to him. From these general precepts still derive a great many of other special rules, that only nature, that is to say reason and equity, suggest to men.
>
> This *natural law* being founded on principles so essential, is constant and invariable ("*Droit de la nature ou Droit Naturel*," 131b).

These invariable principles were labeled "the science of morals" by Boucher d'Argis, in that the authority of natural law has been guided by the "lights" of "reason and religion" all of which God had bestowed upon man. The close connection between the idea of forming human associations and the idea of Natural Law has its source in the Old Testament, the tenets of Roman Law and the principles of Aristotle's *Politics*. Boucher d'Argis traced these sources, in the representative thinkers of the great period of the School of Natural Law, from Grotius and Pufendorf to Hobbes, Locke, and Burlamaqui. Natural Law had become in the seventeenth century an independent and rationalist system and, therefore, he considered these philosophers as "the first" to have "formed a system of natural law" ("*Droit de la Nature*," 132b).[15] Through the popularized translations, editing and writings of the Swiss scholars Jean Barbeyrac and Jean-Jacques Burlamaqui, Pufendorf's influence became discernible within eighteenth-century jurisprudence and philosophical circles. Boucher d'Argis clearly frequented these circles in view of his advice to the readers of the *Encyclopédie* to become familiar with their writings. For "about a century," one scholar noted, "the Grotius and Pufendorf of Barbeyrac represent in France, along with Burlamaqui, the philosophy of law. . . . It is to these authors that one usually refers about 1750 when one wants to treat of political questions." After 1751, "the *Encyclopédie* of Diderot and d'Alembert had contributed equally to popularize the political theories of jurists and to revive them among the French public."[16] Despite Boucher d'Argis' expertise in the philosophy of law, Diderot perceived the need to supplement the former's article with his own contribution, "Natural Law," which was written to stress the role of "humankind" in listening to the "voice of nature." For it is nature's voice that calls individuals to formulate human laws and to regulate human behavior in accordance with its own laws. In the context of these formulations, human agency requires some consideration. "The use of this word," natural law, "is so familiar," he averred,

> that there is almost no one who is not convinced within himself that the thing is evidently known to him. This inferior sentiment is common to the philosopher and to the man who has hardly reflected; with this difference alone that to the question, *what is the law?* the latter immediately lacking in terms and in ideas, returns you to the tribunal of conscience and remains silent. . . .
>
> The interrogated philosopher says, *the law is the foundation or the primary reason of justice*. But what is justice? *It is the obligation to return to everyone that which belongs to him*. But what belongs to one rather than to another in a state of things where everything will be to everyone and where perhaps the distinct idea of obligation still will not exist? . . . It is here that the philosopher begins to feel that all notions of morality, the one of *natural law*, is one of the most important and the most difficult to determine. Also let us believe we have done a great deal in this article, if we succeed in establishing clearly a few principles with the assistance of which one might be able to resolve the most considerable difficulties that one has been in the habit of proposing against the notion of *natural law*.
>
> . . . It is evident that if man is not free, or if his instantaneous determinations, or even his oscillations, being born of some material thing that may be exterior to his soul, his choice is hardly the pure act of an incorporeal substance and of a simple faculty of his substance; there will be neither goodness nor unkindness considered, . . . neither good nor bad morals, neither just nor unjust, neither moral obligation nor law. . . . I do not speak of the voluntary, but of the freedom that one only confuses too ordinarily with the voluntary ("*Droit Naturel*," 115a, b).

The interplay of "will" and "freedom," here, was such that Diderot envisioned human associations to have consisted of individuals who chose to form a social and political unit voluntarily. Natural law, thus, consisted of fundamental rational principles of public human behavior, perpetual and universal, which ought to and often did guide sovereigns in formulating "positive laws" of particular civil societies. Human society was regarded as "being so necessary to man," and as constituting his "natural inclination" to form associations. "From the principle of sociability," he maintained, "all the laws of *society* are derived, as from their source, all of our duties towards other men, as much general as particular. Such is the foundation of all human wisdom, the source of all purely natural virtues, and the general principle of all morals and of all civil *society*." Diderot proceeded to characterize the social group as embodying the corporate idea, "commonweal," and to formulate a theory that the "commonweal" of "civil *society*" established the moral basis for individual behavior:

1. The commonweal ought to be the supreme rule of our behavior, and we ought never to search for our individual advantage, to the detriment of the public advantage; it is that which demands of us the association that God has established between men.
2. The spirit of sociability ought to be universal; human *society* embraces all men with whom one can have dealings, since it is founded on the relations that they all have together, in consequence of their nature and their state. . . .
3. Equality of nature between men, is a principle of which we ought never to lost sight. Within society it is a principle established by philosophy and by religion. . . .
4. Sociability being a reciprocal obligation amongst men, those who by their malice, or their injustice, break the bond of *society*. . . .

   One must return to each one not only the possession that belongs to him, but also the degree of esteem and honor that is due to him, according to his state and his status: because subordination is the bond of *society*, and without that there will be no order in the family, nor in civil government ("*Société*," 253a, b).

Diderot's emphasis on the "subordination" of "individual advantage" to "public advantage," or to the "commonweal," contrasted sharply with the political philosophy of enlightened self-interested individualism attributed to Hobbes and Locke. His stress on "societal bonds" clearly contradicts any tendency towards espousing "anarchism." Indeed, in placing an emphasis upon a reasoned union of individuals within the "commonweal," together with the contrast drawn between "domestic" and "political" authorities, Diderot's position invites comparison with Rousseau, in particular with the latter's article on "Moral and Political Economy."

Diderot's article on "Natural Law" and Rousseau's article on "Moral and Political Economy," appeared in volume V of the *Encyclopédie*; the closeness with which their respective entries were published, and the appearance of the ideas, "commonweal," "general will," in both thinkers prompted speculation as to who influenced whom. Whatever the influence, Rousseau's subsequent usage of the categories "common self," "commonweal," and "general will," carried him in a direction that proceeded beyond Diderot's development of the terms. Though Diderot and Rousseau both wrote about the "general will," and Rousseau about the "moral being," these notions were already evident in the writings of Pufendorf, if not earlier. When Rousseau discussed his conception of "general will," he referred to Diderot's "Natural Law" article as "the source of this grand and luminous principle, [upon] which this article is expanding" ("*Economie ou Oeconomie*," 338b). Diderot introduced the concept of "general will" in the context of the individual's natural state, Rousseau planned to develop the concept further in terms of society and the "body politic."[17]

Diderot's article, "Animal," affirmed the continuity of "living bodies," and the notion of the "imperceptible" progression as to where "animality" "begins and ends." From the perspective of natural history, he refused to set humans apart from the "lower" forms of life. However, in his account of "natural law" and the natural state of human sociability, Diderot espoused the principle that "man is not solely an animal, but an animal who reasons" ("*Droit Naturel*," 116a). Therefore, it is "the issue in which it is a question of the means by which to discover the truth; that the one who refuses to search [for the truth] renounces [his] human quality, and ought to be treated by the rest of mankind as a wild beast." For "once the truth is discovered, anyone who refuses to abide by it, is insensate or spiteful from moral malice" ("*Droit Naturel*," 116a). Once the truth has been discovered, those individuals who learn how to apportion adequately time between human contact and isolation will also be those individuals who renounce self-interested individualism for the sake of the common self. Individualism, he contended, disrupted the foundations necessary for fostering collective human action. Humans possess rationality and sociable qualities, which implies that sociability manifests the natural constitution of human nature, and reason constitutes the common basis of sociability. Human reason and sociable qualities draw individuals together for the purpose of realizing the common interests of society. Upon identifying Rousseau as "our violent reasoner," whose rage he intended to stifle, Diderot denounced the radical "individualism" of "his entire discourse" [*Discours sur l'inégalité*] ("*Droit Naturel*," 116a).

But if we take away from the individual the right to decide the nature of just and unjust, where shall we carry this great question? . . . before the human race: it is to him alone that belongs [the right] to decide, because the good of all is the sole passion that he has. Private individual wills are suspect; they can be good or bad, but the general will is always good: it has never deceived, it will never deceive. If animals had an order nearly equal to ours . . . in a word if they were able to vote in a general assembly, one would have to call them; and they would not plead the cause of *natural law* any more before humanity, but before animality. But animals are separated from us by invariable and eternal barriers; and it is a question here of an order of knowledge and of ideas [that are] characteristic to human kind that emanate from its high position and that incorporate it.

It is to the general will that the individual ought to appeal in order to know how far he ought to be man, citizen, subject, father, child, and when it is fitting for him to live or to die. . . . All that you will conceive, all that you will contemplate, will be good, great, elevated, sublime, if it is of the general and common interest. . . . Say often: I am a man, and I have no other truly inalienable *natural rights* except those of humanity ("*Droit Naturel*," 116a, b).

Diderot's emphasis on the collective human effort to study nature and to search for truth, and on the intuitive inclination toward performing good deeds for others, joins in drawing individuals together in society, and explains how he came to consider Rousseau "the enemy of the human race." The argument advanced by Rousseau, in his *Discours sur l'inégalité*, was in the form of characterizing the natural state as one in which each individual lived in isolation and primarily for him or herself. From this standpoint, he rejected the doctrine of Natural Law and its moral authority. As an effective safeguard against the corrosive aspects of individualism, Diderot invoked the principle of natural law to oppose the "individualism" developed in Rousseau's *Discours sur l'inégalité*, and the combative language, "enemy of the human race" and "violent reasoner," reflects the tenacity with which he challenged the concept.

If you contemplate therefore carefully all the foregoing, you will remain convinced, 1st that the man who only listens to his private individual will, is the enemy of the human race; 2nd that the general will is within each individual a pure act of good judgment which reasons within the silence of passions concerning that which man can demand from his fellow-man, and concerning that which his fellow-man is in the right to demand of him; 3rd that this consideration of the general will of mankind and of the common desire, is the rule of conduct relative from one private individual to another private individual in the same society; from a private individual towards society of which he is a member, and in the society of which he is a member, towards other societies; 4th that submission to the general will is the bond of all societies, without excluding those who are formed through crime. Alas, virtue is so beautiful, that thieves have regard for the reflection even at the bottom of their den! 5th that the laws ought to be made for all, and not for one; otherwise this solitary being would resemble a violent reasoner we have stifled in paragraph V; 6th that, since of the two wills, the one general and the other individual, the general never errs, it is not difficult to see that which would be necessary for the happiness of the human race rests with legislative power, . . . and the authority and infallibility of the general will. 7th that when one would presuppose the notion of species in a perpetual flux, the nature of *natural law* would change, since it would always be relative to the general will and to the common wishes of the whole of mankind: 8th that equity is to justice as cause is to its effect, or that justice can only be something else equity declares: 9th finally that all of these results are evident for the one who reasons, and the one who does not want to reason, renouncing [his] human character, ought to be treated as an unnatural being ("*Droit Naturel*," 116b).

Diderot's pointed criticism of Rousseau's *Second Discourse* did not diminish the latter's efforts to formulate his own notions of the "common self" and the "general will," in the article "Moral and Political Economy." Clearly Rousseau was aware of the controversy, evidenced in his references to the latter's article, "Natural Law," as the "luminous principle" upon which to develop his own contribution to the issue, and to the "brigands" as the "the enemies of virtue in the great society" who "worship the images in their caves" ("*Economie*," 339b, 340a). Apart from the controversy surrounding his earlier essay, Rousseau's "Moral and Political Economy" may be viewed as a transitional work between the "individualism" of the *Discourse on Inequality* and the "collectivism" of the *Social Contract*, in that both conceptions were incorporated in his *Encyclopédie* article. Despite the controversy and the insidious communication between the two thinkers that punctuated their opposing views, the "Natural Law" and "Moral and Political Economy" articles shared perceptions of the "common self" and the "general will" as the fundamental basis of society. These shared perceptions of the collective self demonstrated further how the charges of "anarchism" leveled against the Encyclopedists were unwarranted. However, Diderot's stress upon the "general will" as the natural state of "sociable qualities," and Rousseau's emphasis upon the "general will" as the organized "body" of the civil state is where they parted company.

Rousseau's introductory paragraph to "Economy," in which he defined his understanding of the word, revealed the influence of Aristotle's *Politics*, and, conceivably, Xenophon's *Oikonomikos*. The etymology of the word "economy," or "economics," is Greek, and connotes "household" in the sense of regulating, administering and organizing its members. Furthermore, the broader context of the seventeenth and eighteenth centuries' theoretical writings, that placed particular importance on arguments by analogy between the laws of the family and the laws of political society, resonated in the meaning of the word he offered. His opening remarks pointed to an immediate concern with the analogous relationship between the family and the state, and the treatment of the family as one of the natural bases of the state.

Economie, or *Oeconomie* (*Moral* and *Political*). This word comes from *oikos*, household, and of *nomos*, law, and originally means only the wise and legitimate management of the household, for the joint estate of the entire family. The sense of this term in the process of time has been extended to management of the large family, which is the state. To distinguish these two meanings in the latter case one calls it, *general or political economy*; and in the former case, *domestic or individual economy*. It is only the first case that is the point at issue in this article ("*Economie*," 337a).[18]

In the first book of the *Politics*, *oikia*, the primary human association for the essential needs of life consists of the family, which is based on three natural relations: man and wife, parents and children, masters and both slaves and property. "The family," according to Aristotle, "is the association established by nature for the supply of men's everyday wants, and the members of it are called . . . 'companions of the cupboard,' . . . 'companions of the manager.' "[19]

He proceeded to express further the notion that,

the state is by nature clearly prior to the family and to the individual, since the whole is of necessity prior to the part. . . . The proof that the state is a creation of nature and prior to the individual is that the individual, when isolated is not self-sufficing; and therefore he is like a part in relation to the whole. . . .

Seeing that the state is made up of households, before speaking of the state we must speak of the management of the household. The parts of household management correspond to the persons who compose the household, and a complete household consists of slaves and freemen.[20]

The Greek expression, *oikos*, also emphasized the property component of the household, and *nomos* accented the political in the form of rational management and of the "unwritten laws" of custom.[21] Rousseau also emphasized the politics of the household, in the immediate and extended meaning of the term. In acknowledging the dual connotation of the word, "economy," Rousseau extended his discussion from the domestic to the political realm. In this manner, he came to regard the family and the state as the sole institutions which possessed a basis in Natural Law.

The whole human race, for Rousseau, is not a society, since similarity of human quality alone does not result in forming a bond that unites individual members. Rather a society constitutes a "moral being," emanating from the "true bond" that connects individuals. Rousseau's conceptions of the "common self" and of the "general will" were central to the article, and his notion of "property" as the most basic "sacred" right of society; he speaks of the rights of property as the most sacred of the citizen's rights. The individualist elements in "Moral and Political

Economy" reveal the influence of Locke; the collective elements reflect the influence of Plato, Pufendorf and Diderot.

It is certain that the right of property is the most sacred of all the rights of citizens, and even more important in certain respects than liberty; it may be because it owes more to the preservation of life; it may be because possessions being easier to usurp and more painful to defend than the person, one ought to respect more that which may be ravished more easily; finally, it may be because property is the true foundation of civil society, and the true guarantor of the actions of citizens: because if possessions do not answer for persons, nothing would be so easy as to elude their obligations and to mock their laws ("*Economie*," 344a).

In his *Discours sur l'inégalité*, Rousseau described the transition from the natural state to civil society as one which carried the individual from the condition of liberty to that of slavery through the usurpation of property. He perceived "private property" as the concrete manifestation of self-interest in the individual's natural state. Diderot opposed the self-interested individualism thesis with the notion of the common self as the natural state of individuals. Since Rousseau did not share Diderot's conviction that human societies were natural, he set out to explain how they came to be formed by assenting individuals. In the "Moral and Political Economy," Rousseau proclaimed property, the foundation of civil society, to be essential for the preservation of individual lives and that it must be preserved and protected by the state. In the Greek sense of *oikos*, Rousseau stressed the social significance of property. The head of the household manages and regulates the persons and the property within the group, and thereby exercises power over his possessions. This exercise of authority also served the political function of establishing custom through practice. Indeed, political institutions were established expressly for the purpose of preserving the very foundations of its existence: property. Hence, Rousseau's idea of property, in the "Moral and Political Economy," presupposed the social bonds of "natural society," some form of human association and negotiation prior to civil society, and the subsequent existence of the civil state.

Rousseau's notion of "sacred rights" served the similar function of legislating by custom, as served in the Greek sense of *nomos*, which connotes the "unwritten laws" of custom. The presupposition of the state's existence paved the way for Rousseau to traverse from the "sacred rights" of possessive individualism, which requires civil protection, to the ethos of the "general will," which requires the collectivizing tendency of lawful legislation. The primary concern with this tendency becomes apparent in Rousseau's discussion of the poignant contrasts to be drawn between the family and the state, in that the original family-community evolved into a sovereign civil-political structure. Political communities, as distinct from family communities, were considered as emerging after having constituted a structure of political order based on the conclusion of an agreement among assenting individuals.

When there would be as much a connection between the state and the family as several authors claim, it would not follow that the rules of proper conduct in one of these two societies, may be appropriate for the other: they differ too much in size for power to be administered in the same manner, and there will always be a difference among domestic management, where the father can see everything for himself, and civil government, where the head hardly sees anything save through the eyes of another. In order that things become equal in this regard, it would be necessary that the talents, strength, and all the faculties of the father, increase because of the size of the family, and that the soul of a powerful monarch be as that of an ordinary man, as the size of his empire is to the property of a private individual.

But how would the administration of the state be able to be equal to that of the family from which the foundation is so different? The father being physically stronger than his children, as long as his help is necessary for them, paternal power proceeds with reason to be established through nature. In the large family in which all the members are naturally equal, purely arbitrary political authority about its institution, can only be founded by agreements, nor can the magistrate govern others than by virtue of laws. The duties of the father are dictated to him by natural sentiments. . . .

Through several reasons taken from the nature of things, the father ought to govern in the family. . . . Indeed, the voice of nature is the best counsel that a good father ought to listen to in order to fulfill his duties well. . . . Also nature has made a multitude of good family fathers; but it is doubtful that since the existence of the world, human wisdom has ever made ten good magistrates ("*Economie*," 337a, b, 338a).

After having drawn these various contrasts between the family and the state, Rousseau proceeded to argue that for any analysis of political authority to sustain itself theoretically, "one must distinguish public economy from private individual economy." The readers were enjoined to distinguish further between "public economy," which he calls "management," and "supreme authority," which he calls "sovereignty" and which accounts for the observation that the "state" has "nothing in common with the family." Thus, he perceived the need to proceed beyond patriarchalism as the most important justification of political obligation. Rousseau identified Aristotle's *Politics* and Filmer's *Patriarcha* as two of the "several authors" he believed to be responsible for perpetuating the "ancient error" of equating the head of the household with the head of the state[22] ("*Economie*," 338a).

The "common comparisons" Rousseau eventually made entailed the analogies between the animal and the social organism, between the human body and the body politic. The political authority accompanying paternal power was as natural as the relationship between father and child, and the duties attendant this power dispensed an equity naturally. By contrast, the political authority invested in a government is natural only insofar as the head of state dispenses justice and equality for its citizens through laws. To become natural, in the latter case, laws must become the expression of the community. The organic analogy for a social group highlights the individual parts that form a substantive bond uniting all of its members. The individual organs are separate in function yet mutually dependent. The general principle underlying this conception conveys the understanding of society as a "moral being."

The body politic, taken individually, can be considered as an organic body, living, and similar to the organic body of man. The sovereign body represents the head; the laws and the customs are the brain, mainspring of nerves and the seat of understanding, of the will, and of the senses, of which the judges and the magistrates are the organs; commerce, industry and agriculture, are the mouth and the stomach that prepare the common substance; public finances are the blood of a wise *economy*, in making the functions of the heart, which returns to distribute through the entire body nourishment and life; the citizens are the body and the limbs that make the machine move, live and work, and that one would not know how to injure any part [of the body], immediately the painful impression goes to the brain, if the animal is in a state of health.

The life of one and of the other is the *common self* to all, the reciprocal sensibility, and the inward correspondence of all the parts. Is this communication going to cease, the formal unity to vanish, and the contiguous parts not to belong any more to each other through juxtaposition? Man is dead, or the state is dissolved.

The political body is therefore also a moral being which has one will; and this general will, which always tends to the conservation and to the welfare of each and every part, and which is the source of laws, is for all the members of the state with regard to them and to it, the rule of the just and of the unjust; truth which, to say it in passing, shows how many meanings so many writers have treated of stealing the subtlety prescribed to the children of Lacedaemon [Sparta], to gain their frugal meal, as if everything which arranges the law could not be legitimate. *See the word Law*, the source of this great and luminous principle, from which this article develops ("*Economie*," 338a, b).

The use of the body metaphor essentially implies hierarchy, and Rousseau clearly intends to evoke that implication when he refers to the sovereign as representing "the head" and the citizens as "the body and limbs" of the political body. The sovereign on which Rousseau remarks is a "moral being," and the final normative principle of social life becomes the "general will" of that being. "Moral being" and "general will," then, constitute non-physical entities. The patriarch administered political authority in accordance with "the voice of nature," as his counsel, in regard to representing familial interests and justice. The general will, embodying all of its members, listened to "the voice of the people" which "is in effect the voice of God" or "the law of nature" ("*Economie*," 338a, 339a, b). In identifying the interests of political society with that of the people, Rousseau assigned less importance to the head of the state or the government than to "all of the citizens." This identity of interests positioned him to remark that the rule of justice, then, "in relation to all citizens," as "the will of the state, however general with respect to its members," becomes a principle of "the law of nature." He made this remark based on the observation that "the city of the world becomes the political body of which the law of nature is always the general will, and of which the states and diverse peoples are only individual members" ("*Economie*," 338b). From political society and its members "spring the more universal rules and the safest course on which one may be able to judge a good government from a bad one, and in general, to judge good moral conduct from all human actions" ("*Economie*," 336b). The category "citizen," in this sense, implied an active participation in the public sphere, and especially an active participation in the formulation of laws.

The inclination toward forming societies was a universal human characteristic which demonstrated how assenting individuals with common interests form associations, and how each society develops a general will which regulates the conduct of its members. The general principle underlying this position involves the notion that the state must make liberty possible under the law, and must provide for the material welfare of its citizens by eliminating inequality before the law and in the distribution of wealth. Every political society, he declared, must listen to the "celestial voice" that will affirm the "rights of humanity." For

> the first and most important maxim of legitimate or popular government, that is to say, the one which has for its object the good of the people, and therefore, as I have already said, the object of following the general will in everything; but to follow it one must know it, and above all one must distinguish it well from private individual will beginning with oneself; a distinction always very difficult to do, and for which it belongs solely to the most lofty virtue of giving sufficient understanding. As to will one must

> be free; another difficulty that is hardly of less importance, is to assure at the same time public liberty and the authority of the government. Search for the motives that have supported men joined by their mutual needs in the larger society, to unite more strictly by civil societies; you will hardly find another than the one of assuring the possessions, the life, and the liberty of each member through the protection of all the members: . . . These wonders are the work of the law. It is to the law alone that men owe justice and liberty. It is this salutary organ of the will of all, that reestablishes within the law natural equality between men. It is this celestial voice that dictates to each citizen the precepts of public reason, and teaches him to act according to the maxims of his own judgment, and [dictates to each] not to be at variance with himself. Also it is this voice alone that the chiefs ought to make speak when they govern; because as soon as independently of laws, a man seeks to subject another to his private will, he leaves immediately the civil state, and goes opposite it [the civil state], in the pure state of nature where authority is never prescribed save by necessity ("*Economie*," 339b).

The fundamental rule of "legitimate government," then, consists of the notion that political subjection was primarily ethical and secondarily an issue of law and power. In emphasizing the primacy of the ethical over the legislative, Rousseau makes explicit the assumption that the community comprises the "chief" moralizing agency and thereby represents the highest virtue for its citizens. Rousseau derived the high moral value of the community from Plato. To oppose conventional natural law theorists, who stressed the various capacities of individuals in the state of nature, Rousseau drew from Plato's *Republic* the viewpoint that citizens of a community embody the essential moral "virtue" of the "general will." He took care to emphasize the individual's capacity to grow in reason and in morality. However, the fundamental moral qualities of being human stem from within society, rather than from the "natural" individual. In this manner, political sovereigns become "moral beings" and the best exemplars of the highest virtue, "as virtue is only this conformity of private individual will to the general will." Once political authorities, or "the chiefs" as he called them, strive to reconcile individual with general wills and direct their exercise of power toward the "public good," "according to the law of duty and obligation" instead of "being blinded by their [personal] ambition," they will "make virtue prevail" ("*Economie*" 340b). This second rule of conduct provides a few of the reasons for which Rousseau opposed Diderot's position on the "natural sociability" qualities of the individual. Rather, for Rousseau, sociability derives from the natural inclination of individuals to unite with other humans, an inclination grounded in the rational disposition that seeks the "public good."

The starting point for the political ideas discussed in the articles by Boucher d'Argis, Diderot and Rousseau revealed their concern about the uses to which power was issued. That is, the idea of the origin of human association emanates from the presupposition of the natural association of families as the foundation upon which political communities were organized. This informal association was sustained initially by social impulse and the perception of mutual economic benefits. The natural association becomes a proper political community when its members consent to institute a government in order to promote the "public good," all in accordance with the laws of Nature which constitute God's will. This agreement to agree upon a government requires the free and voluntary consent of all those members who constitute the initial human association. Thus, to be legitimate, political authority must be derived from the consent of the people and must avoid claiming power by the sheer force of violence. The necessary consent of the people limits political power, a limitation which has been defined by natural law. The principles of natural law dictate that no individual, within society, ought to be subject to the arbitrary authority of other individuals, but to rules of conduct known through general laws. Whenever in doubt, sovereigns should be guided by the "celestial voice" which prescribes how one ought to act. Indeed, this "voice" sets limits on the father as "prince" of the family.

In his article "Political Authority," Diderot pursued these issues further and declared that "no man has received from nature the right to rule others. Liberty is a present from Heaven, and each individual of the same species has the right to enjoy it as soon as he possesses reason" ("*Autorité Politique*," 898a). Every father who enjoys "paternal power" cannot legitimately assume "royal authority" through coercion without the consent of the individuals involved.

> . . . If nature has established some *authority*, it is paternal power: but paternal power has its limits; and within the state of nature it [paternal power] will end as soon as the children would be in condition to take care of themselves. All other *authority* comes from another origin than that of nature. When one examines it well, one will always be able to go back to one of these two sources: either the force and the violence of the one who is in possession of it, or the consent of those who have been bound by a contract made or supposedly made among themselves, and the one on whom they have conferred *authority*.
>
> Power which is acquired by violence, is only an usurpation, and only lasts as long as the force of the one who rules prevails over those who submit [to it]; in such a manner that if the latter became in their turn the stronger, and if they throw

off the yoke, they are with as much right and justice as the other who had imposed it on them. The same law that made *authority*, then, unmakes it: it is the law of the stronger. . . .

Power that derives from consent of the people necessarily implies conditions that render legitimate use of it, useful to society, and advantageous to the republic, and that fix and restrict it within limits: because man neither ought nor can give himself entirely and unreservedly to another man; because he has a superior master above all, to whom he alone belongs completely. This is God, in whom power is always immediate over created beings. . . .

True and legitimate power therefore necessarily has limits. . . .

The prince derives from his subjects even the *authority* that he has over them; and his *authority* is limited by the laws of nature and of the state. The laws of nature and of the state are the conditions under which they are submitted, or they are supposed to be submitted to his government. One of these conditions is that only having power of *authority* over them by their choice and their consent, he can never use this *authority* to break the act or the contract by which it has been conferred on him. . . .

Moreover the government, though hereditary within a family, and placed in the hands of a single individual, is not private individual property, but is public property, which accordingly can never be taken from the people, to whom alone it belongs essentially and in full ownership. . . . It is not the state that belongs to the prince [rather], it is the prince who belongs to the state: but it belongs to the prince to govern within the state, because the state has chosen him for that; because he has pledged his word to the people to the administering of affairs, and because these people for their part are bound to obey him in conformity with the laws. . . . In a word, the crown, the government, and public *authority*, are possessions of which the body of the nation is proprietor, and of which the princes are the usufructuaries, ministers, and trustees ("*Autorité Politique*," 898a, b, 899a).

In appropriating "political authority" on behalf of the people, Diderot called into question the politics of possessive royal rule which had become the hallmark of monarchs who rule in accordance with a divine right theory. Louis XIV was such a monarch and his grandson, Louis XV, inherited his "political instructions" as part of his legacy. Louis XIV's reign contributed to this legacy in the form of organizing a government through which all of the French were attached to him, to his own person. In his *Mémoires pour l'instruction du dauphin*, Louis XIV proffered these lessons in referring to himself as king:

. . . all eyes are fixed upon him alone; it is to him alone that all appeals are addressed, he alone receives all respects, he alone is the object of all hopes; nothing is pursued, nothing is expected, nothing is ever done except through him. His good graces are regarded as the sole source of all favors; one hopes to raise oneself only in proportion to as close as one draws nearer to his [the King's] person or to his esteem. All else is fleeting, all else is powerless, all else is fruitless.[23]

To counter this legacy, Diderot put forward the principle that argued for subjecting the legitimate power of the monarch to limitations. The purport of this principle was to supplant "royal authority," claimed by the father and the prince, with human authority. The principles of Stoic moral philosophy, so successful in scientific discourse concerning "man's purpose" in nature, also contributed to undermining the theoretical foundations of absolute monarchy based on the doctrine of divine right. Stoicism embodied an idea of human society that directly challenged the notion of the king as a divine personage: the idea of "human life according to nature," of the identical nature of all individuals, equal in their exercise of reason and choice, all of which combined to develop the concept of a society of individuals who were free, self-governing, legislating for themselves, and also equal before the law. Roman law and the tradition of Natural Law incorporated this Stoic idea and viewed all individuals as having emerged equal from the "creator" of Nature. Namely, all are children of the same father, equally free, equally noble, and all members of one and the same body. Civil or positive law affirmed the principles of nature and thereby focused attention upon that which was considered common to all living creatures, their physical, biological qualities, and to those inequalities which were the consequences of nature. These natural qualities were perceived to be inviolable. De Jaucourt carried the principle another step in making more explicit the doctrine of limited monarchy.

The concept of "liberty" was stressed by all eighteenth-century thinkers including the Encyclopedists. In our earlier discussion of their views of nature and human nature, the concept meant freedom of thought, of judgment and of conscience, including or excluding religious references. In the context of our political discussion, the concept connoted a natural right of freedom, acquired at birth from the "author of Nature."

In his article "Natural Liberty," de Jaucourt defines the category "liberty" employing the latter connotation within the tenets of natural law. "Natural Liberty," he wrote, connotes the

> right that nature gives to all men to dispose of their person and their possessions, in the manner that they judge the most suitable for their happiness, under the limitation that they do it within the terms of natural law, and that they do not abuse it to the detriment of other men. Natural laws are therefore the rule and the measure of this *liberty*; because whatever men may be in the primitive state of nature, some may be in [a state of] independence compared with others, they are all under the subjection of natural laws, from which they ought to direct their actions.

> The first state that man acquires by nature, and that one esteems the most precious of all the property he can possibly possess, is the state of *liberty*; he can neither exchange himself for another, nor sell himself, nor lose himself; because all men are naturally born free, that is to say, they are not subject to the power of a master, and no one has a right of ownership over them.
>
> By virtue of this state, all men have from nature itself, the power to do that which appears good to them, and to dispose to their liking of their actions and of their possessions provided that they do not act against the laws of the government to which they have subjected themselves ("*Liberté Naturelle*," 471a).[24]

Through this understanding of natural liberty, de Jaucourt's idea of a natural right to freedom implied further the rejection of all forms of subjugation to another person. There are no "masters" in his conception of "managing" a household or a state. The natural right to dispose of one's person and one's "liberty," in whatever manner deemed appropriate by the individual, contrasted sharply with the notion of a corporate social order in which individuals were distinguished from one another by hierarchy and privilege. De Jaucourt characterized "natural liberty" as a political order constituted by individual citizens who, no longer subject to the discipline of the monarch, were free to construct their lives as they may choose, living under laws that applied equally to everyone. The concept of equality of rights in the social and political realms, and equality before the law were the logical conclusions drawn by de Jaucourt from the concept of natural liberty. From this perspective, de Jaucourt designated "natural equality" as the principle upon which "natural liberty" becomes possible:

> Natural equality is that which is among all men by the constitution of their nature alone. This *equality* is the principle and the foundation of liberty.
>
> *Natural* or *moral equality* is therefore founded on the constitution of human nature common to all men, who are born, grow, subsist, and die in the same manner.
>
> Since human nature finds the same within all men, it is clear that according to natural law, each ought to esteem and to treat the many others as human beings as himself and naturally equal, that is to say that they are men as well as he ("*Egalité Naturelle*," 415a).

The pivotal point of Divine Right Theory, systematically developed in the writings of Bishop Bossuet and passed on in Louis XIV's "political instructions" to his heirs, was the Aristotelian assumption that individuals are by nature unequal in rank, and that there is, therefore, a natural hierarchy among individuals. This presupposition became untenable once challenged by the assumption of natural law that the natural rank of all individuals is *ipso facto* equal. Politically, this implied that together with all other rank orders the hierarchies of society could no longer claim continuous stability. For if individuals are equal by nature, then social inequalities cannot be established by nature or by God; and if inequalities were not established thusly, they, in turn, are subject to change. The "author of Nature" bestowed upon human nature the gift of life, liberty, reason and the ability to exercise judgment.

The second half of the eighteenth century was the period in which the sense of the variety of all life and the sense of its inherent meaningfulness, in the form of experimentation, mathematical analysis, and natural history, combined to produce works, such as the *Encyclopédie*, in which the universal principles of reason were invoked to order the natural world and the broad range of human experience. The deliberate application of a method of analysis to several different areas of knowledge became the systematic device by which the Encyclopedists organized natural and social realities. The mathematical and natural history ideals of the *Encyclopédie* promoted principles of nature that provided individuals with "chains" of judgments, reasoning, "chains" of "living bodies," nature, and knowledge that connected the physical world to the world of individuals. Secular and "revolutionary" as the *Encyclopédie* may be, with respect to method and approach to human knowledge, the principles of Nature and human nature formulated by the Encyclopedists continued to find their bases in God's ordinance. These principles of nature were closely connected to the fundamental conception of Reason as the common element in all humans. The premise, Reason appears in every individual, derived the conclusion that Reason must be acknowledged as Natural, if not Divine, Law. Natural Law provided the foundation upon which all human associations were constructed and, thus, became identical with moral principles. Through this form of reasoning, the Encyclopedists appealed to the Laws of Nature and to Nature's God through which moral postulates were made to posit a "common humanity" in natural and social realities. Natural philosophy carried moral philosophy with it. The philosophical espousal of the ideal of unifying knowledge, nature, history and human history, engendered the political ideal of natural equality. Both ideals of which were founded upon the notion of a natural order which operated in accordance with laws that were accessible to all who exercise their rational powers. Hence, the contributors to the *Encyclopédie* bequeathed to eighteenth-century thought a compendium of knowledge, methods of analysis, and the normative responsibility to discern correct from incorrect judgment.

Wesleyan University

NOTES

* "Aphorisms concerning the Interpretation of Nature and the Kingdom of Man," in *The Works of Francis Bacon*, collected and ed. by James Spedding, Robert Leslie Ellis, Douglas Denon Heath, 15 vols. (Boston: Houghton, Mifflin and Company, 1900), 8: 67–68.

[1] *L'Encyclopédie ou dictionnaire raisonné des sciences, des arts et des métiers*, ed. by Diderot and d'Alembert, 17 vols. (Paris: Chez Briasson, David, Le Breton, Durund, 1751–1765). Reprinted (New York: Readex Microprint Corporation, 1969) 5 vols.

[2] *Arrest Du Conseil d'Etat Du Roy*: Qui ordonne que les deux premiers volumes de l'ouvrage intitulé, *Encyclopédie ou Dictionnaire raisonné des Sciences, Arts et Métiers*, par une Société de gens de Lettres, seront & demeureront supprimés. Extrait des Registres du Conseil d'Etat, Du 7 février 1752. Salle des manuscrits, Bibliothèque Nationale.

[3] D'Alembert, *Eléments de Philosophie*, in *Oeuvres Complètes de d'Alembert*, 5 vols. (Paris: A. Belin, 1821), I: 122. *Encyclopedia Britannica*, or a *Dictionary of Arts and Sciences and Miscellaneous Literature*, Supplement to the Third Edition, 2 vols. (Edinburgh: Thomas Boner, 1801), I: III–IV.

[4] Francis Bacon, *The Works of Francis Bacon*, 8: 131; *New Organon*, Part II, XCV.

[5] René Descartes, *Les Principes de la Philosophie* in *Oeuvres de Descartes*, 4th ed., French trans. by Charles Adams and Paul Tannery, 11 vols. (Paris: Léopold Cerf, 1897–1909), second part, 9: 83–84.

[6] *Ibid.*, *Discours de la Méthode*, 6: 1–2.

[7] *Ibid.*, *Rules for Direction of the Mind* (Regulae ad Directionem Ingenii), Rule 4, 10: 371–372.

[8] "Du Système du Monde dans les principes de la gravitation universelle," in *Histoire de l'Académie Royale des Sciences: Année MDCCXLV* (Paris: de l'Imprimerie Royale, 1749), p. 329. D'Alembert also read a paper at that time on "Méthode générale pour déterminer les orbites et les mouvemens de toutes les Planètes, en ayant égard à leur action mutuelle," pp. 365–390. Although the papers were not published until 1749, both Clairaut's and d'Alembert's papers were read "à l'Assemblée publique du 15 novembre 1747," pp. 329, 365. By way of contrast to Clairaut's signal of a "great revolution," see Colin Maclaurein's *An Account of Sir Isaac Newton's Philosophical Discoveries, in Four Books* (London: Patrick Murdoch, 1748), pp. 39, 42, in which he suggests a circular sequence concerning revolution, when one witnesses "the history of learning thro' its various revolutions in the later ages," and observes that "after a long interval, learning has returned to them; and the period which commenced upon the revolution we have mentioned, has already continued for some hundred years."

[9] D'Alembert, *Eléments de philosophie*, I: 3–6.

[10] Descartes, *Opuscules de 1619–1621* in *Oeuvres* ed. by Charles Adam and Paul Tannery, 13 vols. (Paris: Librairie Philosophique J. Vrin, 1974), 10: 255, 215. See also

Joseph F. Scott, *The Scientific Work of René Descartes* (1596–1650) with forward by H. W. Turnbull (London: Taylor & Francis, Ltd., 1952), p. 7.

[11] Denis Diderot, *Oeuvres Complètes*, ed. by Roger Lewinter, 20 vols. (Paris: Le Club Français du Livre, 1970), 19 février 1758, 3: 630.

[12] George Louis LeClerc Buffon, *L'Histoire Naturelle*, "De l'Homme," in *Oeuvres Complètes de Buffon*, 4th ed., 6 vols. (Paris: A. Ledoux, 1845), 3: 224.

[13] Buffon, *L'Histoire Naturelle* in *Oeuvres Complètes*, I: 3.

[14] Diderot derived his positive characterizations of eclecticism from Jakob Brucker's *Historiae Critica Philosophiae*, in which Brucker defined the "eclectic method of philosophizing," as having been "approved by intelligent men and practiced by philosophers of the greatest ability." Brucker identified philosophers such as Giordano Bruno, Francis Bacon, Campanella, Hobbes, Descartes, Leibnitz and Christian Thomasius to be "men who renewed the universal eclectic philosophy" in the seventeenth and eighteenth centuries. See Jacobi Bruckeri, *Historia Critica Philosophiae a Tempore Resuscitatarum in Occidente Literarum ad Nostra Tempora*, second ed., 4 vols. (Leipzig: Weidemanni et Reichii, 1766), 4: part I, 190; part 2, 3–543. Compare Diderot's definition with Brucker's, 4: part 2, 4. For a discussion of Diderot's reliance on Brucker, also see Jacques Proust, *Diderot et l'Encyclopédie*, pp. 247–284, 548–555.

[15] Boucher d'Argis drew upon Burlamaqui's theories, the fourth edition of which had been published in Geneva in 1747. See Jean-Jacques Burlamaqui, *Principes du droit naturel*, 2 vols. (Genève: Barrillot et fils, 1767), *passim.*

[16] Robert Derathé, *Jean-Jacques Rousseau et la science politique de son temps* (Paris: Presses Universitaires de France, 1950), pp. 32–33. For the influence of Pufendorf in Europe and in America, see Leonard Krieger, *The Politics of Discretion: Pufendorf and the Acceptance of Natural Law* (Chicago: University of Chicago Press, 1965), pp. 255–269.

[17] Compare Diderot's concept of "sociability" with Pufendorf who considered "sociability" to constitute "the fundamental law of natural right." See, *Le Droit de la nature et des gens, ou système général de la morale, de la jurisprudence, et de la politique*, trans. from Latin by Jean Barbeyrac, 2 vols. (Amsterdam: Chez Gerard Kuyper, 1706), 1: 178. Concerning the issue of whether Rousseau borrowed the idea of "general will" from Diderot, see René Hubert, *Rousseau et l'Encyclopédie: Essai sur la formation des idées politiques de Rousseau (1742–1756)* (Paris: J. Gamber, 1928), pp. 43–59; *The Political Writings of Jean-Jacques Rousseau*, ed. with an introd. by C. E. Vaughan, 2 vols. (Cambridge: The University Press, 1915), I: 424–425; Jacques Proust, *Diderot et l'Encyclopédie* (Paris: Armand Colin, 1962), pp. 387–393; and Arthur M. Wilson, *Diderot* (New York: Oxford U. Press, 1972), pp. 233–234. On the earlier usage of "general will" and "moral person" in Pufendorf, see Otto Gierke, *Natural Law and the Theory of Society, 1500–1800*, trans. with an introd. by Ernest Barker (Cambridge: The University Press, 1950), pp. XIII, 97, 118–135.

[18] Though commentators refer to this article as "Political Economy," Rousseau defined his understanding and purpose, in the introductory paragraph, to entail a discussion of "Moral and Political Economy." Boulanger wrote the entry, "Political Economy." By the middle of the eighteenth century, the connotation of the word in its modern sense was evident. The concept political economy, as theory of the art or practical science of managing the resources of individuals and of their government, so as to promote material prosperity, was later formulated and made popular by Adam Smith in 1763. Adam Smith

was a student of Francis Hutcheson who published in Latin his *Short Introduction to Moral Philosophy*, in 1742, which was translated five years later into English. The chapters on "The Principles of Oeconomics and Politics" and "Elements of the Law of Nature" discussed the duties of parents and children, masters and slaves, marriage and divorce and succession to property rights among other topics. In *The Wealth of Nations* Adam Smith proclaimed that "the great object of the political economy of every country is to increase the riches and power of that country." The latter eighteenth-century formulation, in which concern for the "economic" aspect of the state has been expressed, was anticipated in Rousseau's article on moral and political economy. See Francis Hutcheson, *Short Introduction to Moral Philosophy*, in *Three Books: Containing the Elements of Ethicks and the Law of Nature*, 2 vols. transl. from the Latin, third ed. (Glasgow: Robert and Andrew Foulis, 1764), 2: 273–324; Adam Smith, *An Inquiry into the Nature and Causes of the Wealth of Nations*, 2 vols. (London: W. Straham and T. Cadell, 1776), I: 453.

[19] *Politics* in *Introduction to Aristotle*, sec. ed., edited and intro. by Richard McKeon (Chicago: The University of Chicago Press, 1973), p. 597; 252b 10–11; cf. pp. 595–620; 1252a 1–12, 1260b (22).

[20] *Ibid.*, pp. 598–9; 1253a, 19, 26–29, 40–42, 253b 1–4.

[21] In Sophocles' *Antigone*, Antigone raised the question of political legitimacy and familial obligation when she cited the higher authority of the "eternal" "Unwritten laws" of custom to challenge Creon's edict prohibiting the burial of her brother. Cf. Sophocles, *Antigone*, second Ode, lines 440–450.

[22] See Aristotle, *Politics*, especially the first book, *ibid.*, pp. 595–620; and Robert Filmer, *Patriarcha and Other Political Works*, ed. with an introd. by Peter Laslett (Oxford: Basil Blackwell, 1949).

[23] *Oeuvres de Louis XIV, Mémoires historiques et politiques*, 6 vols. (Würtz 1806), 2: 26.

[24] Compare John Locke's *Two Treatises of Government* where he maintains that the natural liberty of the individual necessarily connects to the individual's knowledge of the law of nature, similar to the manner in which the freedom of an Englishman consists in his "liberty to dispose of his Actions and Possessions" in accordance with the laws of England, Chapter VI, para. 59 and 63, pp. 325, 327.

WALTER H. CAPPS

DISCERNING HUSSERL'S PHILOSOPHY IN HAVEL'S PROPOSALS

The thesis of this paper is uncomplicated and straightforward, namely, that a central tenet in the philosophical phenomenology of Edmund Husserl became an ingredient in the philosophical outlook that guided the now famous "Velvet Revolution" that occurred in Prague in 1989. Though the paper concerns itself with political change of rather dramatic proportions, it is not about politics, first of all, but about continental philosophy, and, specifically, about that portion of it that came to prominence in the writings of certain Czech thinkers, and not least, Vaclav Havel, now President of the Czech Republic.

In the essay entitled "Politics and Conscience" (1984), Havel pays direct tribute to Husserl, and cites specifically the latter's development of the concept of *Lebenswelt*. In this brief citation Havel employs paraphrases such as "the natural world" and "the world of lived experience" to undergird the argument that Husserl's concept provides a reliable vantage point from which to approach and investigate "the spiritual framework of modern civilization and the source of its present crisis." Here Havel identifies children, working people, and peasants as being "far more rooted in what some philosophers call 'the natural world,' or *Lebenswelt*, than most modern adults." He explains:

> They have grown alienated from the world of their actual personal experience, the world which has its morning and its evening, its *down* (the earth) and its *up* (the heavens), where the sun rises daily in the east, traverses the sky and sets in the west, and where concepts like "at home" and "in foreign parts," good and evil, beauty and ugliness, near and far, duty and rights, still mean something living and definite.

It is important to recognize that these words were issued in 1984, following the author's second imprisonment. He was arrested and jailed for four months for Charter 77 activities in 1977. In 1979 he was convicted again, and sentenced to four and a half years in prison, but received a suspended sentence owing to his poor health, and, most likely, in response to an international protest campaign. Clearly, his reflections on "lived experience" were developed as contrast and antidote to the conception of reality that forced his own imprisonment as well as

A-T. Tymieniecka (ed.), Analecta Husserliana, Vol. LIV, 259–266.

the situation of his country as being under the domination of totalitarian rule.

The same theme is enunciated in the essay entitled "What I Believe" wherein Havel levels a criticism against two dogmas. The first is the prevailing "systematically pure market economics," and the second is prevailing Marxist ideology. Both can be treated in the same way, for they represent and illustrate the same mental disposition to assume that life devolves from theory, and that whatever does not conform to fundamental theoretical precepts, by definition, deserves to be rejected. Havel charges the advocates of these dogmas with assuming that "operating from theory is essentially smarter than operating from a knowledge of life." And when elucidating this latter, more reliable knowledge, the Czech essayist-playwright-turned-politician holds up the reliability of "our understanding of individual human beings and the moral and social sensitivity that comes from such understanding." Indeed, in summary fashion, Havel states that "social life is not a machine built to any set of plans known to us." This, he explains, is why "new theories are constantly being fashioned." And this time, through the paraphrase for *Lebenswelt* is "the flow of life which is always taking us by surprise."

In numerous places within his essays, Havel examines "the collapse of an ideology," and the consequences pertaining thereto. To his mind, the collapse of one ideology is not to be followed by the creation, construction, or imposition of a successor ideology, presumed to be better equipped than the one it replaces to deal with social, cultural, economic, political, and intellectual challenges and complexities. Rather, the collapse of an ideology is indicative of the collapse of ideological presumptions. Havel asserts that "the world of ideologies and doctrines is on the way out for good," for the post-ideological age will be formed by ideas that have their roots and sources, once again, in the world of lived experience, or that fundamental human environment within which "the flow of life" is most fundamentally honored.

I have suggested that it is not difficult to find such contentions in Havel's writings. They are simply present, explicitly or implicitly, in virtually everything he has written, perhaps most poignantly in his famous *Letters to Olga*. It is also not difficult to discover the same kind of contentions within other Czech literati of the modern and contemporary era. Indeed, throughout his impressive work, the Czech novelist Milan Kundera, well known for such books as *The Book of Laughter and Forgetting*, *The Unbearable Lightness of Being*, and (my favorite title) *Life is Elsewhere*, also pays tribute to Husserl and to Husserlian concepts. For example, in his "The Depreciated Legacy of Cervantes" (1983), which stands as introduction and orientation to the essays included in his *The Art of the Novel* (1986), Kundera simply states that he is responding to the problematic Husserl addressed in his writings on "the crisis of European humanity." Kundera proceeds to make a gallant effort to place an exposition of *Lebenswelt* within the framework of a portrayal of the roots of the modern era, principally in Galileo and Descartes. In so doing, Kundera describes the novel as being "grounded in the relativity and ambiguity of things human" and, thus, as being "incompatible with the totalitarian universe." He explains:

> The world of one single Truth and the relative, ambiguous world of the novel are molded of entirely different substances. Totalitarian Truth excludes relativity, doubt, questioning; it can never accommodate what I would call the *spirit of the novel* (pp. 13–14).

Yet, it is perhaps in his *Jerusalem Address* (entitled "Man Thinks, God Laughs," when subsequently published in *The New York Review of Books* in 1985) that Kundera offers the most striking formulation of his contention:

> For if European culture seems under threat today, if the threat from within and without hangs over what is most precious about it – its respect for the individual, for his original thought, and for his right to an inviolable private life – then, I believe, that precious essence of the European spirit is being held safe as in a treasure chest inside the history of the novel, the wisdom of the novel.

In less formal philosophical terms, precedent for taking the life-world as being normative can be found in perhaps the most popular piece of Czech literature in the twentieth century, namely, Jaroslav Hasek's *The Good Soldier Schweik* (1930), which is an attempt to define heroism in terms directly contrary to the assumptions that would call Napoleon, for example, a great man. Hasek describes Schweik as a modest, unpretentious man, who troubles no one, expects not to be troubled by anyone else, goes his own way unassumingly, rather shabbily dressed, who does nothing to get his name in lights or in the papers, and yet is a "heroic, dauntless man who was the talk of all citizens in the Kingdom of Bohemia when they were under Austrian rule, and whose glory will not pass away even now that we have a republic." Admittedly, it would be a large stretch to identify the disposition of "good soldier Schweik" with the approved side of the contrast Havel draws when comparing *Lebenswelt* with the view of the world that is deduced from systematic, aprioristic ideological invention. Yet the example of Schweik as hero in contrast to models of previously prevailing heroic expectation can be likened to the devotion Havel exhibits for the world of ordinary human experience. Schweik is an ordinary human being who is fully in touch with the flow of life, even when confronted with the realities of ideological conflict and military hostility. The temper of Hasek's portrayal of the heroic soldier is in keeping with the spirit of Karel Capek's precedent-setting *An Ordinary Life* (1934), which Rene Wellek described as "one of the most successful attempts at a philosophical novel in any language." In Hasek and Capek is also something of the disposition of "the inversion formula," exhibited in Ludvik Vaculik's *The Axe* (1966) and Ivan Klima's *An Hour of Silence* (1963), wherein the professed dogmas of socialist realism are seen to produce consequences that contradict the ethical premises on which they claim to be based. Perhaps the best known and most dramatic examples of this genre are the writings of Franz Kafka, reinforced for Czech writers and readers by the searing memories of the sharp juxtapositions of intention and outcome of World War I. In this respect, our intention is simply to sketch in some of the intellectual background factors that pervade prominent Czech plays, novels, and essays, which, even for Vaclav Havel, serve as the primary vehicle for the communication of philosophical ideas.

A more direct and explicit reference to Husserl, and to the fundamentality of *Lebenswelt*, can be found in the writings of the Czech philosopher Jan Patocka (1970–1977), to whom Havel pays tribute on numerous occasions. It was Patocka, who was immersed in the writings of Husserl and Martin Heidegger, who became one of three spokespersons for Charter 77, the statement that was promulgated Prague in 1976 to be supported by additional signatures of scholars, educators, and musicians who were calling for a more tolerant and open Czechoslovakia. In *Disturbing the Peace* (1986), Havel mentions that he "hungrily devoured" the writings of Professor Patocka. In "Six Asides About Culture" (1984), Havel calls Patocka Czechoslovakia's "most important philosopher." Perhaps his most famous essay, "The power of the Powerless" (1978), was dedicated to Patocka, who died – Havel and others call it a martyr's death – shortly after being interrogated by police after serving as spokesperson for Charter 77. Indeed, it is highly possible that Havel's exposure to Husserl and Heidegger came more through Patocka's writings and lectures than through any study and analysis of primary texts. In this connection, it is important to add that

Patocka impressed upon Havel the truth that "the real test of a man is not how well he plays the role he has invented for himself, but how well he plays the role that destiny assigned to him" (*Disturbing the Peace*, p. 72). Again, the equation is not self-evident, but no large stretch is required to discern the contrast between the workings of invention and destiny, and to perceive therein the sharp contrasts between world as product of aprioristic ideological construction and world as inhering in lived experience. For Vaclav Havel, Jan Patocka stands as an embodiment of the teaching that impersonal power can only be resisted and confronted by the one thing everyone has at their disposal, namely, their own humanity. Patocka extended the concept *Lebenswelt* so as to provide a resilient philosophical foundation for the civic activism reflected in Charter 77.

The rather astounding aspect of Havel's thought is that he employs the world of lived experience as the basis or foundation of transcendent reality. Recently, when addressing an audience at the Law School of Stanford University, Havel spoke of "unconscious experience" and of "archetypes and archetypal visions." His contention is that cultures formed thousands of years ago, quite independently of one another, nevertheless employ the same basic archetypes. This leads him to assert, or at least to suggest, that "there exist deep and fundamental experiences shared by the entire human race, and that traces of such experiences can be found in one another." The second time through, this same thought is expanded so as to acknowledge that there is "prescientific knowledge" such that "the whole history of the cosmos, and especially of life, is mysteriously recorded in the inner workings of all human beings." On the third take, this thought is extended and formalized metaphysically to read: ". . . after thousands of years, people of different epochs and cultures feel that they are somehow parts and partakers of the same integral Being, carrying within themselves a piece of the infinity of that Being." And, on the final pass through the idea is phrased as follows, namely, that "all cultures assume the existence of something that might be called the 'Memory of Being,' in which everything is constantly recorded." Havel's point is that the guarantees of human dignity, freedom and personal responsibility lie not in systems of thought or programs of action, but "in man's relationship to that which transcends him," which he translates as "the respect of man for that which transcends him, without which he would not be and of which he is an integral part. . . ." From these and other statements, it is clear that Vaclav Havel firmly believes that meaningful transcendence – which is coupled with the exercising of a "higher responsibility" – is inextricably tied to the world of actual, lived experience. His understanding of "Being" is perhaps most lucidly explained in *Letters to Olga*, specifically Letter 76, wherein he writes:

> Behind all phenomena and discrete entities in the world, we may observe, intimate, or experience existentially in various ways something like a general "order of Being." The essence and meaning of this order are veiled in mystery; it is as much an enigma as the Sphinx, it always speaks to us differently and always, I suppose, in ways that we ourselves are open to, in ways, to put it simply, that we can hear.

Thus, even when the roots and sources of transcendence are under consideration, *Lebenswelt* is accorded a fundamentality that is unquestioned and unqualified.

The scope of this paper does not allow me to extend the thesis into additional realms of inquiry or application. But I must point out that the association of the exercising of "higher responsibility" (in Havel's political philosophy) with his comments on transcendence (in his philosophy of being) has some direct correlation with Husserl's exposition of transcendental subjectivity. I stated at the outset of this paper that my exposition of themes from phenomenology is about philosophy and not about politics. It is apparent, however, that Husserlian philosophical substance – probably as mediated by Jan Patocka, J. B. Kozak, the Masaryk-Radl school of philosophy in Prague, and their associates, disciples and students – found its way into a collective political reform movement that eventually produced the now famous "Velvet Revolution" of 1989. Given the prominence of the inversion formula, congruent with a focus on ordinary human life, it is also apparent that the intention of this impressive literary production was to encourage a kind of spiritual victory within adverse social and political situations wherein questions concerning the nature of human freedom are raised with extraordinary force. My own view is that an extension of *Lebenswelt* sensibilities into this larger, more extensive individual and social context is not inconsistent with Edmund Husserl's understanding of his own aspirations. It was, after all, in the third and final phase of his career, in Freiburg (from 1916 to 1929), while the Great War was occurring, that he radicalized (or, if you will, de-academicized) phenomenological idealism, and placed it in the service of the renewal of life. It was during the same period that he, a man of Jewish parentage living in Nazi Germany, experienced profound personal alienation by virtue of totalitarianism's rule. All of this encouraged him to explore even further the intersubjective and historical dimensions of experience. But the tendency was there from the beginning. He did bring closure to his *Cartesian Meditations,* we recall, with St. Augustine's exhortation: "Do not go abroad, but turn within, for in interiority dwells truth" – counsel that is harmonious with Ivan Klima's observation that "the real struggle, be it between good and evil, life in truth or life as a lie, life as universal order or being in nothingness, takes place as Kafka saw it: within every one of us." Thus, when summarizing the purpose and intention of his vocation, Husserl, who, after all, was born in Moravia, and, of course, was thoroughly familiar with the life and work of the Protestant Reformer John Hus, said, "All I claim is the right to speak according to my best lights – primarily to myself and correspondingly to others – as one who has lived through a philosophical existence in all its seriousness." The same, for many of the same reasons, could be said, and, indeed, has been said, by Vaclav Havel.

University of California, Santa Barbara

REFERENCES

French, A., *Czech Writers and Politics: 1945–1969* (New York: Columbia University Press, 1982).

Gellner, Ernest, *Conditions of Liberty: Civil Society and Its Rivals* (New York: Penguin Books, 1994).

Goetz-Stankiewicz, Marketa (ed.), *Drama Contemporary: Czechoslovakia* (New York: Performing Arts Journal Publications, 1985).

Harkins, William E., *Karel Capek* (New York: Columbia University Press, 1962).

Hasek, Jaroslav, *The Good Soldier Schweik* (London: Penguin Books, 1930).

Havel, Vaclav, *Disturbing the Peace,* translated by Paul Wilson (New York: Random House, 1991).

Havel, Vaclav, *Letters to Olga,* translated by Paul Wilson (New York: Random House, 1988).

Havel, Vaclav, *Living in Truth,* edited by Jan Vladislav (London: Faber and Faber, 1986).

Havel, Vaclav, *Open Letters: Selected Writings, 1965–1990,* selected and edited by Paul Wilson (New York: Random House, 1985).

Havel, Vaclav, *The Power of the Powerless,* edited by John Keane (Armonk, New York: M. E. Sharpe, Inc., 1985).

Havel, Vaclav, *The Spiritual Roots of Democracy,* address to Stanford University Law School, 1995.

Havel, Vaclav, *Summer Meditations,* translated by Paul Wilson (New York: Random House, 1992).

Havel, Vaclav, *Toward a Civil Society: Selected Speeches and Writings, 1990–1994*, edited by Paul Wilson and others (Prague: Lidove Noviny, 1995).

Husserl, Edmund, *Crisis of European Sciences and Transcendental Phenomenology: An Introduction to Phenomenological Philosophy*, translated by David Carr (Evanston: Northwestern University Press, 1970).

Husserl, Edmund, *Ideas* (New York: Collier, 1962).

Husserl, Edmund, *Ideas Pertaining to a Pure Phenomenology and to a Phenomenological Philosophy* (Netherlands: Martinus Nijhoff, 1983).

Klima, Ivan, *An Hour of Silence* (New York: Vintage Books, 1963).

Klima, Ivan, "Post-Wall World," in *Index on Censorship* (November/December, 1994), pp. 93–100.

Klima, Ivan, *The Spirit of Prague*, translated by Paul Wilson (New York: Granta Books, 1994).

Klima, Ivan, "Writing from the Empire behind the Wall," Introduction to *Description of a Struggle*, edited by Michael March (New York: Vintage Books, 1994), pp. xix–xxiv.

Kundera, Milan, *The Art of the Novel*, translated by Linda Asher (London: Faber and Faber, 1988).

Kundera, Milan, *The Book of Laughter and Forgetting* (New York: Alfred A. Knopf, 1981).

Kundera, Milan, "Man Thinks, God Laughs," in *The New York Review of Books*, June 13, 1985.

Liehm, Antonin and Peter Kussi (eds.), *The Writing on the Wall: An Anthology of Contemporary Czech Literature* (New York: L Karz-Cohl Publishing Company, 1983).

March, Michael (ed.), *Description of a Struggle: Contemporary Eastern European Writing* (New York: Vintage Books, 1994).

Mestrovic, Stjepan G., *Habits of the Balkan Heart: Social Character and the Fall of Communism* (College Station: Texas A&M Press, 1993).

Patocka, Jan, *Le Monde naturel comme problème philosophique* (Phaenomenologica Series No. 68 (Netherlands: Martinus Nijhoff, 1976).

Ringen, Stein and Claire Wallace (eds.), *Societies in Transition: East-Central Europe Today* (Prague Papers on Social Responses to Transformation, Vol. I) (Prague: Central European Unversity, 1993).

Rosenberg, Tina, *The Haunted Land: Facing Europe's Ghosts After Communism* (New York: Random House, 1995).

PART FIVE

THE CLASSIC PHENOMENOLOGICAL THINKERS CONVERGING TOWARD THE HUMAN LIFE PERSPECTIVE

Antonio Zirión Quijano and Robert Sweeney.

ROBERT SWEENEY

SCHELER'S PERSPECTIVISM AND LIFE

My aim here is to explore the relevance of Scheler's perspectivism to the theme of life in his thought. From time to time I have reexamined Scheler's theory under the influence of the new forms of relativism grouped under the catch-all term, "post-modernism." What has struck me is the importance of trying to clarify and rethink this situation so that in academic and other public discourses we might avoid misconstruing or even demonizing various positions. Scheler's position, while not always clear, constitutes an early effort to think through the problem of relativism. The current vogue comes primarily, it would appear, from Nietzsche's usage where perspectivism represents the equivalent of a radical, skeptical (and for some, pernicious) relativism. Perspectivism is opposed, by Nietzsche, to a single, changeless meaning "in-itself" as in repeated passages like the following: "Is meaning not necessarily relative meaning and perspective?" or ". . . it (knowledge) has no meaning behind it but countless meanings, – 'perspectivism.'"[1] The implication is that we are cut off from reality by our own perspective on it. It is amplified by linguistic positions in which, e.g., truth is described as an "army of metaphors." Nietzsche's perspectivism may indeed, as Danto and others have pointed out, entail a positive metaphysical position – one that describes the world as consisting of points of origin for perspectives. Even if this renders the relativism inconsistent, the intention of it is clear and its nihilistic implications compelling. Nietzsche's heritage remains predominant today among the "neo Nietzscheans" despite the fact that a few philosophers, including several French phenomenologists – notably Merleau-Ponty and Ricoeur – have used perspectivism in a modified sense closer to Scheler's, a sense that rejects a radical relativism but maintains a relativity.[2]

To evaluate such approaches is the first aim of this paper, that is, to present Scheler's use of expressions involving "perspective" and "perspectivism" and then examine some of the possible models for interpreting its basic thrust. A conclusion about which interpretation is superior (if any) will depend on the model that can best be fitted to his approach taken generally. In turn, this model will be applied to life.

A-T. Tymieniecka (ed.), Analecta Husserliana, Vol. LIV, 269–279.

What is noteworthy, first of all, in Scheler's usage, is the fact that he applies perspectivism broadly not only to the epistemic (Nietzsche's main area) but also to axiology as well as philosophy of religion – at least judging from the *Formalismus*, the major work of Scheler's middle period. Even so, perspective is not a predominant term in this work or elsewhere, occurring as it does almost exclusively in the section on "Ethos"; rather, "relativity," "levels" or "strata" are the predominant terms. The overall point of this section is to demonstrate the absolute status of value in general by distinguishing it from ethos and the other levels of value perspective found subtended to ethos. That is, an ethos is the determinative component in a culture (cf. ideology) with respect to value; through it, the culture expresses its value commitments in terms of attitudes, beliefs, etc. It sees the basic realm of values from a certain angle of vision and embodies that vision. This constitutes, therefore, Scheler claims, a relativity of value but not a relativism.[3]

But there are other levels through which absolute value(s) is (are) refracted: these are variations in: the ethics of a time, in the morals, i.e., types of institutions and goods; in practical morality – what people actually do as contrasted to what they profess; and finally in mores and customs. Each of these levels varies in the degree to which it is distant from the highest level of the scale; for example, the last-named (custom) exhibits more variability than the first (ethos). Ethos actually receives most of Scheler's discussion since it serves as a foundation for the other levels. Ethics of a time (second level, right after ethos) is a corollary of the ethos and thus cannot transcend it although in its "scientific" version – as distinguished from its natural language form – one can compare applied ethics with the ethos in order to refine and correct it accordingly.[4]

The third level of variations – that of *types* – covers those forms of behavior usually emphasized by moral relativists to make their case. For example, theft, adultery, and murder have been constantly cited as categories marked by constant change because of which the same conduct can be designated good at one time and evil at another, and vice versa. Typically, sufficient attention is not paid to the fact that the change here is in institutions, not the values themselves (which have their own scale and relativity). The essence of murder, that is, what gives a unity to all the different possible definitions of murder, remains a disvalue and therefore evil, despite the long history of human sacrifice, infanticide and abortion. Murder was not predicated of these acts because no personal life was recognized as given. Either it was felt that personhood was found only in the tribe as a whole, rather than in its members, or in the members of one tribe rather than of another.[5]

Another area where perspectivism was applied by Scheler was that of religion. Against the sociologism of Durkheim and Marx, who asserted that the social group or the economic interest group determines the thought and religion of a given people, Scheler claimed that only the selection of particular aspects of that religion was so determined. For example, the concept of God is so rich that no single nation or period can exhaust it. Indeed, Scheler recognizes that different nations attribute aspects of their national character to God; in the case of the Israelites, God was considered the God of mountains, herds and battles in that order.[6]

The cognitive or epistemological aspect of perspectivism is dealt with sketchily by Scheler in his sociology of knowledge. Here the variabilities are not set into historical levels (vertically or hierarchically) but simply juxtaposed historically. But they are compared to each other on the basis of their proximity to or distance from a spiritual center (standard, ideal, model). Scheler uses "anticipatory schemata" – conceptual frameworks in and through which images of reality are captured and arranged in different societies according to the social circumstances and historical situations, the direction of interests and the preference of values. Each 'anticipatory scheme' originates from a functionalization of a certain sector of essence-perception and from a certain form of ethos, i.e., a living system of positive and negative ranking of values" (e.g., in the middle ages, knowledge was sought in a schema of contemplation – a knowledge that desired to perceive the essences of reality. But the modern world-view replaced this with a knowledge that seeks power and domination over nature. Thus, although different societies have different sets of truths and values, still they all "spring from the perception of the one ontic realm of ideas and value orderings.")[7]

Thus it would appear that the underlying assumption of this "perspectivism" is that an intuition of absolute value yields the point of reference with respect to which all the variations can be judged as perspectival, as partial. This in turn raises the question of whether the "perception," or feeling of value can really be an intuition in Scheler's sense of a "phenomenological experience" in which there is a "coinciding of the meant and the given." Such an intuition would imply the overcoming of a perspective or transcending one's historical condition, which possibility is denied by the theory of perspectivism itself. In other words, the question is readily raised as to the validity of the knowledge of value and being that transcends this frame of reference, this perspective. Scheler's answer would seem to lie in some form of intuition such as what he calls phenomenological knowledge. This version of Husserl's intuition of essence (*Wesensschau*), but one involving Scheler's own version of the reduction based on a suspension of moral or affective distortion, presumably must be undertaken explicitly and methodically. And it would also take place in different eras, presumably those that have an appropriate methodology, such as Plato's among the Greeks. But again the question returns as to how this intuition (even if, as Scheler claims, it is "factual") can in any way share in the situatedness of the perspectival view. As I read him, Scheler virtually ignores the issue. We might even consider Scheler's interpretation of Platonism (even as he rejects it) as the actual model of his position, i.e., a highly transitory, fluctuating (temporal) mode of experience/knowledge alongside an eternal, transcendent, intuitive mode, especially since Scheler also espouses the Platonic theme of an essential asceticism.

Another solution to such difficulties might appear to be available in Scheler's theory of "models" or "value-person-types." It is the model who has primary access to the world of values, and who effects changes in the ethos and the other perspectival levels: "The law of all moral growth (and decline) of values [is that] moral growth (and decline) occurs through the effectiveness of models or counter-models in the form of the person."[8] Optically considered, Scheler explains, the model is a "determinate and structured value-quality whose efficacy of the model is found in its exemplarity. This efficacy is actually experienced not as a duty but as a powerful attraction. Models draw the person; we do not go toward them by our own volition; rather, our response is pre-logical and precedes choice."[9] The attraction of the model is not a blind constraint, like the power of suggestion; rather, it carries with it a fundamental awareness of being obligatory and right. Thus the relation of "discipleship" or of the follower to the model is one of fidelity "founded on love for the constituent experienced in the formation of his being as a moral person"; we do not imitate the model but respond to its "creative power."[10] An underlying distinction here is that between the model as ideal and as concrete: to remain a model it must be ideal, but to be efficacious it must be concretized. "The model is contemplated more or less adequately . . . through the person who is supposed to fulfil the function of exemplar." A hierarchy of such value-person-types derived from the hierarchy of value modalities runs as follows: saint, genius, hero, leader and connoisseur (artist) of the sensual. Historical figures illustrate the different levels: St. Francis, that of saint; Augustine, a combination of saint and hero; Frederick II, a mélange of hero and genius, etc. Scheler warns, however, that we must be careful not to hypostatize these models into historical figures, since this would be a source of rigid and false traditionalism which places the values of the past above those of the present or future.

It is not clear that a real solution lies here to the paradox mentioned above, namely, that absolute value is claimed to coexist with historical relativity, since the "guiding-model" apparently does not himself intuit the world of value but simply approximates it (somewhat like Hegel's world-historical individual) and in so doing creates change in the ethos. That is, Scheler seems to suggest that, while transcending the ethos, he (the moral genius) does not actually perceive the world of values as such. In other words, a strictly dualistic or absolutistic interpretation (more or less Platonic), while it might seem to cover the main thrust of Scheler's position, would miss some of its modifications, particularly its effort at concretizing value in order to take history seriously, as well as the role of the future as representing a "framework of expectations", or that the movement into the future has a potential for transcending the framework that launched it. Still the parallelism of two worlds remains.

Another possible model for solving the paradox might be found in Hegel's "absolute knowledge" in which history, process, is at the core of Being itself. In this approach, each limited perspective is overcome, at least for a moment, by reason of the dialectical movement to absolute knowledge. In other words, for Hegel, the overcoming of the one-sidedness of differing perspectives can be found in the claim of an ultimately non-perspectival point of view. But Scheler explicitly rejects Hegel's position (despite, we can surmise, considerable influence from it) on the grounds of an implicit pantheism.[11]

If, in another strategy, we return to Nietzsche's model already mentioned, we might see his description of the "will to power" as an expression of an instinctual thrust that combines with that of other instincts to form a total thrust – a single individual perspective. The net effect, however, is not really a synthesis but merely an aggregate of the various instincts. Transferring this position to the social/historical, we would be talking about how different perspectives contest each other and in the process accidentally align themselves with other perspectives to form a momentary coalition. But there would be no progress here and certainly no real "transgression." Nietzsche's "horizons" are quite isolated and narrow.[12]

A more helpful model might be derived from Merleau-Ponty's discussion of the theme of perspectivism (even if not *expressis verbis*). We can think, for example, of his treatments of perception and thought as intersubjective, or even (or especially) in tandem with his discussion of the moral order, the social world. In this model the different perspectives do overlap at some point so that "narrowness" is not the last word: "the perception of the other founds morality by realizing the paradox of an *alter ego*, of a common situation, by placing my perspectives . . . in the visual field of another and all the others."[13] In other words, there is a transgression, a partial breaking out of the narrowness of perspective in such a way as to overlap on the field of the neighboring perspective without in any way absorbing it. A similar model is implied, I believe, in Ricoeur's passing references to perspective in a number of areas, including both the perceptual and the axiological. In Ricoeur's case, of course, it has become explicitly hermeneutical and thus incorporates a number of components from Heidegger and Gadamer including the hermeneutic circle, effective history and so forth. Despite the hermeneutic insistence on finitude, Kantian limitation, etc., Ricoeur might still be said to add to this model a note of optimism by reason of the component of hope that he sees as energizing the effort at dialogue and achieving "fusions of horizons." This suggests, to me at least, the possibility of a kind of convergence of perspectives as a culminating point – not a synthesis – but one that is neither guaranteed nor automatic.[14]

Of relevance on the specifically social level is the way Ricoeur has treated the perspectivism implicit in ideology. I find his concept of ideology – closely tailored to Marx's – similar to that of ethos (although clearly not identical) at least in the sense that it represents a pervasive cultural milieu or mentality that influences all action taking place within it. At the same time, in Ricoeur's analysis of ideology, there is transgression of the ideological perspective but only through utopias – which are also perspectival.

Now, is it forcing things to find Scheler's position in this last model – at least in its general sense of perspectival overlap? I believe that would be the case if the Platonist interpretation were the only valid one. But there are some suggestions of a kind of perspectival overlap in Scheler's language, and beyond that I think that is the basic point anyway, since one can't demonstrate an intuition of an absolute value hierarchy.

SCHELER'S PHILOSOPHY OF LIFE

To discern Scheler's theory of life, one must extract it from various sources; even though it is a "genuine essence," it is not given any simple definition. Indeed, in places it is situated within a more fundamental metaphysical framework, viz. *Alleben*, which Frings describes as "a life-agent determining vital-centers and positing "force-centers"; it differentiates itself into bundles of transcendent objective appearances. Since it applies to both inorganic and organic nature, *Alleben* represents a metaphysical unity in matter that is non-substantive and non-dualistic.[15]

Whereas the inorganic force-centers display motion from without, the vital centers display spontaneous motion and rhythmic growth as a simple thrust (urge-forward) into an undifferentiated environment. Beyond the vegetative level, these centers exhibit sensation, spontaneous locomotion, and distinct drives and instincts. Most important here are the instincts understood as abilities to emotionally pre-feel (*vorfühlen*) values of life; they follow a fixed inflexible course tied to an environment, are inborn and hereditary, and are triggered by sense-experience. With the higher animals, the instincts begin to be displaced by practical intelligence – a primitive form of insight and sense-governed mastering of "environment-alien" situations. Here we find cunning, the rudiments of choice (e.g., of mates), a kind of prudence and a competitiveness. It should be understood that every life center is both an organized body, and, if only to an elementary degree, a psychic being, a soul.[16] If he speaks in *Man's Place in Nature* of "the identity of body and soul," it is because the physiological vital process and the psychological vital process are ontologically identical. Thus the terms "physiological" and "psychological" are "but two ways of looking at one and the same process of life."[17]

But while the human being shares psychical life as practical intelligence with the higher animals, it is different from the animal not merely by reason of a greater complexity, but by a spirituality that makes life independent of drives and of attachment to the environment. Humans are thus world-open – responsive to a world of objects in their thisness (*Sosein*), whereas higher animals experience only sense qualities.[18]

Such differentiations are multiplied by Scheler within the metaphysical position of his last work, a dualism of drive and spirit. To explicate the details of this question would be far beyond the reach of this session – and beyond my capacities, in any case, since the thorny question of the relation between early and late Scheler is a major complicating factor. Suffice it say that the shift is a real one: from the position in *On the Eternal in Man* of a distant *Ens a Se* vis-à-vis a declining humanity, to the position in *Man's Place in Nature* according to which developing Spirit and Drive are opposed in a collaborative way – one in which an optimism accompanies the description of human development.[19]

In this polarization that accounts for the developmental process of reality, Drive (*Drang*) gives power or energy, Spirit (*Geist*) gives direction; God is spirit at its apex; no synthesis culminates – only greater integration or interpenetration: "Spirit, originally impotent, and the demonic drive, originally blind to all spiritual ideas and values, may fuse in the growing process of ideation, or spiritualization, in the sublimation of the drives, and in the simultaneous actualization, or vitalization, of the spirit."[20]

Life, then, here is the most salient manifestation of drive, especially in the form of human drives. But, as described above, life in the human extends only to the level of the psyche: spirit is more or less independent of psyche to the degree that it (we) has achieved integration. This manifests itself in terms of ideation, knowledge of essence, objectification and – most important for Scheler – the capacity for negation, the ability to say no to urges, to drive. Sublimation applies to spirit here not, as in Freud, in some mysterious libidinal transaction, but as designating the "natural process by which energies of a lower sphere of being, in the course of evolution, are made available to higher forms of being and becoming."

But of course, much of our understanding of life has changed since Scheler's day (as well-informed as he was at the time), under the influence of newer versions of evolution, the life-sciences and DNA/genome theory, neurophilosophy, etc. Still, irrespective of the period, Scheler's position might seem to be a reductionist vitalism, except that Scheler opposed the vitalisms of his day – Klages, Spengler, Jung, *et al.* – by his counter position that spirit represents a counteracting, correcting, moderating principle that directs but does not control.

Now, returning to the perspectivism I have claimed to discern throughout Scheler's thought, but especially in the theory of models as interpreted through Ricoeur's hermeneutics, one might understand Scheler as follows: Life manifests itself through levels or strata based on vital centers, but these vital centers can also be seen as various stages of development that are thus equivalent to perspectives – not just points of force, but zero points of origin for fields of meaning. In other words, my suggestion is that there is, implicit in, intrinsic to, the levels, a set of perspectives on life – angles of vision we can take without fully comprehending the overall essence. These stages are somewhat like value-person-types ("guiding models") in that they are stratified according to value levels – animal life, for example, is of higher value than vegetal. But in addition they are also temporal in that they have appeared in different eras. They function, that is, like the horizons of different epochs even as they overlap. But, of course, perspectives cannot be equated with levels – only analogized.

Such stages might seem to constitute simply the chapters of a "natural history," – except that there cannot be such, *strictu sensu*. That is, as Ricoeur points out, there is no cosmic history as such, since history can only be understood phenomenologically as an existential narrative that involves discordance and concordance of time, mimesis (1, 2 and 3), "spaces of experience" and "horizons of expectation." Across the gap of the aporia of phenomenological vs. cosmic time there are only "connectors" that bridge but do not eliminate the gap. These bridges are: dates (calendar), generations, and traces – each of which shares components of human and cosmic time.[21] Specifically, on the vital level in Scheler's approach, one could say that the generations and their datings allow us to see life in its evolutionary stages (vegetative through animal) as emergent properties detectable, for example, through traces in the fossil record, i.e., presences that refer to what is absent, as deciphered by science. Such stages do not coalesce in the sense of forming a seamless whole that is identifiable as the *Alleben*. But they are like the horizons of a text that can overlap with the perspective of the reader, i.e., the observer/researcher – constituting, thus, a fusion of horizons – the hermeneutic mark of validity and eventually of truth. A full manifestation of life is not possible, therefore; it only displays itself on a chronological map on which we project a future but with no outright predictability, and no essence except, perhaps, at its culminating stage, at its end.

But there is also, as suggested, the personal dimension of life that cannot be reduced to the impersonal or the global or the biological or the evolutionary. It is what Marcel highlighted as "my life" and what, in Ricoeur's terms, can be considered "a story in search of an author" that has its own play of traditionality (sedimentation) and innovation. It is a pre-narrative capacity that is continually reinterpreted in terms of narratives read and internalized.[22] Consequently, my life is at the center of my understanding of life globally, yet it is not written by me; rather, it is a series of small stories that are caught up in life generally, that are dependent on life understood globally as a matrix, a gift, a resource – but also as threat (viruses), a danger (scarcity), etc. The perspectivism of life, then, begins from the narrowness of my experience and the difficulty of understanding and reaching out to other forms – to diversity in the realm of the living. And it extends to the most elaborate reconstructions of natural development – to evolutionary emergences as we and Scheler refer to it.

Spirit could also be reinterpreted in this context. Instead of being a polar opposite of life, it could be understood as its self-transcendence when the psyche gets beyond its narrowness to collaborate with others in the search for truth, or to enhance life – for example, in the scientific community (all occurring in a rhythm of unity and plurality, of dispersion of self in narratives and re-identification in reception and reflection). The source of this inspiration might not be identifiable in philosophy, and in any case is not prescribable or controllable – "the spirit breathes where it will." But, with Scheler, we can make the case for the spirit of love as the main energizing inspiration, and, with Ricoeur, one could make a case for spirit as simply "hope to be in the truth" and, as he puts it, "that you also are in the truth even though your conviction differs from mine."[23]

The importance for life of the recognition of spirit, for Scheler, is that the value of life presupposes the founding value of spirit. That is, in contrast to the erroneous but widespread view that all values are relative to life – which would mean, Scheler insists, that life itself would then be a value-indifferent fact – the value of life presupposes a valuing spirit and the pregivenness of spiritual values. Thus, in Scheler's terms, we can be said to have levels of life and of its vital values that reach me through time as perspectives on its different aspects; and in Ricoeur's terms only in a narrative context.

Life is not a unity, after all, but unifying in terms of a hope of understanding, integration, reconciliation.

John Carroll University
Cleveland, Ohio

NOTES

[1] Friedrich Nietzsche, *The Will to Power*, trans. by W. Kaufmann and R. J. Hollingdale, ed. by W. Kaufmann (New York: Vintage Books, 1968), p. 267.
[2] E.g., Maurice Merleau-Ponty, *The Phenomenology of Perception*, trans. C. Smith (London: Routledge Paul, 1962) p. 353; Paul Ricoeur, *Temps et récit, III, Le temps raconte* (Paris: Seuil, 1985) p. 319.
[3] Max Scheler, *Formalism in Ethics and Non-Formal Ethics of Value*, trans. by M. Frings and R. Funk (Evanston: Northwestern U. Press, 1973), p. 302.
[4] *Ibid.*, pp. 307–308.
[5] *Ibid.*, pp. 309–317.
[6] *Ibid.*, p. 292.
[7] Max Scheler, *Problems of a Sociology of Knowledge*, trans. M. Frings, ed. K. Stikkers (London: Routledge & Kegan Paul, 1983), p. 154.
[8] Max Scheler, *Formalism*, p. 583.
[9] *Ibid.*, pp. 578–580.
[10] *Ibid.*, pp. 585–587.
[11] *Ibid.*, p. 503.
[12] E.g., in Friedrich Nietzsche, *On the Advantage and Disadvantage of History for Life*, trans. P. Preuss (Indianapolis: Hackett Publishing, Inc., 1980), p. 11.
[13] Maurice Merleau-Ponty, *The Primacy of Perception*, ed. J. Edie (Evanston: Northwestern U. Press, 1964), p. 18.
[14] See e.g., Paul Ricoeur, *Time & Narrative*, trans. K. MClaughlin, D. Pellauer (Chicago: University of Chicago Press, 1988), p. 220.
[15] See M. Frings, *Max Scheler* (Pittsburgh: Duquesne U. Press, 1963), pp. 33–34.
[16] *Ibid.*
[17] Max Scheler, *Man's Place in Nature*, trans. H. Meyerhoff (NY: Noonday Press, 1961), p. 74.
[18] *Ibid.*, p. 51.
[19] *Ibid.*, p. 71. Max Scheler, *On the Eternal in Man.*
[20] *Ibid.*, p. 78.
[21] *Ibid.*, p. 68.
[22] Paul Ricoeur, "Life: A Story in Search of a Narrative," *Facts and Values*, ed. by M. C. Doeser and J. N. Kray (Dordrecht: Martinus Nijhoff, 1986), pp. 121–132.
[23] Paul Ricoeur, "De l'Esprit," *Revue Philosophique de Louvain*, No. 2–3, Mai-Août 1994, p. 252.

MANFRED S. FRINGS

MAX SCHELER: AN APHORISM OF IMPORT

It is the purpose of this paper to present in summary form essential points of the central notions of *Drang* and *Geist* as Max Scheler used them throughout his later philosophy between 1922 and 1928. We shall do this on the basis of a hitherto unknown posthumous aphorism.

When the bulk of Max Scheler's posthumous manuscripts of his projected *Philosophical Anthropology* was published for the first time in 1987 in *Gesammelte Werke* (GW), Vol. XII, the mass of material did not allow all of it to be included. This has not been seen to have been a shortcoming, however, because the published material does provide the reader with a comprehensive view of Scheler's project over which he died on May 19, 1928 at the age of fifty-four.

Already three years before his demise, Scheler had voiced on several occasions to his wife that he had feelings of oncoming death. No surprise that many manuscripts of the *Philosophical Anthropology* deal also with the phenomenon of death. Scheler entertained strong intentions to organize his manuscripts consisting of thousands of pages. Other works, including his *Metaphysics* (now GW, Vol. 11), were also in the making simultaneously with the *Philosophical Anthropology*. It was a life-long habit of Scheler's to work simultaneously on several books. But amidst his frequent avalanches of insights and involvement in practical matters of the day, he failed to bring the material into any order. He told his wife, Maria, that he was quite at peace with himself knowing she would at some time take this immense technical work into her hands.

It was not only Scheler's habit of incessant writing and reading books that prevented him from said organizing intentions. He was not an organized man himself either, in contrast to his contemporaries Heidegger and Husserl. At the spur of the moment he would jot down notes on pages torn out from books, on tickets, menus, on pieces of sheets or in his numerous note-books whose scribbles asked for special efforts to be decoded from his various styles of his handwriting. Scheler even mixed up ordinary lettertypes with the old German script (that Heidegger consistently used all his life and that was officially abolished in Germany in 1947).

A-T. Tymieniecka (ed.), Analecta Husserliana, Vol. LIV, 281–294.

Scheler's disorder teamed up also with another trait of his. In his eulogy on him, the Romance philologist Ernt Robert Curtius, author of the incomparable book, *European Literature and the Latin Middle Ages*, said, "Scheler loved to give of himself lavishly to others." Scheler was neither an egocentric nor a vain person. He was no "Herr Professor," as it were, – the typical status phrase common at German universities then and later.

Let a few remarks follow here concerning the general ambiance of the times in which Scheler jotted down later notes and aphorisms.

In 1923, a deteriorating physical condition made Scheler apply for a leave of absence from the University of Cologne. His physician and colleague Dr. Moritz had urged him to take a break. The university had hired him in 1918 to head the *Research Institute of Social Sciences* and to lecture philosophy. But Cologne University's president, Christian Eckert, was obdurate and rejected the application: he informed the Mayor of Cologne, the later Chancellor of West Germany, Konrad Adenauer, who had a say in such matters, that there were too many professors that did not take their duty of teaching seriously. But officials of the Ministry of Culture disagreed and granted Scheler the leave all the same. Ever since, Adenauer and Eckert held a grudge against him. This, however, was not the only reason for their animosities.

During the twenties, Scheler was at the peak of his career. He had gained a distinguished international reputation. Numerous invitations from abroad to lecture on his forthcoming *Philosophical Anthropology*, on his philosophy of religion and sociology of knowledge were extended to him. Scheler did not feel flattered about all this, nor did he flaunt his honors. All the same, his renown during the twenties was met with envy by bureaucrats, factional figures and perhaps colleagues.

Another such application to the administration of Cologne University pertained to a 1924 invitation from Japan. There is a longer story to this than we can report here.

As a strict Catholic, Adenauer believed that Scheler was not enough of one. He held that Scheler had really no place in Cologne's centuries-old Catholic tradition. Anything that came to the mayor's desk concerning Scheler had a ring of suspicion about it right from the beginning. In Japan, Scheler was to lecture at the University of Sendai and to give talks in Tokyo and Kyoto. The then German ambassador to Japan, Wilhelm Solf, wrote Adenauer about the political and economical significance for Germany that would ensue from Max Scheler being in Japan. Solf also suggested that after Scheler's visit Einstein and other luminaries be encouraged to visit Japan. Such moves, the ambassador stressed, would counterbalance the growing efforts pursued by the French government to expand its own influence in Japan. The German Ministry shared the ambassador's view. It granted Max Scheler the leave and, in addition, he was to receive full salary for the two years' stay abroad (including also invitations from China, Russia, and the United States).

The Ministry's ruling was too much for Herr Eckert. He began to amass material for a campaign against Scheler. In a letter to Adenauer, he downplayed Sendai University as being just a provincial school and did not mention to him the other invitations that had been extended to Scheler. The City of Cologne, president Eckert argued, could not afford financial extravagances of paying full salary to persons not even living in the City. It must be stressed at this point, that money had never been one of Scheler's concerns. Whatever the money made, he spent it lavishly also. What mattered to him more was his deteriorating health condition. As he had written in his application, he needed tranquility and peace of mind, an escape from the mêlée of politics. He believed he would find this in Japan.

From his early Munich years (1906–1910) on, his restless lifestyle got Scheler, inadvertently or not, into personal or public troubles. During the twenties, he was an adamant foe of educational trends in Germany that one could compare, *cum grano salis*, to present-day fashions of speculative and political "multiculturalism" and "interdisciplinary studies," except that a number of German universities, including Cologne, promoted these fashions by instituting special chairs for them. Of course, depending on who was offered such a chair, it could easily be misused for the promotion of an ideology. The chairs were referred to as *Weltanschauungsprofessuren* at the time (world-view professorships). Even the theologian Romano Guardini held one at Berlin University, although he did not misuse it for political ends. Speaking out sharply against the fashion of teaching multicultural generalities passing as higher education, Scheler alienated bureaucrats and administrators alike: he openly promoted elitism in higher education and diplomatic service, saying, "There is no more serious mistake than to hold democracy and an elite to be mutually exclusive. . . ."[1]

So much for the story.[2] My purpose in mentioning it is to draw the reader's attention to the very atmosphere that suffused Scheler's surroundings during his last years. It constituted to a substantial part of the German style of roaring twenties, of the post-World War I defeatist mentality and the political and economic turmoil that eventually led to Hitler's takeover in 1933, five years after Scheler's demise. In contrast to many philosophers at the time, Scheler did not cease, already then, to warn the public against both Fascism and Communism.

It was in troubled social and political conditions that Scheler jotted down aphorisms, as if he had no time to elaborate on these, or they befell him by a fluke.

In the following I will offer an explication of an aphorism that, upon first reading it, strikes one by surprise. This is because Scheler, every so often, argued that the human being or "man" (*der Mensch*) can never be defined. He rejected all past definitions of man for this reason, spanning those of "homo sapiens," "created man," or the more modern definitions like "evolutionary tool-man," man as "Dasein," and the egological definitions that had suffused the course of philosophy since Descartes. Contrary to Scheler's refusal to define man, the original Greek meaning of the word, *aphorismos*, does mean "definition." An aphorism is a terse saying that aphorizes a general truth, i.e., in an aphorism something is defined in the shortest terms as in "time will tell."

Before discussing the aphorism we have chosen, let one more point be made with regard to definitions. We indicated earlier that Scheler (and others like Nietzsche) would sometimes have an intuition at the spur of the moment. In light of the term aphorism qua definition, there are two types of definitions (which would indicate a possible dilemma with the meaning of "definition" itself).

A definition can be logical and it can be intuitional. It is the logical definitions that philosophers have in general found to be insufficient or redundant because they can turn out to be tautological. Let us refer to an example. If one wants to define "equality" by saying that if A is B, and if B is C, then (in all cases) C is A, such a definition is analytical and cannot be questioned. Note, however, that this definition begins with the word "if." In an aphorism, however, one does not use the if . . . then relation. One does not say "if there is time it will tell," but simply "time will tell."

By contrast then, an intuitional "definition," i.e., an aphorism, is not conditioned by the word "if." "Time will tell" is such an aphorism. It says something that is presented as a whole and *at once* without logical deduction, reduction or inferences. An intuitional definition, therefore, presents *itself* as a whole to the mind which, of course, does not make it necessarily true. One can also have inadequate and false intuitions. To be sure, if one wants to define the nature of aphorisms itself, one can also argue that an aphorism contains either a half-truth or a truth and a half.[3]

But the intuition itself and as a phenomenon by itself is neither false nor true. For what is logically true or false shows in both cases in an intuition. I do not need to prove that $11 + 2 = 1$ is false. It is seen at the very spur of the moment in intuition to be false without the "if . . . then" relation. But stop for a moment. $11 + 2 = 1$ can also be logically true when modulo 12 is used, that is, if one chooses a mathematical formalism having only 12 and no more numbers. The terms of logical syntax change accordingly. In the case given, one has to revert to number 1 because in this formalism there is neither a number 13 nor more than twelve numbers. There exists an infinite number of different mathematical formalisms, depending on a modulo chosen. Our everyday numerical formalism may be practical for keeping our checkbooks balanced, for designing homes, for observing mileage, the stock market, or for any computer use: it itself is neither mathematically nor logically privileged over any other formalism. This could indicate that an intuited aphorism should not be subordinated to a logical definition either.

Now the aphorism:

In Scheler's vernacular, the aphorism selected runs as follows:

Der Mensch ist der Einigungsort von Drang und Geist.

We translate it as:

Man is the meeting-place of impulsion and spirit.[4]

The aphorism defines man as a meeting-place of two terms: impulsion and spirit. They need explication because prior to the publication of the posthumous manuscripts contained in GW, Vol. XI–XV, they had not infrequently been misconstrued.

What, then, are impulsion and spirit?

I. IMPULSION

The German term for impulsion, *Drang*, can also be translated as "urge." But "urge" would more relate to urges that animals and human beings have, such as urges felt in hunger, thirst, desires, etc. Although Scheler's notion of *Drang* pertains also to such physiological and biological factors, its meaning goes far beyond them: *Drang* is an active principle of energy that posits reality because its capacity consists in the ability to resist. Preliminarily, one can aphorize this by saying, "Reality is resistance." This theory of the givenness of reality is by no means new. In various versions, it appears in Duns Scotus, A. Schopenhauer, W. Dilthey, F. P. Maine de Biran and others whom Scheler sometimes refers to.

The physical appearance of reality and its occurrences in sensations is given only *after* a central experience of resistance in our lives has occurred. For this reason, the givenness of reality is always concurrent with *suffering* the resistance of the world.[5]

As a vital self-activity, Scheler does not identify impulsion to be a biological principle that would serve, for instance, a philosophy of life (*Lebensphilosophie*). Impulsion may be applicable to biological factors, but in itself it is not dependent on them. We will describe below how Scheler assigned impulsion to all levels of being (and even including the Divine that we can not touch upon in our context).

It is at this point that a number of details of the energy of impulsion should be pointed out. In order to do so, let us briefly remind ourselves first of the more familiar concept of energy used in science.

There are main and several other subsidiary forms of energy that science deals with. The main forms of energy occur as light, motion, electromagnetism, potency, and heat. For our purposes of separating them from impulsion's energy, let an operating definition of physical energy suffice here: *Physical energy possesses the capacity to do work.* (A more precise, scientific definition of physical energy explains it as that which diminishes when work is done by an amount equal to the work done.)

All forms of energy will, according to the second thermo-dynamic law, eventually dissolve into heat which will no longer convert into other energies or vice versa. When this will occur in trillions of years, there will be neither motion nor work. That is, the universe will end up in "heat-death." In its comparison to impulsion, it is important to note that physical energy can be measured. We handle and manipulate physical energy from the outside and manage it every day. Filling up our gas tank, we fill it with energy of the sun. The larger part of this heat energy remains unused in the engine's work while we are driving our cars. Heat distributes over the engine, hood, etc. That quantity of heat is lost. Our cars use approximately only 20 percent or so of the potential heat contained in the gas. A race car has a somewhat better ratio. In all forms of work occurring in the universe, heat gets lost.

The energy of impulsion is fundamentally different from measurable physical energy. This is why it can easily be misconstrued. The definition of the energy of impulsion would run: *Impulsion is the capacity of growth and, as such, it is an immeasurable capacity to posit reality.*

It lies in the nature of the growth to be the process of the self-growth of something that continuously forms itself and undergoes continuous modifications. Self-growth and its modifications occur already in molecules, more noticeably so in the growth of organs and organisms and in the mind and individual development as well as in social conditions and general history.

In every fleeting moment of growth ever new phases emerge and pass away to give room for new ones to emerge and pass away, etc. That is, self-growth and self-decline are intercontained. What grows declines simultaneously in growing and vice versa. For this reason, Scheler attributed to the essence of growth and its source, impulsion, a continuous "*Werden und Entwerden*," i.e., "becoming and unbecoming."

A most apparent form of becoming and unbecoming lies in the process of all *aging*. Aging sets in with birth or: no birth without aging. Scheler provided a phenomenological structure of all processes of aging also with regard to the experience of time in the various stages of aging, childhood, maturity and progressed aging.[6]

That the capacity to grow is the foundation of the givenness of all reality through impulsion has two components. In this, Scheler set himself off from other thinkers of his time like K. Jaspers, M. Heidegger and E. Husserl as well as from all modern and present-day representatives of French thought. For his principle of resistance, anchored in the capacity of individual and universal impulsion, is also the origin of *space and time*. Space and time, Scheler vigorously tried to show, have their origin in the self-activity of impulsion but are not yet separated in it. In short, without the self-activity of impulsion and life-centers there is neither space nor time. They separate only when individual impulsion disjoins in drives. This is shown first in the essay "Idealism and Realism."[7] Later he dealt with this in said posthumous manuscripts. In the latter he showed that the energy of impulsion forms a four-dimensional "expanse" (*das Auseinanderein*) whose effects can be traced from drives on to the mind. In perception, for instance, we see that two parallel railroad tracks meet at the horizon (i.e., not being parallel in the field of perception), like it is the case in our non-Euclidean universe. We must keep in mind that impulsion is both in individual organisms and a metaphysical principle of world-reality.

Note that we also experience a resistance of something that is absent. When one feels hungry, one feels food missing which makes material food all the more real at those moments. Resistance can also be felt in the lack of something mental as in the lack of an idea or of unfulfilled wishes and desires. Hence, resistance does not only pertain to physical presence, say, when the weight of a bag resists our muscles while we are waling. Indeed, toward the end of his life, Scheler stressed the functions of resistance qua reality to the extent that he spoke of "World-Resistance" as an essential component of human existence. World resists against all practical and mental activity. In terms of this concept, Scheler was in disagreement with Heidegger's conceptualization of *Dasein*, arguing that if there was no world-resistance such that everything were plainly "there," i.e. without being an experienced *difference* among all entities at any moment of time, such a Dasein-World may be true, said Scheler, but could never be real. Using the fairy-tale image of the *Land of the Plenty* where everything thought, desired, willed, etc., is there at once, such a world is thinkable, but it could never have reality. But at every split moment, entities and the world are *different already* from what we expect or think them to be. It is an old phenomenological datum that perceiving or thinking an entity is *different* at least in the sequences of time it is there for us, including its noematic midpoint which – although it is said to remain identical with itself – also flows in the flux of time.

II. SPIRIT

The above remarks on impulsion allow necessary conclusions with regard to the nature of spirit (or mind, consciousness, depending on how one understands the term spirit in each case)

Since impulsion posits reality by dint of its capacity to resist, spirit does not and, as we shall see, cannot posit the reality of anything or account for existence as has most often been at least implied. In Kant's respective category, for example, it is reason that posits reality. In Scheler's concept of spirit, all that spirit can do is to point and be related to whatever meanings or to what traditionally is called "whatness." Hence, spirit cannot effectuate existence and realizations in the practice of whatever ideas it has.

While Scheler agreed with Husserl that spirit (qua consciousness) is correlated to meanings and noemata, he disagreed emphatically with him that reality is tantamount to having a place in time. Spirit alone does not make anything in the world real. If spirit devises ideas of a plan, say, a social program to reduce unemployment, it is, on Scheler's premise, utterly dependent on whether or not social realities at the time allow the plan to become doable in practice. If, on the other hand, spirit would have a capacity of realization of its own, one would live in a totally different world than we do. Many things that in our present world cannot and do not work would do so under the premise that spirit had realizing capacity. One would have to also expect such spirit with realizing powers, in theory at least, to be able to bring into existence nonsensical objects like a square-circle. True, the square-circle has a noematic meaning (of contradiction) in consciousness, but no consciousness could posit its reality by itself. Hence the well-known expression of Scheler's that spirit is, in its nature and by itself, impotent (*ohnmaechtig*) to realize and bring by itself anything into existence. (As we indicated earlier, this applied also to Scheler's later conceptualization of a "becoming" Deity. It is because of his theory of reality qua resistance that Scheler abandoned the concept of a Creator-God also. No God can create anything either without the function of impulsion.)

The argument may be shared by anyone who agrees that the essence of spirit relates only to meanings and whatness. But one would have to emphatically disagree with the argument if it were to deny in addition that spirit in practice is not involved in bringing anything into existence, i.e., to look at the nature of spirit as if it were locked up for good in whatness alone, an idea sometimes shared in more or less extreme forms of idealism in the history of philosophy by thinkers like Fichte or Berkeley. But as we stated earlier, Scheler had always been at odds with any kind of idealism.

The question for us now is this: how does Scheler explain the realization of ideas in our practical world? We can already begin to see that an answer to this question is implicit in the wording of the aphorism under discussion: man is the "meeting-place" of impulsion (positing reality) and spirit (inner relatedness to whatness only).

What is this meeting-place? Its explication can be furnished in two ways: 1. in terms of Scheler's sociology of knowledge and 2. in terms of philosophical anthropology. Let us take a look at both ways.

In his *Probleme einer Soziologie des Wissens*[8] he draws a distinction between areas in which ideas are already realized in practice and areas in which ideas first occur.

The former areas manifest factors that realize ideas, so-called "realizing factors" (*Realfaktoren*). Realizing factors are, for instance, race, distribution of population, geo-politics, and economics. All realizing factors are grounded, in turn, in the three main drives: propagation, striving for power, and nutrition. Hence, all realizing factors are anchored in impulsion because impulsion is also shown to be the "unity" of all drives of which Scheler assumed there to be some twenty-four.

The other areas are referred to as "ideal factors." Among them are art, law, science, religion and philosophy.

Scheler discovered general laws obtaining among both of these factors: 1. the realizing factors interact among each other, 2. the ideal factors interact among each other, and 3. both realizing and ideal factors interact among each other. Concerning the sociological explication, one must look at the third law of interactions. It points to an interpenetration of spirit and impulsion. One can show, for instance, that ideas interact with realizing factors in history when religion interacts with a particular race or population (Islam, Jewish Religion or Buddhism). Whatever the instances of this kind are, the historical course that all realizing factors take are to high degrees independent of human action. They have laws among themselves that are mostly beyond human controls. We often do not know why populations and races die out and what leads to their extinction. Nor can we pinpoint why and how social conditions change or also become extinct. Real factors just take their own courses.

Let us illustrate this a bit further. Geographical real factors like water waves and winds show this. A wind-surfer or a pilot has to maneuver his craft through *self-governing* currents of waves, winds and turbulences that are largely beyond their control but they must maneuver them in order to realize their ideas of how to steer their craft through them and their ideas of destination. Already such elementary illustrations of the "meeting-place" and interactions between ideal and realizing factors show that there appear to be, as we stated, unyielding limits of controlling the real factors. As a rule, these limits determine the progress of technology. Technology, can never get control over the courses of all real factors; indeed, it controls them only to a rather small degree. How often can we foresee unanticipated courses that realizing factors take. Not often at all. In economics, this is a well known fact. We can at best offer good guesses and have either success (sometimes just by chance) or are gravely mistaken. This occurs not only in economics but with all real factors, especially those of political and social prognostications that Scheler was highly interested in.

The points made about the sociology of knowledge may help us to understand the central word of our aphorism: "meeting place." We did not translate it as a "meeting point," because the word "point" would be misleading: if two equal lines converge in a point that divides them each into half their length, this midpoint has a fixed position; i.e., geometrically, it does not allow for flexibility, shifts or adjustments to the dividing position it has. "Meeting place," on the other hand, allows shifts between ideal and real factors to occur. Ideas have to be amended and rectified to better fit and conform to the course that realities are taking. In a manner of speaking, the course that real factors take is always "ahead" of being grasped by spirit.

The state of affairs of said flexibility is implicit in the very title of one of Scheler's most popular texts: *Die Stellung des Menschen im Kosmos.*[9]

The text was, however, only a "very condensed summary," as Scheler himself stated, of a speech he delivered on the occasion of a conference in Darmstadt in April, 1927. The conference had the title *"Mensch und Erde"* (Man and Earth). C. G. Jung was also one of the contributors. How much import Scheler put into the clarification of the relationship between impulsion and spirit is seen by the length of his speech: it lasted about four hours. Nonetheless, the audience was reported to have been spellbound. Although he was hesitant to write the summary, in the end Scheler honored the requests coming from both organizers and audience. The text is, indeed, very condensed when one compares it to the posthumous manuscripts of the above mentioned *Philosophical Anthropology.*

The translator of the English text rightly rendered the title as "*Man's Place in Nature*," but the word "nature" for German "*Kosmos*" in the title was ill-chosen. Perhaps, a more commensurate translation of the title might have been, "Man's Cosmic Place." Be this as it may, in the German title, Scheler used the word "*Die Stellung*," a word that allows for the above shifts of ideal and realizing factors among each other. It allows for rectification and adjustment. He did not use the word "*Die Stelle*" because in German this word pertains precisely to a fixed position or location. "*Die Stellung*," however, implies flexibility of place and position. The German word can also mean "opinion," "social position," "coming out in favor; or against," etc., all connoting possible change.

As a better and more concise summary than *Die Stellung des Menschen im Kosmos* concerning his theories of impulsion and spirit, Scheler offered us his hitherto untranslated 1926 lecture *Leib und Seele* (Lived Body and Soul) whose manuscripts also belonged to his *Philosophical Anthropology*. At the very end of it, Scheler summarized for his audience in terms of a succinct pictorial analogy what he himself admitted must have been a difficult subject for anyone to comprehend in its import and consequences. We make use of this particular analogy because it, rather than other texts, appears to elucidate concisely the meaning of our aphorism.[10]

The analogy is taken from a performance of a symphony. Scheler starts out by telling us again, that spirit, by itself, is impotent because it can not bring any of its ideas into existence by itself. Unless a composer's ideas of a symphony are realized by an orchestra and its conductor, his ideas and imagination keep dwelling in his spirit. While they are thus idling in his mind, so to speak, they are asking for realizing factors like a performance to enter into a function with them and thereby become real.

In the analogy, Scheler designates the conductor as *the* realizing factor, but one that cannot do without those of the orchestra and instruments either. Other realizing factors are the real estate and the building of the concert hall (all of them also referring to economic factors). In short, the realization of the ideas of the symphony is dependent on performance. Without it, they remain idle and inactive. They are, as Scheler also puts it, only "in potentia."

In the analogy of unrealized musical ideas and a musical performance of them, Scheler takes the conductor to be "impulsion" serving as the principle opposite idling ideas of spirit. The "conductor" is the vital energy that propels the growth of the symphony forward during the performance. One can see that the conductor qua individual impulsion also feeds emotions, auditory and visual perceptions, imagination and even acts of the minds in both orchestra and audience.

Up to this point, the analogy shows two moments of impulsion and spirit that we have not yet mentioned. Concerning the conductor qua impulsion, there must be a "longing" in him to realize the symphony. Scheler tells us that both individual and universal impulsion possess such longing. He calls it impulsion's *Sucht* toward spiritualization. The meaning of *Sucht* is a very strong hungering and yearning for realization. The German word *Sucht* (related to the English "seeking") comes close in German to "addiction," a word we want to avoid here. On the other hand, spirit, too, possesses this *Sucht*. It, too, has this yearning, but for its realization.

Hence, ideas meet with the conductor, and impulsion with ideas.

Man is the meeting-place of impulsion and spirit because only in man does impulsion become "conscious of itself" through its yearning for spiritualization; and only in man does idle spirit become "conscious of itself" through its yearning for realization.

What is the analogy saying about the orchestra? The players of the instruments stand for drives. Without the conductor, they would soon be out of whack and lose rhythmic unity. In this sense, the contention that impulsion must also be the "unity" of drives should be understood. Without this unifying function drives, too, would take erratic courses. In the analogy, one should add, the conductor cannot realize the symphony either without the players in the orchestra.

And what about the instruments? They stand for physical, mechanical and chemical factors as well as for sound waves and the air transmitting them. In short, they stand for all necessary inorganic materials taking part in the realization of the symphony's ideas. Although they seem to be of lesser import than impulsion, they are an essential part of Scheler's philosophy of science running through all of his later papers. Let us mention one point of this in our context.

Scheler identified impulsion with "force centers" that set up the physical "fields" of physical energy. The modern physical field theory has a distinct place in his philosophy. The field theory rests, however, on *impulsion which, by dint of its capacity to bring about all reality it, too, brings about the reality of fields*. Here and there, Scheler even speaks of "atoms" in impulsion, very likely referring to their manifestations as waves. We said at the beginning of this paper that he identified impulsion with a four-dimensional expanse. Now we can see that he had in mind that this expanse is equivalent to said force centers in impulsion. Impulsion therefore had for Scheler two inseparable sides that are comparable to a concave-convex structure: impulsion is also underneath the reality of inanimate matter. The vital side of impulsion "picks it up" for its own constitution. Impulsion, therefore, possess both force centers and vital centers. The force centers constitute space (fields), vital centers time (aging).[11]

A third center in Scheler's *Philosophical Anthropology* and *Metaphysics* is the center of the human person: "*das Personzentrum.*"

It is always "form" of spirit and consciousness. Or, without the form of personhood, there is no spirit or consciousness. Neither a "pure spirit" nor a "pure consciousness" have a place in Scheler's philosophy.

It had been mentioned that Scheler at the end of the lecture told his students that what had been presented must have been difficult to grasp. This difficulty referred to his statement that all the analogy stood for was not to be thought in the sequence of the points made. Rather, the dynamic, mutual penetration of impulsion and spirit in the human being had to be intuited as a whole and at once. The above aphorism, too, requires the same arduous effort of the intuition mentioned earlier.

It is at this point that our attempt to explicate the above aphorism reaches its end. This is because at this juncture the puzzle of Scheler's philosophy of an unfinished Deity becoming real only through the real factor of human history would have to be tackled. This thematic is, in a way, a challenge to contemporary theology and philosophy of religion but lies beyond what our aphorism appears to have defined.

Max Scheler Archives
Albuquerque, New Mexico

NOTES

[1] Max Scheler, GW, Vol. IX, p. 145.

[2] For further information consult: Heimbuchel, Bernd, and Papst, Klaus, *Kölner Universitätsgeschichte*, Band II, *Das neunzehnte und zwanzigste Jahrhundert* (Köln Wien: Bohlau Verlag, 1988), 530–578.

[3] Kraus, Karl, *Sprüche und Widersprüche* (Dicta and Contradictions), 1909.

[4] The aphorism has the Archive No: B II, 9. (CD, IV, 1).

[5] Scheler elaborated on resistance qua reality first in "Idealismus-Realismus," GW, IX, pp. 208–236 (English tr. by David Lachterman in: Max Scheler, *Selected Philosophical Essays*, Evanston: Northwestern University Press, 1973, pp. 288–356). GW, IX contains further posthumous sections of "Idealismus-Realismus" (including one on M. Heidegger) that were not available to the translator at the time.

[6] For instance, GW, IX, p. 229; GW, pp. 19–21.

[7] GW, IX, pp. 222, 228.

[8] GW, VIII, pp. 9–190. English translation by M. S. Frings (with an introduction by K. Stikkers) as *Problems of a Sociology of Knowledge* (London, Boston and Henley: Routledge and Kegan Paul, 1980).

[9] GW, IX, pp. 7–71. English translation (with an introduction) by H. Meyerhoff (New York: Noonday, 1961).

[10] GW, Vol. XII, pp. 148–150.

[11] GW, 11, pp. 160, 181.

MAIJA KŪLE

THEODOR CELMS: FORERUNNER OF THE PHENOMENOLOGY OF LIFE

Prof. Theodor Celms (1893–1989), the most prominent Latvian philosopher of his day, in the twenties and thirties conducted phenomenological investigations and published significant philosophical works in Latvian and German. His philosophical heritage is *Der phänomenologische Idealismus Husserls* (Riga: 1928), *Vom Wesen der Philosophie* (Regensburg: 1930), *Lebensumgebung und Lebensprojektion* (Leipzig: 1933), and *Subjekt und Subjektivierung. Studien über das subjektive Sein* (Riga: 1943). All these works have now been republished in Germany, under the title *Der phänomenologische Idealismus Husserls und andere Schriften* (Frankfurt a.d.M.). What attracts our attention now is that Celms has intuitively aimed at a phenomenology of life.

In 1922–1925 Celms went to Germany and took courses in philosophy conducted by Rickert and Husserl. Husserl recognized him as one of his best pupils in phenomenology. At the University of Freiburg he obtained a doctorate in philosophy. Later he became a research assistant at the *Deutsche Literaturzeitung für Kritik der internationalen Wissenschaft*. His main philosophical book on Husserl was translated into Spanish and published in Madrid, 1931. This work has not lost its significance to this day. Garland Publishers of New York recognized its importance and published it in 1979, but that edition is no longer available. Celms became famous as one of the deepest critics of Husserl's transcendental phenomenology, as one who tried to find a way out of phenomenology's discrepancies.

In the thirties Celms wrote reviews in German on Heidegger's and Scheler's philosophies and published volumes in Latvian – *Tagadnes problēmas* (The Problems of Today, Riga: 1934), and *Patiesība und šķitums* (Truth and Appearance, Riga: 1939), as well as separate articles in papers, magazines and encyclopedias. The themes of Man, subject, life, consciousness, culture, society, occupied a most prominent place in Celms' philosophical articles and lectures at the University of Latvia.

At the end of the Second World War, Celms emigrated to Germany and then moved to the USA (1949). He worked at Augustana College in Rock Island, Illinois until his retirement in 1963. In the USA he

A-T. Tymieniecka (ed.), Analecta Husserliana, Vol. LIV, 295–302.

continued philosophical studies and wrote the book *Phänomen und Wirklichkeit des Ichs. Studien über das subjektive Sein* (The Phenomenon and Reality of the I. Studies of Subjective Being). This work is not available, not having been published yet. Therefore our understanding of Celms' philosophy must be based mainly on the first part of his philosophical activity, which is connected with his life in Latvia and Germany. The publication of Celms' manuscripts will enable the development of a more thorough assessment of the ideas he conceived.

Celms worked on the problems of human nature, the relationship between man's personality and the culture of mankind, man's volitional life. These touch upon the themes of psychology and sociology. In his works Celms follows the traditions of critical realism, neo-Kantianism and phenomenology and tries to develop some new aspects of phenomenology.

In the European history of philosophy Celms is little known. But he has been well noted as belonging to Husserl's phenomenological school. Waldenfels mentioned him together with such figures as Roman Ingarden, Ortega y Gasset, Jan Patočka, Marvin Farber, Hajime Tanabe, and Shuzo Kuki. He was mentioned also as a Latvian phenomenologist in Herbert Spiegelberg's history of phenomenology. Thomas Seebohm in his book *Der Bedingungen der Möglichkeit der transzendental Philosophie* writes that Folwart, in analysing Celms' views on phenomenology, had noticed that Celms failed to express the thought that phenomenology is nothing but eidetic psychology. This thought has been clearly voiced in German philosophical literature by Ryle and Erlich, who point out that phenomenology was an aprioristic psychology.

The way in which Celms criticizes Husserl's phenomenology, places him close to the Munich-Göttingen school. Celms was closely connected with Pfänder, to whom he dedicated his work *Lebensumgebung und Lebensprojektion* in 1933. Pfänder published an excellent review of Celms' book *Der phänomenologische Idealismus Husserls*. At the same time, however, Celms occupied a peculiar position which distinguishes him from Husserl and the Munich-Gottingen school.

To my mind, it is worthwhile, therefore, to view his heritage as the starting point of the phenomenology of life. From the present point of view, when the standpoints of phenomenology of life have taken shape, Celms' contribution is more obvious than it was at the time when phenomenology was taking its first steps.

Celms calls Husserl's phenomenological idealism life philosophy and compares Husserl with Bergson. Transcendental idealism chooses consciousness, I, subject, spirit as the starting point of thinking. The former distinction between idealism and realism gives place to a new distinction between I-philosophy (*Ich Philosophie*), and a World-philosophy (*Weltphilosophie*) which has no I at its centre. Husserl, to Celms' mind, belongs to the I-philosophy which completely subjugates being to spiritualization (*Durchgeistigung*). If different forms of transcendental philosophy view pure consciousness as transindividual consciousness, then Husserl, according to Celms, poses the problem of the transience of experience brimming with individual consciousness. That Husserl interprets the world pulse as the pulse of I-life proves that his idealism is close to life philosophy.

In studying Husserl, Celms pays attention to how the concept of life functions in phenomenology. Pure life is the basis for reduction in Husserl's phenomenology, Celms acknowledges. Life is grasped in phenomenological reflection. In describing the model of Husserl's reflexive consciousness, Celms stresses that life does not consist of objects but of experiences ("besteht mein Leben nicht aus Objekten, sondern aus Erlebnissen"). We perceive experiences as the infinity of an immanent observation process.

Celms often compares Husserl's approach with Kant calling to mind the limitations of the philosophy of transcendentalism. The world in Husserl's teachings is an idea motivated by interconnected experiences, which resembles Kant's approach as the world is attributed existence in itself. Husserl regards the stream of experiences as absolute being. In Celms' opinion, Husserl's phenomenology is one-sided rationalism. However, to turn to life's actualities, as Celms noted as far back as 1928, despite absolute a priori regularities (*Regelung*), this leads to the perpetually irrational.

Celms is one of the first in European philosophy to notice the likelihood of a transition in phenomenology from rationalism to irrationalism. He notes that in Husserl's teaching everything in the long run leads to an absolute a priori defined, immanent development, which can also be called immanent evolution. All the objective, potential and real worlds are outer manifestations of the immanent evolution of life's actualities, or, in other words, the intentional illumination of an immanent evolution, writes Celms. This describes the word evolution, *créatrice*. Though Husserl's and Bergson's philosophies are different, nevertheless they possess surprising similarity. Celms discerns in both the philosophers a turning to the phenomenological actuality of pure life as duration in immanent time. Husserl's satiated pure duration (*erfüllte reine Dauer*) is analogous to Bergson's *durée réelle*.

It should be noted here that Celms' view of phenomenology is truly prescient. He pays attention not only to the concept of life but also to the concepts of evolution and creative factors. As is well known, in the second half of our century these are the concepts accentuated by the phenomenology of life. A-T. Tymieniecka writes this about the phenomenological understanding of life:

> This world of life, however, should not itself be understood to be identical with the life-world conception of Husserl and Merleau-Ponty, who saw in it a fictitious ideal of a primogenital constitutive structure of subjective-objective life operations representing the originary milieu of man. In the descent they make following its genesis alongside and in human constitutive operations, they overlooked the evolutive nature of the constitutive forms alongside and in the evolution of life itself. I have already from the outset of these inquiries, challenged the validity of the Husserlian life-world conception by introducing the crucial factor in the evolution of its forms: THE SELF-INDIVIDUALIZATION-IN-EXISTENCE OF THE LIVING BEING OF WHICH THE CREATIVE ACT IS THE VORTEX.[1]

Obviously, in this conception the decisive role is played by understanding life evolution as a process of self-individualization. As in Bergson's system, the creative aspect of life is accentuated. If phenomenology wants to disentangle itself from excessive subjectivization, from sinking into spiritual metaphysics for which Celms reproaches it, then contemporary phenomenology of life is a variant moving in the opposite direction. However, phenomenology in turning to the theory of evolution, to life's connection not only with man's subjectivity but with everything-that-is-alive, may yet lead to naturalism.

In Husserl's later works the life-world concept acquires priority and thus has an enormous impact on the development of phenomenology. Celms was unable to fully follow the development of Husserl's ideas, to read his manuscripts and the unpublished works. That is why in writing about Husserl (in an article "Edmund Husserl" in Latvian in 1931) he refers mainly to *Logical Investigations* and *Ideas I* and mentions *Vorlesungen zur Phänomenologie des inneren Zeitbewusstseins* (1928), *Formale und transzendentale Logik* (1929) and *Nachwort zu Ideen* (1930) noting, however, that he was not acquainted with *Ideen II, III*. In an addendum of 1939 Celms acknowledges that *Cartesian Meditations* do not present anything new but confirm once again how close Husserl is to Descartes. Celms also mentions the publication of *Die Krisis der europäischen Wissenschaften und die transcendentale Phänomenologie* in Belgrade; however, in this work he mainly sees Husserl's conviction that contemporary philosophy is inevitably to come to phenomenology.

Taking into account Celms' incomplete familiarity with the development of Husserl's phenomenology, one must admit that in working out the concepts of life-world and life Celms does not copy Husserl, but retaining the original motive coming from phenomenology, follows his own path. This is testified to by his work *Lebensumgebung und Lebensprojektion* (Leipzig: 1933). One of the main problems in interpreting life is the question of how to join the phenomenological standpoint about pure life, life subjected to reduction and the actual concrete and empirical reality of life. In other words, how to link the abstract philosophical approach to life and its concrete psychological view?

Interpreting Husserl's teaching, Celms notes that the natural trend places in the foreground "I and the surrounding world" that actually is the psychological I which finds itself in the world. The pure I in the natural trend is not to be found in principle. However, as everything that is to be found in the natural trend is a form of the experience of pure consciousness, then natural life is in the long run also pure life, the purity remaining unconscious. The reduction of psychological consciousness envisaged by phenomenology means the refinement of that same (not only other) life from its psychological form to its pure form. Strictly speaking, psychological life is not at all reduced. What is reduced is life's psycholcgical form, writes Celms.

Die Reduktion des psychologischen Bewusstseins bedeutet also nur die Überführung eines und desselben Lebens aus der psychologischen Form in die reine. Streng genommen wird also nicht das volle psychologische Leben, sondern nur die psychologische Form des Lebens reduziert.)[2]

So, in Celms' understanding, life is regarded as a unifying basis that acquires manifold forms which can be reduced. There are four forms of life: the direct manifestation of life, the naturalistically reflected form, the phenomenologically reflected form and the phenomenologically reduced form. The former two are psychological, the latter two are phenomenological forms. Husserl acknowledges an important principle: there is the possibility of transition from the psychological to the phenomenological life forms (and vice versa).

Obviously, such a stance does not draw apart but rather draws together phenomenology and psychology. Celms, in his work *The Subject of Psychological Investigation* in Latvian makes full use of phenomenological method in psychology. The basic idea expressed by Celms in his psychological studies is that from phenomenology we can take only the reflexive method. He considers the phenomenological philosophy in its entirety to be intuitive idealism.

Doubtless, Husserl's phenomenology provides a basis for a psychological interpretation. For instance, how does the pure I appear in relation to one's own I? As soon as the I starts losing logical universality, the originally given identity, as soon as it is filled with acts, the stream of experience, life, it gets more and more psychologized and becomes one's own I notwithstanding the strict demand for observing the phenomenological reduction. Husserl views the I as a personality and that is why he often speaks of phenomenology being linked with personalism. Husserl himself sees parallels with psychology and notes that what a phenomenologist does is similar (parallel) to what a psychologist does in turning to inner experience and the experience of one's self. However, alongside the transcendental phenomenological stand, the I reaches the last (deepest) point of experience and cognition in which the I becomes a disinterested observer of his worldly-natural I and I-world. This I-world is only a part or a layer of the transcendental life.

Consequently, in Husserl's phenomenology the notions "life," "to live" function in their transcendental meaning and imply pure structures.

Celms with his standpoints, in his turn, strives to go deeper into the concrete, psychological manifestations of life forms without denying the significance of the fundamental principles of philosophy in order to free himself from the naive trend. In the work *Vom Wesen der Philosophie* he notes that life from the philosophical point of view is life elucidated in the light of ultimate principles. The fullness of life comes as superficial naivety is replaced by the fundamental principles of philosophy. The function of philosophy, to his mind, is "Die Welt- und Lebenstatsachen müssen in ihren Bedingungen und diese wieder in den Grundbedingungen gerettet werden."[3] (Emphasis added.)

In his work *Lebensumgebung und Lebensprojektion* Celms works out a theory for viewing life in philosophy. It is significant that, in fact, he follows phenomenology, yet nowhere does he refer to Husserl.

In relation to life there are two distinct views – that of the natural sciences, and the philosophical one. The view of the natural sciences is based on the reality inherent in man's biological life. The view of life as a stream of psychological experiences is quite different. As long as Man is an organism, it should be regarded as a biological life unit (*Lebenseinheiten*). In the stream of experience Man is intentional. That is why in the end life may be viewed as physically real, psychically real and intentional, writes Celms. In the first instance it is characterized by the spatio-temporal, in the second by the temporal and in the third by intentional parameters. Celms reviews the body environment and the I environment. The body is connected with the physical and biological environment, and the I with pure psychic life, formed by intentional activities. The I environment is the embodiment of cognized objects. The body environment, in its turn, is the embodiment of realities which are independent of consciousness. Human life, however, has to be understood as a psycho-physical I within which are combined the forms of the I and the body life environment. By life projection, Celms really means the life experiences which relate to the world as it is "real." Objects from all objectivities of intentional experience are understood to be real by man. "Das bewusstseinsmässige Sicheinbeziehen in eine real gehaltene Welt werden wir im weiteren die Lebensprojektion nennen,"[4] Celms writes. Thus, we can deduce that I includes I in its relationships with the non-I. There exists an inner view of life and an outer one. Consequently, one should distinguish causality (which refers to outer life) from motivation (which applies to inner life). Don Quixote's example can make it clear. From the outer point of view he is a dreamer fighting imaginary windmills. However, from the point of view of his inner life, the windmills are quite real to him. From the outer point of view he is a fool, and from the inner point of view a sage. Celms also speaks about distance from the organically physical world. In a psychic sense, that I environment possesses distance from other I lives. The strange I is a part of the I environment. While grappling with this question, Celms returned to the theme of intersubjectivity investigated by Husserl.

Theodor Celms comes to the conclusion that Husserl's phenomenology cannot cope with the question of intersubjectivity as in the course of its development it only manages to substitute pluralistic solipsism for monistic solipsism. To save the situation, we need to recognize the pregiven harmony described by Leibnitz. However, in Celms' opinion, that would be the creation of a new metaphysics.

Celms' attitude to phenomenology is formed in relation to how and to what extent the teaching grasps reality independent of consciousness and subjectivity. In his main work Celms shows that the theme of reality is not adequately substantiated and calls Husserl's teaching solipsism and spiritual metaphysics. In his works not specially devoted to analysing Husserl, however, Celms to a great extent relies on the phenomenological teaching. That is evident from the characterization of the subject matter of the psychological investigation, in his stance on what philosophy is and in his interpretation of culture striving for a broader substantiation outside consciousness. Celms turns to the concept of life. He strives to give life not so much transcendental, phenomenologically reduced, pure forms, but rather a content based on concrete realities and psychological motivation. His explanation of life has a lot parallels with phenomenology. Celms works out concepts of I and the body environment which are similar to concepts of Husserl's in his teaching about life-world and a phenomenologically viewed body. His teaching on motivation in contradistinction to causality has a phenomenological drift, too. Thus, Celms' attempts to find solutions to the problems of phenomenology, on the one hand, are closely connected with the possibilities provided by Husserl. On the other hand, however, Celms sketches in a shift the connection of phenomenology with the life, a theme which is further developed by phenomenology of life at the end of the twentieth century.

Institute of Philosophy and Sociology
Latvian Academy of Sciences
Riga

NOTES

[1] A-T. Tymieniecka, *Logos and Life*. Book 1: *Creative Experience and the Critique of Reason*, Analecta Husserliana, Vol. **XXIV** (Dordrecht: Kluwer Academic Publishers, 1988), pp. 308–309.

[2] Celms, T. *Der phänomenologische Idealismus Husserls and andere Schriften*, S. 152.

[3] *Ibid.*, p. 216.

[4] *Ibid.*, p. 236.

JIM I. UNAH

HEIDEGGER'S PHENOMENOLOGICAL ORIENTATION

> Phenomenology is our way of access to what is to be the theme of ontology . . . only as phenomenology is ontology possible.
>
> – Martin Heidegger*

Ontology is the name of the science which gave the initial impetus to Heidegger's inquiry into being. And the being question crystallised into Heidegger's career preoccupation. Heidegger's renewal of the problem of being is made more radically different from conventional ontology,[1] more comprehensible and more globally significant by the phenomenological attitude. Ontology researches into being. But not every research into being is informed by the phenomenological attitude. The phenomenological attitude is the orientational habit of letting things come into light as they are without forcing them into our ready-made, artificial, conceptual straitjackets. Ontology makes possible the exhibition of the general structures of beings. The phenomenological attitude enables beings to come into light as they are without any preconceptions. Being itself is the light which makes possible the coming into light of beings. Being itself is the horizon which enables beings to be, the opening which makes beings come into openness. Hence "only as phenomenology is ontology possibie."

Now if ontology is possible only as phenomenology, are we to take it that all forms of ontology are phenomenological in character? And if all forms of ontology are phenomenological in character in the sense in which Heidegger accepts phenomenology, what is special about Heidegger's Being-talk? Only Heidegger can answer these questions satisfactorily. We should, therefore, in the phenomenological spirit, allow Heidegger to speak for himself. And he speaks quite elaborately, untiringly, eloquently about these matters right from the time he receives Brentano's *On the Manifold Sense of Being in Aristotle* (1862) to his last days when, for him, authentic thinking becomes the thinking of Being.

A-T. Tymieniecka (ed.), Analecta Husserliana, Vol. LIV, 303–330.

BRENTANO'S INFLUENCE

In an academic correspondence prefacing William Richardson's massive work[2] on his philosophy, Martin Heidegger acknowledged his debt (as an ontologist) to Brentano (1838–1917) whose doctoral dissertation stimulated his thinking on Being.

> ... the first philosophical text through which I worked my way again and again from 1907 on, was Franz Brentano's dissertation: *On the Manifold Sense of Being in Aristotle* (1862). On the title page of his work, Brentano quotes Aristotle's phrase: ". . . A being becomes manifest (sic) with regard to its Being in many ways." Latent in this phrase is the question that determined the way of my thought: what is the pervasive, simple, unified determination of being that permeates all of its multiple meanings? This question raised another: What, then, does Being mean?[3]

Thus, by a single stroke, Brentano's work did two things for Heidegger: it roused his interest in the Being question, and introduced him to Aristotle's[4] theory of Being. According to Richardson the problem of Being came into focus when Heidegger began to meditate with Brentano on what the word "being" means for Aristotle. Henceforth, a small but powerful word that enjoys an unusual presence became a passion for Heidegger. Every time we open our mouth, says Heidegger, we utter it. This word that is small and powerful but which enjoys an unusual presence, is the word "is." Although if finds itself in practically all our utterances, it is not turned asunder by its everywhereness. It retains its marvellous unity. According to Heidegger, what guarantees the unity of the word "is" irrespective of its everywhereness, that is, what makes possible all that there "is," must be Being itself. But then what about Being itself? What is its meaning? This pattern of questioning stimulated by Brentano's thesis brought Heidegger into the mainstream of ancient Greek philosophy.

To be sure, the Being issue in Aristotle is a metaphysical issue. Aristotle's thesis of a "going-beyond" physics (things of nature) to the transcendence which characterises metaphysics is analogous to the question of Being which demands "going-beyond" beings as beings to that which makes them be, to Being itself. The Being question in Aristotle is a metaphysical question pure and simple because both metaphysics and the Being question involve a passing beyond, a transcendence of the things of sense, the physical or beings by reason of which the things of sense, the physical or beings as beings emerge into light as beings.

Thus, a being becomes manifest in many ways by reason of the emergent power or light of Being.

But Being itself, by reason of which beings come into light or come to be, from the point of view of metaphysics remains obscure. That is why, says Heidegger, Aristotle was not the first of Greek thinkers to construe the Being question in the metaphysical sense. It is rather Plato, by the metaphor of the Cave in which he makes a distinction between the world of experience or sense as a world of shadow and illusion and the real world as one of reason and ideas, who was the founder of the metaphysical enterprise. Here, Plato says that the philosopher, the man of wisdom, is one who can transcend or go beyond the world of sensible things, of shadows and illusion to the world of reason and ideas. This notion of transcendence or going beyond, as we noted earlier, is the subject matter of metaphysics. And if according to Plato ideas constitute the light by reason of which beings come to be, what is the light which makes it possible for ideas to shine and bring beings out of their concealedness? Heidegger says that it is not within the purview of metaphysical thinking to raise and answer this question. In the same way, if we are to pose the question in the Aristotelian fashion: "what does it mean to inquire after being qua being?" the answer, like the question, would remain unalterably metaphysical which means a going beyond the physical in such a way that we come back to affirm the "why" of the physical, a going beyond beings as beings to the reason by which beings become manifest. In his explication of Heidegger, Richardson says ". . . metaphysics means the transcendence of beings to their Being in such a way that beings are thereby considered as beings."[5] It is this sense of Being that is predominant in the thinking of ancient Greek philosophers. And it is this same sense of Being, which is essentially metaphysical, that leads to the oblivion of Being.

So although the sense of Being as the light which makes possible the coming into light of beings (i.e., Being itself) is inherent in the thoughts of Plato and Aristotle, the metaphysical sense of Being (i.e., Being as a being) such as Being as ideas, Being as Pure Act, Being as Absolute Concept or Being as Will to Power, etc., has dominated the history of Western philosophy. That is why every metaphysics is a thinking of Being as beings, or an investigation of the essence as such. But the inquiry which goes into the root of metaphysics in order to consider Being itself precisely as it is, is called the "metaphysics of metaphysics" by Kant or the "overcoming of metaphysics" by Heidegger. This is how the Being question stimulated by Brentano's dissertation inspired Heidegger to meditate on the essence of metaphysics which culminated in a kind of super metaphysics.

There is, however, another dimension to Brentano's influence in the development of Heidegger. This concerns the question of the nature of consciousness through which *Dasein* apprehends beings and at the same time maintains a vague understanding of Being. This aspect of Brentano came down to Heidegger through Edmund Husserl, whose influence on Heidegger we shall consider in a moment. First, if we must go beyond metaphysics in order to overcome it or investigate pure Being, we have to do so through a higher or a kind of super ontology which Heidegger designates as a "fundamental ontology," which simply means the analysis of the being of man – who is fond of asking metaphysical questions and raising the problem of Being. Any authentic analysis of the being of man will require an analysis of his mode of cognition. Heidegger's analysis of man's[6] mode of cognition makes him a beneficiary of Brentano's theory of consciousness or mental act.

In Brentano's descriptive psychology, all mental states, which he calls "psychological phenomena," are intentional in character. By this Brentano means that every mental act has its intentional object. There is an indissoluble link or relationship obtaining between psychological phenomena and their objects which is discoverable by what Brentano calls "immediate experience" or "inner intuition." The mental or psychic sphere is a self-transcending intentionality. Nothing within consciousness can be understood save on the basis of its directionality: consciousness is always directed at something. This is how Brentano designates "intentionality" as the mark of the psychological. Quoting Brentano's *Psychology from the Empirical Standpoint*, David Carr writes:

> Every psychic phenomenon is characterised by what the scholastics of the Middle Ages called the intentional (or perhaps mental) inexistence of an object, and what we, though not without some ambiguity of expression, call the relation to a content, the direction towards an object (which is not to be understood as a reality), or immanent objectivity. Each contains something as an object within itself, though not in each case in the same way. In representation something is represented, in judgement something is acknowledged or rejected, in love something is loved. . . . This intentional inexistence is characteristic exclusively of psychic phenomena. No physical phenomena exhibit something similar.[7]

By "intentional inexistence" or "relation to a content" or "direction towards an object" of any psychic or mental phenomena Brentano means that thinking is necessarily thinking about something; whether that "something" exists in concrete form or not is outside the question. When, for example, we say that James is thinking of the River Niger we do not, in the intentionalist view, imply that such a River Niger exists or that it does not exist. What we mean, rather, is that somebody (James) is thinking of a thing or an object, or that a thing or an object occupies the thought of somebody (i.e., James). In the same way, when we deny the statement and say that James is not thinking of the River Niger, we do not mean, following the intentionalist thesis, that the River Niger exists or that it does not exist. We simply mean that the object "River Niger" does not occupy the thought of a certain James.

Brentano uses the term "immanent objectivity" to designate, not that which is outside the mind as concrete material thing, but that which is the object of thought. By "object of thought" he means "what it is that the thought is about, whether or not there is anything outside the mind corresponding to the thought."[8] Brentano further distinguishes between "immanent object" and the "object of thought" which he says are not identical. What we think about, according to him, is the "object" or "thing" and not the "object of thought." Brentano says that if, in our thought, we contemplate a horse, our thought has as its "immanent object" – not a contemplated horse, but a horse. And strictly speaking, only the horse – not a contemplated horse – can be called an object.[9] This limits intentionality to the psychological, and rejects the absurd idea that one could be thinking of thought.

Now the thing or object that we think about need not exist in concrete terms. Somebody may think of a thing or an object in his mind even though that very thing or object does not exist. ". . . I allowed myself the term 'immanent object' in order to say, not that the object exists, but that it is an object whether or not there is anything that corresponds to it."[10] And elsewhere, Brentano says it is impossible to have an idea without having an idea of something. The term "something" signifies a thing. "If the thinking is a compound thought, then things are what one has compounded."[11] To say that when one thinks, one necessarily thinks of something is different from saying that the thought of a thing, say a tree, is the same as thinking of the existence or non-existence of such a tree. The former is a first order act and it is intentional in character. The latter – the question of the existence or non-existence of a thing – is a second order activity and it is not a mark of the psychological, not intentional in character, but rather involves the issue of "content of judgement."[12] Now what does it take for a sentence to be intentional in character?

According to Roderick Chisholm's reformulation of Brentano's theory of mental act, a statement is intentional:

> If it uses a substantival expression – a name or a description – in such a way that neither the sentence nor its contradictory implies either that there is or that there is not anything to which the substantival expression truly applies.[13]

The gist of this criterion[14] of intentionality is that when we think, when we imagine, when we hope for, etc. – all of which are mental acts – we think of something, we imagine something, we hope for something without implying that what is thought of or imagined either exists or does not exist or that what is hoped for could be realised or not. And the point at issue here is that the human mind, for all intents and purposes, cannot be conscious of nothing where "Nothing" is understood to mean "not anything"[15] or "total nothing" and that whatever the mind thinks about must be something connected to the world in one way or another. This concept of "Nothing" is central to Heidegger's analysis of Being. In a sense, Brentano's theory of intentionality has nourished Heidegger's concept of "Nothing" which in turn facilitated his inquiry into Being. It is, however, to Edmund Husserl – one of the most resourceful and influential pupils of Brentano – that Heidegger owes his phenomenological culture.

HUSSERL'S INFLUENCE

In "My Way to Phenomenology"[16] we are immediately greeted by the fact that it was essentially under the tutorial supervision of Edmund Husserl that Heidegger acquired his phenomenological culture. Although Brentano's dissertation determined Husserl's thought and his own, it was from Husserl's *Logical Investigations* that Heidegger "expected a decisive aid in the questions stimulated" by Brentano's dissertation. But as he was searching in the wrong direction, Heidegger's efforts proved abortive. This was a perplexity for Heidegger. How was thinking's manner of procedure which called itself "phenomenology" to be carried out? This question, says Heidegger, was exacerbated by the "ambiguity which Husserl's work showed at first glance."

On the one hand, Husserl's *Logical Investigations* was an assault on psychologism, since it contended in the first volume (1900) that the "doctrine of thought and knowledge cannot be based on psychology." On the other hand, the work was a psychology since the second volume "contains the description of the acts of consciousness essential for the constitution of knowledge." Now the point Heidegger makes is this: any phenomenological description of acts of consciousness is a psychology, and it puts one in a position of "psychologism" and any contention that "thought and knowledge cannot be based on psychology" is a repudiation of psychologism. Husserl's work, the *Logical Investigations* (Volumes One and Two) seems to contain both elements. Heidegger says that if this huge error cannot be attributed to Husserl's works, it devolves on us to show how phenomenology is different from logic and psychology. But how is one to make such clear-cut distinctions between phenomenology and psychology unaided?

Husserl's edited work, the *Yearbook for Philosophy and Phenomenological Investigation* published in 1913 by Max Niemeyer,[17] brought in a ray of hope. According to Heidegger, the *Yearbook* showed that "pure phenomenology" is distinctive in the sense that the "subjectivity of the knowing, acting and valuing subject is posited as transcendental." In pure or transcendental phenomenology "experience of consciousness" remains the central theme. But added to this is the thorough and systematic exhibition of the "structures of acts of experience together with the investigation of the objects experienced in those acts with regard to their objectivity."[18]

Much to his relief, Heidegger discovered when he "met Husserl personally in his workshop" that the act of philosophical thinking called "phenomenology" could not be attained merely be reading philosophical literature.[19]

> My perplexity decreased slowly, my confusion dissolved laboriously. . . . As I myself practised phenomenological seeing, teaching and learning in Husserl's proximity. . . .
>
> The distinction which is worked out there between sensuous and categorical intuition revealed itself to me in its scope for the determination of the manifold meaning of being.[20]

So the problem of Being became clearer for Heidegger with his phenomenological orientation under Husserl particularly after the publication of the last chapter of the *Logical Investigations*. It dawned on Heidegger that what one finds in "the phenomenology of the acts of consciousness as the self-manifestation of phenomena is thought more originally by Aristotle and in all Greek thinking and existence as *aletheia*, as the unconcealedness of what-is-present, its being revealed, its showing itself."[21] Now that the key principle of phenomenology is *Zu den Sachen* – "back to the things themselves," Heidegger asks:

> Whence and how is it determined what must be experienced as the things themselves? . . . Is it consciousness and its objectivity or is it the Being of beings in its unconcealedness and concealment?
>
> Thus I was brought to the path of the question of Being illumined by the phenomenological attitude. . . .[22]

If in phenomenological education one is cultured to patiently allow the phenomenon manifest to itself as it is, and if in ontology one learns to exhibit the profiles or general structures of experience (which in a sense allows for the object of experience to show itself), then, it is impossible to carry out a genuine ontological research without adopting the phenomenological attitude.[23] The point to make here, however, is that the influence of Brentano and Husserl (master and pupil) conjoin in the emergence of Heidegger. It is now time to focus attention on Heidegger's concept of the phenomenon.

HEIDEGGER'S CONCEPT OF THE PHENOMENON

The term "phenomenon" goes back to the Greek expression whose derivative verb signifies "to show itself." So the word "phenomenon" has the connotation of the self-validating, the self-revealing, the self-manifesting. In other words, a phenomenon is that which validates itself by its manifestness, that which shows itself as it is, or that which reveals itself in accordance with its profiles. In Heidegger's own words, ". . . we must keep in mind that the expression 'phenomenon' signifies that which shows itself in itself, the manifest."[24] Similarly, Heidegger takes the word "phenomena" as the totality of "what lies in the light of day or can be brought to the light." This question of being exposed to light which characterizes the word "phenomena" points to another important fact – that depending on the type of access we have to it, an entity can show itself from itself as something which "in itself it is not."[25] When an entity shows itself as something which "in itself it is not" the showing becomes"seeming" for the entity exhibits itself as something else. Thus the term "phenomenon" means also "that which looks like something, that which is semblant 'semblance' . . ." The *phenomenon as semblance* relates to "something good which looks, but in actuality is not, [like] what it gives itself out to be." It is important to keep these two senses of the term "phenomenon" in mind for both the sense of the *phenomenon as manifest* and the sense of the *phenomenon as semblance* are structurally interconnected.[26]

Heidegger says that both the *phenomenon as manifest* and the *phenomenon as semblance* are structurally interconnected because it is only when an entity has the character or potentiality to be something that it "makes a pretension of showing itself as something which it is not: only then can it merely look so and so." Even when we talk of the *phenomenon as semblance* the original meaning of the phenomenon as manifest is already implied, which shows that the former is founded upon the latter. Hence we shall allot the term "phenomenon" to this positive and primordial signification . . . and distinguish "phenomenon" from "semblance," which is the privative modification of "phenomenon" as thus defined.[27]

But this distinction is not to be understood on the analogy of the distinction between appearance and mere appearance. Instead, Heidegger likens the distinction which he makes between *phenomenon as manifest* and *phenomenon as semblance* to what happens when one speaks of the "symptoms of a disease" with a view to showing that the all-important concept of the phenomenon is different from the question of "appearance" and "mere appearance."

> Here one has in mind certain occurrences in the body which show themselves and which, in showing themselves as thus showing themselves "indicate . . . something which does not show itself, the emergence of such occurrences, their showing – themselves, goes together with the Being-Present-at-hand of disturbances which do not show themselves. Thus appearance of something," does not mean showing itself; it means rather the announcing-itself by . . . something which does not show itself, but which announces itself through something which does not show itself. Appearing is a not-showing itself. But the "not" which we find here is by no means to be confused with the privative "not" which he used in defining the structure of semblance. What appears does not show itself; and anything which fails to show itself, is also something which can never seem.[28]

Even though we can say that "appearing" is not a showing-itself in the sense in which the phenomenal being can show itself, appearing, we must concede, is possible only on the basis of a showing-itself of something. The showing-itself which brings about the appearing, is not the appearing itself. "Appearing is an announcing-itself . . . through something that shows itself." Heidegger says that we simply muddle up the concept of the phenomenon and the word appearance if we talk of the latter (appearance) as "something wherein something appears without being itself an appearance." If this muddle is left unclarified, Heidegger says, the word "appearance" and the expression "appear" would get used in two ways: In the first place, that wherein something announces itself, and therefore does not show itself; and secondly, in the words . . . "without being itself an 'appearance,' appearance would stand for the showing-itself."

> But this showing itself belongs essentially to the "wherein" in which something announces itself. According to this, phenomena are never appearances, though on the other hand every appearance is dependent on phenomena.[29]

The point Heidegger makes in all this is that on account of the bewildering number of uses into which the words "appearance," "appearing" and "appear" can be put, it would be a miscalculation to define the term "phenomenon" through the vehicle of a conception of appearance unless of course we can clearly distinguish the manifold sense of appearance. The first double significance of appearance takes "appearing" to be announcing-itself, as not showing itself, on the one hand, and "that which does the announcing . . . that which in its showing-itself indicates something which does not show itself on the other." Of course, there is the sense of appearing which links it with the genuine concept of the phenomenon "as showing-itself."

Furthermore, we can legitimately talk of appearing as the emanating, the emerging from that which is in itself non-manifest; that which emanates from the non-manifest such that the latter really gets thought of as something which essentially never ever manifests itself. Heidegger puts it succinctly:

> When that which does the announcing is taken this way, "appearance" is tantamount to a "bringing-forth" or "something brought forth," but something which does not make up the real Being of what brings it forth: here we have an appearance in the sense of "mere appearance." That which does the announcing and is brought forth does, of course, show itself, and in such a way that, as an emanation of what it announces, it keeps this very thing constantly veiled in itself.[30]

This double sense of appearance as the announcing itself or a bringing-forth which veils what is thus announced or brought forth is explicit in Kantian usage of the term. In Kant's view, says Heidegger, "appearances" are, in the first instance, the "objects of empirical intuition," they are what show themselves in such intuition.[31] In the second place, "appearance" carries the meaning of the "phenomenon" in the original sense of the word "as emanation of something which hides itself in that appearance – an emanation which announces." Thus the "phenomenon," that which shows itself in itself, can become "appearance" in virtue of its announcing itself through something which shows itself. In this case, however, the phenomenon will acquire the privative character of semblance, just as appearance can become mere semblance. Now the difference between the phenomenon proper and appearance in the Kantian sense is this: the phenomenon describes the distinctive way in which we can encounter something, but appearance denotes "a reference-relationship" between what is thus encountered and the announcing or bringing forth itself which also qualifies as an entity to be encountered. For ". . . what does the referring (or the announcing) can fulfil its possible function only if it shows itself and is thus a 'phenomenon.' By thus illustrating that both appearance and semblance are founded upon the phenomenon, albeit in differing ways, Heidegger shows that the Kantian conception of the phenomenon can be used as "prolegomena" in ushering in the genuine phenomenological conception of the phenomenon.

We shall summarize Heidegger's concept of the phenomenon (in the phenomenological sense) by saying that the "phenomenon" is that which shows itself exactly as it is. But this "showing itself in itself" does not mean that all possibilities of that which shows itself or that which manifests itself are thereby exhausted. What is shown by the phenomenal being or the entity is exactly what it is at that point in time. Hence any phenomenological conception of the phenomenon must take account of the temporal character of Being. The sense of the phenomenon as that which "shows itself in itself" is the positive and original meaning of the term. This original and positive meaning of the phenomenon is what Heidegger calls the *phenomenon as manifest.*

But since it is also possible for an entity to show itself as that which in itself it is not, whereupon the showing becomes a seeming, we can say that the phenomenon has the character of resemblance. In this case an entity merely looks like what it is not. Now this is the privative signification of the phenomenon, and it is what Heidegger calls the *phenomenon as semblance.*

So an entity (the phenomenon) has both positive and negative characteristics. These characteristics – the positive and privative significations of the phenomenon – are ontologically or structurally coterminal. First, an entity shows itself in itself, exactly as it is. Secondly, the same entity has the potentiality to show itself as something which in itself it is not. On the one hand, the entity shows itself properly. On the other hand, the same entity can show itself improperly. Now the proper showing and the improper showing of the entity – the positive and the negative significations of the entity – belong together. When the phenomenon is thus understood as both that which shows itself exactly as it is and that which can show itself as what it is not, we would have been radically poised to do an ontological research, and the way would have been paved for us to appreciate the nature of "Nothing" in Heidegger's concept of Being.

We can now see why Heidegger's concept of the phenomenon contrasts sharply with those of certain philosophers who understand the phenomenon through the instrumentality of the theory of appearance. The problem with various conceptions of appearance is that none of them could ever lead to the primordial meaning of the *phenomenon as manifest*. Every traditional conception of appearance leaves out the proper exhibition of the entity as it is in itself. Only a phenomenological conception of the phenomenon can exhibit the entity properly in accordance with its basic structures. As a means of illustrating the deficiency of the traditional notion of the phenomenon which arises from certain conceptions of appearance, may we, by way of digression, take a brief and cursory excursion into the history of the concept "phenomenon" and emphasize the difference between previous approaches to it and Heidegger's.

HISTORY OF THE PHENOMENON

The plural form of the concept "phenomenon" is "phenomena," a technical philosophical term which has been of diverse usage since the inception of Western thought. Conventionally, there is a certain amount of dualism associated with the term "phenomena." Anthony Quinton distinguishes "two main ways in which phenomena are contrasted with other things" which carry this dualistic association. The first is the straight forward, scientific way of differentiating public, material objects from mental constructions or theoretical entities,[32] or of distinguishing facts from fiction. The second way is the traditional philosophical habit of characterizing phenomena as

> ordinarily private or subjective entities; impressions, sense-data, ideas; the things that really appear to or are directly perceived by, percipients, as contrasted with the public, objec-

tive, material things which are the presumed source, in some way or other, of them. In each case phenomena, as what is actually observed or what actually appears to a perceiver, are set against observable, non-appearing things which are somehow inferred, supposed or conjectured on the basis of the phenomena.[33]

If we collapse the two ways identified by Quinton of contrasting phenomena with other things in favour of the latter (the traditional philosophical way) we can show that the history of philosophy until the advent of the phenomenological movement is a history which makes phenomena the announcers or advocates or harbingers of what in themselves they are not. In Plato, phenomena taken as appearances are imperfect reflections of their genuine forms in the supersensible world. Thus the objective, material table is not a genuine table but simply an announcer of or a pointer to "real" table. In other words, an objective, material table is not a real table but merely announces what it is not. And so phenomena in the Platonist sense have been taken as characterizing the tangible, the transient, or the ephemeral aspects of reality. Following Plato, J. H. Lambert describes phenomena as the "theory of illusion."[34] In Platonism, therefore, the "thing" cannot speak for itself; the phenomenon does not show itself as it is. Instead, it shows itself as it is not.

In Kant, the phenomenon is understood in terms of appearance. Quinton says that Kant regards appearance and phenomena as synonymous: ". . . appearances, so far as they are thought of as objects according to the unity of the categories, are called phenomena."[35] For Kant, continues Quinton,

a phenomenon is an articulated system of appearances unified by the categories in the form of an object. Appearances are ingredients of phenomena, the properties, perhaps, that enter into the constitution of a concrete individual.[36]

Apart from this conception of the phenomenon as appearance, Kant insinuates that the entity which appears and is thus subjected to the discipline of the categories of human understanding is not the thing-in-itself. The thing-in-itself is not the phenomenon. The phenomenon is simply the town-crier, the announcer or the messenger of the Noumenon. The Noumenon is the thing-in-itself which, though completely veiled from our perceptual apparatus, keeps sending out in a centrifugal manner, entities to be transformed by the dynamic agency of the human mind. The phenomenon, then, is the announcer of that-which-is-veiled, the phenomenon announces itself. As it were, the phenomenon, for Kant, is the appearing of the non-appearing, the speaking of the ineffable, the comprehension of the incomprehensible. In phenomenology proper, the phenomenon does not speak for the dumb. It speaks for itself. On this score, Kant's conception of the phenomenon, though philosophically formal, falls short of a phenomenological expectation.

The same applies to Hegel's concept of the phenomenon. For Hegel, phenomena are mere excretions of the Giant Spirit (the Absolute) which seeks to purge itself of all contaminating impurities, through the triadic motion of position, opposition and synthesis. The phenomenon, in this case, is an emanation of that-which-announces. In Hegel, the phenomenon is not the manifest, not that which "shows itself in itself," but that which is manifested, that which is shown or brought forth by the absolute idea. And phenomenology for Hegel would be the science of appearances as bestowed by the Universal Idea. Thus does the Hegelian notion of phenomena fail to carry in its shell any thorough going phenomenological conception of the phenomenon.

Now whether in Platonism or Kantianism or in Hegelianism, the traditional conceptions of phenomena and phenomenology can be summarised thus:

(i) that things have a double nature which are not normally accessible to our ordinary perceptual apparatus (Platonist).
(ii) that our perceptions of things are different from what things are in themselves (Kantian);
(iii) that things are mere ejaculations of a self-actualizing Universal Reason (Hegelian);
(iv) that there is a distinction between reality and appearance, and
(v) that phenomenology is the study of the transient, ephemeral, outward features of human experience.

None of these notions of phenomena and phenomenology has any title to legitimacy in genuine phenomenological research. They furnish us merely with a "prolegomenon" (an introduction) to the study of Husserlian and Heideggerian phenomenology. In particular, they contrast sharply with Heidegger's concept of the phenomenon. Whereas Heidegger holds the phenomenon to be essentially (i.e., in the primordial, positive sense of the term) that which shows itself as it is in itself in profiles – which does not entertain the notion of a world-behind-the-scene – the traditional concept of the phenomenon from Plato down to Hegel understands the phenomenon as a shadow of the "in itself," as a pointer to a world-behind-the-scene.

Now we should shift attention to Husserl and Heidegger. It is important to mention here that while Husserl and Heidegger share a number of things in common as it affects the concept of the phenomenon, they also differ in certain respects. Let us highlight a few of these similarities and differences between master and pupil.

HUSSERL AND HEIDEGGER ON THE PHENOMENON

In determining the similarities and differences of Husserl and Heidegger on the concept of the phenomenon, two questions are of interest: The one relates to what the two philosophers take to be the outstanding phenomenon of phenomenological inquiry. The other revolves around what they regard as the nature of consciousness. Husserl states explicitly that consciousness is the absolute starting point of any genuine phenomenological inquiry. And following Brentano, Husserl designates consciousness as intentionality.

> Like perception every intentional experience – and this is indeed the fundamental mark of all intentionality – has "its intentional object," . . . its objective meaning. Or to repeat the same in other words: To have meaning, or to have something "in mind," is the cardinal feature of all consciousness, that on account of which it is not only experience generally but meaningful, . . .[37]

The doctrine of intentionality is a description of the nature of consciousness. It signifies the fact that consciousness is directional, "that it is given in experience as an outward-moving vector."[38]

Consciousness is like a stream, or a current that flows between subject and object. The object of my consciousness, say a hexagon or a U-boat, "is something meant, constituted, in short, intended by me."[39] In Husserl's words, "the stream of consciousness is permeated by the fact that consciousness relates itself to object. This relation is an essential characteristic of every act of consciousness."[40] Such is the nature of consciousness that "True being, . . . whether real or ideal, has significance only as a particular correlate of my own intentionality."[41] Consciousness is always consciousness of something intended by the subject. "And one of the hallmarks for my being as a conscious being is that I am free and can spontaneously bring about certain states of affairs: I am, in other words, an agent."[42] Thus does intentionality bestow freedom and selfhood. And this discovery that thinking is free is another cardinal feature of Husserlian phenomenology. Now if thought is free and if whatever is thought of or intended by the subject must necessarily be a being, then, we can say that critical *a priori* research (i.e., phenomenology) paves the way for an inquiry into "Nothing," since the mind cannot think of "absolute Nothing."

Heidegger comes in here. He, like Husserl, takes intentionality to be the primitive fact of phenomenology. Although he expounds no elaborate theory of consciousness, Heidegger's analysis of the role of the imagination[43] in the achievement of the ontological synthesis bears some family resemblance with Husserl's theory of consciousness. In Husserl as well as in Heidegger the established ground of all forms of knowledge and all ontologies is the intentional operation of consciousness. Whatever theory we hold about the world and whatever meaning we assign to things are a function of man's mental activities. For Husserl, without the intentional operation of consciousness, world-constituting activity would be impossible. And for Heidegger, without a certain comprehension of Being by the human being we could not even utter being, that is, we could not say of a thing that it is.[44] *Dasein* for Heidegger, at least, in *Being and Time* is the existing, thinking subject – the locus of truth and untruth, of meaning and meaninglessness. Thus, the world as it is presently constituted, understood and assigned meanings is the product of the intentional operation of consciousness. There is nothing "that is" that was never intended or meant by the existing, thinking subject. This position is held by both Husserl and Heidegger. Heidegger also subscribes to the intentionalist view that the phenomenological subject and its objects are correlative, that is, that objectivity has meaning only in relation to subjectivity. Like Husserl, Heidegger holds that things are not intelligible *a priori*. They are meaningful only for us. The two philosophers maintain that it is not possible to confront an object which cannot be intended or meant by a subject. The lesson to learn from this is that whatever has meaning for or is intelligible to the subject must be connected to the world in one way or the other. And whatever is presented to the subject as an object in the world, must, in some sense be real about the world. This also prohibits speaking of what occurs to someone else as an object of thought as total nothing.

Another area of comparison between Husserl and Heidegger is the nature of the object of a phenomenological inquiry. The outstanding phenomenon of phenomenological thought (for Husserl and Heidegger) has possible modes which are not exhaustible in the given. Every particular profile of a phenomenological object refers to (not necessarily present) the object as a whole. The present "given" of the object points to its possible other givens such that the object as a whole – the intended object – unifies every other profile as given in the many acts of perception. Each perception points to other perceptions and "perception becomes a process of fulfilment." Commenting on this question of profiles or aspects which are unified in the many acts of perception, Richard Schmitt says:

> The understood object is never fully given. It is given only in profiles or aspects, and we understand every given profile as pointing to other profiles. Our understanding of the object does not stop at the actually given, but extends to possible other aspects of the given, to its "horizons," in Husserl's language. We understand the object also in so far as it is not now given. The intelligibility of the object therefore does not end with what is given, but extends to what we make of the given. Hence the intelligibility of the object is not altogether a property of the object independent of our awareness of it, but is intelligibility for us.[45]

In other words, phenomenological objects are what they give to us, what we make of what they give to us, what they can possibly give to us, and what we can make of what they can possibly give to us. This classical phenomenological understanding of the object is in complete accord with Heidegger's analysis of the "essent" as the "thrown," that is, as that which shows itself exactly as it is in terms of the unity of its profiles.

But whereas Husserl takes the outstanding phenomenon to be that which is universally and necessarily posited by consciousness or that which could yield universal essences through the intentional operation of consciousness or, better still, through a phenomenological reduction, Heidegger connects the outstanding phenomenon with the concept of Being. Let Husserl and Heidegger speak for themselves on this matter. Husserl states:

> . . . in speaking about investigating the objects and modes of cognition, we always mean investigation into essences, which, in the sphere of the absolutely given, exhibits in a general way the ultimate meaning, the possibility, the essence of the objectivity of cognition and of the cognition of objects.[46]

And Heidegger retorts:

> What occurs for the phenomenology of the acts of consciousness as the self-manifestation of phenomena is thought more originally by Aristotle and in all Greek thinking and existence as *aletheia*, as the unconcealedness of what-is-present, its being revealed, its showing itself.
>
> . . . Whence and how is it determined what must be experienced as "the things themselves" in accordance with the principle of phenomenology? Is it consciousness and its objectivity or is it the Being of beings in its unconcealedness and concealment?[47]

It is for thus insisting that the Being of entities in its self-communication rather than *cogitationes* and their universal essences constitute the outstanding phenomenon of phenomenological research that Husserl denounced Heidegger. But the humble opinion of this researcher is that master and pupil need not quarrel on what should constitute the object of phenomenological research, since their stated aims are different. For whereas Husserl's phenomenological purpose is epistemological[48] and while he insists that scientific philosophy consists in determining a theory of knowledge as the foundation of all science, Heidegger's phenomenological aim is ontological and he insists that the primary problem of philosophy is the problem of Being – the problem of exhibiting the general profiles of the world as it is.

Be that as it may, Heidegger's understanding of phenomenology (which we shall discuss in a moment) as the letting be of beings or letting what is show itself clearly to our unbiased, unprejudiced mind is reminiscent of Husserl's call for a return to pre-reflective consciousness – "to the things themselves."

HEIDEGGER'S UNDERSTANDING OF PHENOMENOLOGY

Heidegger distinguishes the genuine phenomenological conception of the phenomenon from the formal conception of the phenomenon. In the genuine phenomenological sense, the "phenomenon" is that which shows itself as it is in itself, the *manifest*. Even in this genuine sense, the phenomenon is also capable of showing itself as that which in itself it is not – as *semblance*. Both the phenomenon as *manifest* and the phenomenon as *semblance* signify the positive and privative senses of the phenomenon in the genuine phenomenological context, and they are different from the formal conception of the phenomenon as Appearance.

Next, Heidegger examines the different uses to which the word "Logos" has been put, and insists that the interpretation of "Logos" as discourse where discourse is understood to mean that which "lets something be seen" is the genuine sense of the word. Thus, an interpretation of the genuine senses of the terms "phenomenon" and "logos" enables Heidegger to enunciate a preliminary conception of phenomenology. According to him, when we construe the words "phenomenon" and "logos" in their genuine senses, we see that there is a close affinity between them, "we are struck by an inner relationship between the things meant by these terms." And we find it legitimate and appropriate to understand "phenomenology" as meaning "to let that which shows itself be seen from itself." And this, according to Heidegger, amounts to expressing the maxim: "To the things themselves."[49]

On account of the foregoing, Heidegger insists that it would be improper to construe phenomenology as one construes such sciences as "theology," "sociology" and so on.

Those terms designate the objects of their respective sciences according to the subject matter which they comprise at the time . . . "phenomenology" neither designates the object of its researches, nor characterises the subject matter thus comprised. The word merely informs us of the "how" with which what is to be treated in this science gets exhibited and handled. To have a science "of" phenomena means to grasp its object in such a way that everything about them which is up for discussion must be treated by exhibiting it directly and demonstrating it directly. . . .[50]

Even in ordinary parlance and in the formal conception of the "phenomenon," "any exhibiting of an entity as it shows itself, may be called phenomenology with formal justification." The point now is to deformalize the official conception of the phenomenon into the phenomenological one, and show how the latter is different from the ordinary conception of it. The way to go about this is to determine what it is that must be called the "phenomenon" in a distinctive sense and what it is that by its very nature is always the theme whenever something gets exhibited explicitly.[51] According to Heidegger, what necessarily becomes the theme whenever something, an entity, is exhibited properly, explicitly, is something that does not show itself at all most of the time:

. . . it is something that lies hidden, in contrast to that which proximally and for the most part does show itself: but at the same time it is something that belongs to what thus shows itself, and it belongs to it so essentially as to constitute its meaning and its ground.

Yet that which remains hidden in an egregious sense, or which relapses and gets covered up again, or which shows itself only "in disguise" is not this entity or that but rather the Being of entities as our previous observation have shown. This being can be covered up so extensively that it becomes forgotten and no question arises about it or about its meaning. Thus that which demands that it becomes a phenomenon, and which demands this in a distinctive sense and in terms of its ownmost content as a thing, is what phenomenology has taken into its grasp thematically as its object.[52]

In other words, Heidegger is saying that what is presupposed by every inquiry is the Being of entities. And what is forgotten and never made to show itself clearly as it is, not even by the formal conception of the phenomenon, is the Being of entities. Thus the Being of entities is that which, though inherent in every conceptual preoccupation, is nevertheless covered up or made to lie hidden by the essent as such, by the thing which is the object of inquiry. Be that as it may,

In the phenomenological conception of the "phenomenon" what one has in mind as that which shows itself is the Being of entities, its meaning, its modifications and derivatives.[53]

But Heidegger cautions that the showing-itself of the Being of entities is not just any kind of showing-itself, not like the showing-itself of other entities, and not like any such thing as appearing. In no way can the Being of entities ever be anything such that "behind it" stands something else which does not appear.

Thus, for Heidegger, the pure datum of phenomenological inquiry – that which lies hidden, which does not show itself in itself but which belongs to that which shows itself so much so as to constitute its meaning and its ground, and which cries out to be shown, to become a phenomenon – is the Being of all beings. And whenever we successfully exhibit an entity properly, explicitly, as it is in itself, it is this Being of entities that always becomes the theme of discussion. Phenomenology may well inquire after other entities. But every such inquiry usually ends up in callous obliviousness to Being.

Even in the concrete work of phenomenology itself there lurks the possibility that what has been primordially "within our grasp" may become hardened so that we can no longer grasp it.[54]

Hence the primary task is to wrest the way in which the Being of entities and its structures are encountered from the objects of phenomenology.[55]

Now, the question of the being of entities is an ontological question. By making Being the object of phenomenological inquiry is Heidegger saying that ontology is synonymous with phenomenology? And if all forms of ontology are phenomenological in character, what is novel in Heidegger's analysis of Being?

PHENOMENOLOGY AS ONTOLOGY

We raised this question of the novelty of Heidegger's phenomenological ontology at the very beginning of this paper. Heidegger specifically insists that ontology is possible only as phenomenology and that whenever, in the phenomenological conception of the "phenomenon," one talks of that which shows itself, one simply has in mind the "Being of entities, its meaning, its modifications and derivatives." In phenomenology it is the "phenomenal structures" of entities that are displayed.

That which is given and explicable in the way the phenomenon is encountered is called "phenomenal." . . . Everything which belongs to the species of exhibiting and explicating and which goes to make up the way of conceiving demanded by this research, is called "phenomenological."[56]

So because "phenomena," in the phenomenological sense, are always concomitant elements of Being, and Being is always the Being of some entity, any exhibiting of the general profiles of anything, any entity, is ontology, and the manner in which such exhibition is carried out such that one sees what is thus exhibited as it is in itself, is phenomenology. In other words, Heidegger is saying that for us to explicate and exhibit the general structures of the world as it is – ontology – we have to employ a method of inquiry whose subject matter is to make manifest any object whatsoever. And the only method qualified to do this is the phenomenological method. As he rightly says, in terms of subject matter,

phenomenology is the science of the Being of entities-ontology. . . . Ontology and phenomenology are not two distinct philosophical disciplines among others. These terms characterise philosophy itself with regard to its object and its way of treating that object. Philosophy is universal phenomenological ontology. . . .[57]

William Richardson says that on Heidegger's own terms, "phenomenology lets-be-seen the Being of beings. Now to let-be-seen . . . to investigate thematically, the Being of beings is the classical function of Ontology."[58] If ontology is the thematic investigation of the Being of beings and phenomenology, the letting-be-seen-of-the-Being of beings, Heidegger's insistence that there is close affinity between phenomenology and ontology is perfectly in order.

What is novel about Heidegger's phenomenological ontology is that his letting-be-seen of the Being of beings had to proceed through the "hermeneutic" of *Dasein*'s finite essence – that is, the ontological analysis of "there being" or what, loosely speaking might be called the being of man. Phenomenology becomes the phenomenology of *Dasein* – an interpretation of the basic structures of *Dasein* as a precondition for the ontological study of other entities – and ontology becomes a fundamental ontology which focuses primarily and mainly on the being of man with a view to renewing the question of the meaning of Being in general. The point, however, remains to be made: that Heidegger regards ontology as phenomenology does not argue the case that all forms of ontology are phenomenological in character. For one thing, not all forms of ontology proceed through the existential analytic of *Dasein*. Heidegger's phenomenological ontology is unique in this sense. If every form of ontology is not necessarily informed by the phenomenological attitude, is there any sense in which we can distinguish ontology from phenomenology?

THE DIFFERENCE BETWEEN PHENOMENOLOGY AND ONTOLOGY

If one thinks that because Heidegger insists that Being is the outstanding phenomenon of phenomenological inquiry, and that phenomenology is essentially ontology, and therefore conclude that there is no difference between phenomenology and ontology of any kind; then one would be in error. For Heidegger has specifically stated that there is a prevalent notion of ontology (the traditional notion) which treats it (ontology) merely as "an academic classification" or a "branch of the philosophical system."

> The word "ontology" was first coined in the seventeenth century. It marks the development of the traditional doctrine of the essent into a discipline of philosophy and a branch of the philosophical system. . . . And it is in this sense that the word "ontology" is used today.[59]

Ontology in the traditional sense is a special branch of philosophy or a special branch of metaphysics called *metaphysica generalis* as distinct from *metaphysica specialis*.[60] In this same sense (the traditional sense), ontology is a science of being, an inquiry into the essent as such. The function of traditional ontology is to describe the essent as such, "what is what is?" Now a description of "what is," the essent as such, is different from an effort – a humble effort – to make Being show itself clearly as it is. In the former, in the description of "what is," we always dress being in a ready-made conceptual attire. In the latter, in the Heideggerian sense, we maintain an attitude of patience. We open up to the openness of Being. Again in the former (in the traditional) sense of ontology, Being is not a phenomenon in the genuine phenomenological sense. Being is something to be subjected to our ready-made conceptual strait jackets. But in the latter (in the Heideggerian) sense, Being is a phenomenon in the genuine phenomenological conception of it. So whereas ontology in the traditional sense signifies a description of being on the basis of our ready-made conceptual tools whereupon we determine for a being what it is, ontology in the Heideggerian conception of it is the humble endeavour to let Being disclose itself as it is, whereupon we do not compel Being to become something we desire, but allow it to speak for itself.

Thus, ontology, we may say, given the Heideggerian dimension, is a science that investigates the concept of Being on the basis of two different modes of questioning as informed by two radically dissimilar attitudes. Traditional ontology investigates Being as an entity and poses the question, "Why are there essents rather than nothing?" Heidegger's ontology investigates Being, not as an entity, but as a process – a process that on its terms constitutes a phenomenon in the genuine phenomenological conception of the phenomenon as that which primarily and mainly does not show itself, but which belongs to and asserts itself so much on that which shows itself as to constitute its meaning and its ground – and poses the primordial question "How does it stand with Being?"

But Heidegger would not have any association of his ontology with the traditional brand, which he considers a distortion, and in righteous indignation he wishes strongly that he could dispense with the use of the term in future, for "Two modes of questioning which, as we now see clearly, are worlds apart, should not bear the same name."[61]

Now, given our characterization of traditional ontology, we can state unequivocally that it is different from phenomenology, and that not every form of ontology is informed by the phenomenological attitude. We state also that not every conception of Being meets the standard requirements of the genuine phenomenological conception of the phenomenon. We say that Heidegger's Being qualifies as an outstanding phenomenon in the phenomenological sense. Hence also his ontology (concept of Being) qualifies as phenomenology.

THE PHENOMENOLOGICAL CHARACTER OF HEIDEGGER'S PHILOSOPHY

As ontology is phenomenology for Heidegger, so is phenomenology philosophy for him. As he makes no distinction between ontology and phenomenology, so does he not make any clear-cut distinction between phenomenology and philosophy. According to him both phenomenology and ontology characterize philosophy with regard to its object and its way of treating that object. Philosophy, he declares, is universal phenomenological ontology.[62]

The primary task of philosophy, according to Heidegger, is the rethinking of the problem of Being which the tradition of Western philosophy has subjected to callous forgottenness. So the real problem of philosophy is the problem of Being, and philosophy is the original thinking of Being. Hence the primary task of philosophy is ontological, not epistemological. In performing this essential task of rethinking Being, philosophy employs the procedural tool of phenomenology which lets-be-seen the Being of beings. Otherwise stated, philosophy (original thinking) inquires after Being (ontology) through the procedure of thinking (phenomenology) which guarantees the thorough X-raying (exhibition) of the subject of thought.

Philosophy, which Heidegger calls "universal phenomenological ontology," points to no special direction or tendency. The phenomenological spirit in philosophy is to let things stand out in their own terms and according to their own merits. This spirit is predominant throughout the three phases of Heidegger's thought – the stages of *Dasein, Being* and *Ereignis*. In the first phase, often described as the phase of *Dasein*, Heidegger embarks on a rigorously critical, consistent, careful and exhaustive analysis of the Being of entities through the ontological telescope of human essence. In this phase, according to Vincent Vycinas, we are introduced, not to anthropology, but to a truly ontological philosophy,[63] one in which phenomenology becomes a hermeneutic, an interpretation of man's historical "being there" with a view to exhibiting the Being of entities. Heidegger's *Daseinsanalytik* portrays *Dasein*'s relationship to *sein* – loosely speaking, man's relationship to Being. Vincent Vycinas says something on this:

> Heidegger's first phase of thought is the thought of *Dasein*, namely, of man in his ultimate essence. *Dasein* primarily is not man, but Being approaching man to reveal itself. *Sein und Zeit*, the basic work of the first phase of Heidegger's philosophy, is an established bridge leading from the common man's standpoint to the road of philosophy.[64]

The phenomenological attitude is also held in perfect balance in the second phase of Heidegger's thought. In this phase, Heidegger applies *epoche* to the history of Western ontology. He does this by taking an excursion into the history of philosophy. Each example of a historical philosophy that he cites is laid bare to the foundation, thus making it possible for us to see "clearly and distinctly" how each philosophy in the Western tradition has woefully omitted the ontological difference, that is, the difference between Being and beings. The consequence of such omission is far-reaching. Being has come to be thought of as a being, thereby subjugating the former to a hardened forgottenness. Such callous obviousness to Being arising from the omission of the ontological difference has led to an unfortunate treatment of things as mere objects which in turn leads to nihilism. Now that things are reduced to objects and as objects disintegrate under our vengeful manipulations, only a full-blooded phenomenological attitude can bring us to regain the essence of Being.

Accordingly, in the third phase of his thought on the *Ereignis* – or what Vycinas calls the phase of "earth and gods" – Heidegger, in a kind of detached Olympian majesty reminiscent of the Husserlian transcendental ego that he rejects, meditates on the essence of Being. Thinking becomes thanking. Philosophy becomes listening to the voice of Being. And we see a thoroughly humble and respectful approach to things. According to Vycinas,

> Heidegger tries to approach the "thingness" of a thing by a phenomenological method, namely, by letting things rest in themselves, i.e. appear the way they are in the light of Being.
>
> In a phenomenological description, the thinker sustains himself from any a priori interpretation of the described thing and lets the thing appear as not placed in one or another theory.[65]

This means returning "to the things themselves" in a humble, respectful and thankful attitude. Such a humble, respectful, prayerful and thankful approach to the treatment of things (Heidegger's fashion) can only be informed by the phenomenological spirit. On this count, we stress here that Heidegger is, first and foremost, a phenomenological ontologist. We stress this point irrespective of what certain critics of Heidegger say about his alleged unfaithfulness to the phenomenological movement.

We should, perhaps, by way of digression, mention that Heidegger has been accused, among other things, of bastardizing phenomenology, of making an anthropology of it, and abused for not swallowing Husserl hook, line and sinker. Herbert Spiegelberg, for example, states that "Heidegger never was a phenomenologist in the strictest sense defined by Husserl's subjectivist transcendentalism."[66] A view of this sort usually commits its author far beyond what he himself had imagined. Did Heidegger in any way suggest that he was giving a verbatim account of Husserl's phenomenology, or that he was an uncritical disciple of Husserl? Is it not fundamentally unphenomenological to contend that

Heidegger should have conceived of phenomenology exactly as Husserl did? Are we not forbidden in advance, given the phenomenologist's criterion of intentionality, from speaking in such dogmatic fashion?

Phenomenology as a concept can be shown to be a mental construction, that is, something meant or intended by the subject. If it is, then, no one, given the phenomenological criterion of intentionality, can describe what occurs to him or what is presented to his mind other than the experiencing subject. By the same token, Heidegger has described phenomenology precisely as it has occurred to him, not as it occurred to Husserl.

> Heidegger sees the phenomenological method as a means to think away from traditional philosophy, that is, from the subject-object relation, thus allowing things to manifest themselves as they are without any predisposition. For Heidegger what ultimately manifests itself is Being in its differentiation from beings.[67]

And while acknowledging his debt to Husserl, Heidegger insists that what is crucial in phenomenology does not lie in its actuality as a philosophical movement, but in our ability to seize upon it as a possibility with which to realize our projects since possibilities are higher than actualities.[68] On this count, Heidegger is basically a phenomenologist of unbounded imagination!

University of Lagos

NOTES

* Martin Heidegger, *Being and Time*, trans. J. Macquarrie and E. Robinson (Oxford: Basil Blackwell, 1962), p. 60.

[1] There is this essential difference: conventional ontology thinks of Being as beings (i.e., as objects); phenomenological ontology (i.e., Heidegger's) thinks of Being as pure Being (as a process).

[2] The work in question is William Richardson's *Heidegger: Through Phenomenology to Thought* (The Hague: Martinus Nijhoff, 1963).

[3] *Ibid.*, p. x.

[4] *Ibid.*, pp. 3–4.

[5] *Ibid.*, p. 5.

[6] We should note that Heidegger avoids talking of or using the word "man." He would rather talk of *Dasein* or "there-being" which cannot be reduced to a being with a fixed essence. So, only in a very non-Heideggerian loose sense can we use the word "man" interchangeably with *Dasein*.

[7] Quoted by David Carr in "Intentionality," in *Phenomenology and Philosophical Understanding*, ed. Pivevic (Bristol: Cambridge University Press, 1975), pp. 19, 20,

from *Franz Psychologic Brentano, vom Empirischen Standpunkte* (Leipzig: Von Dunker and Humblot, 1874), Vol. 1, pp. 115f.

[8] Franz Brentano, *The True and the Evident*, O. Kraus and R. M. Chisholm (London: Routledge and Kegan Paul, 1966), p. 77.

[9] *Ibid.*, p. 77.

[10] *Ibid.*, p. 78.

[11] *Ibid.*, p. 83.

[12] *Ibid.*, p. 83.

[13] R. M. Chisholm, *Intentionality, Mind and Language*, ed. A. Marras (Chicago: University of Illinois Press 1972), p. 32.

[14] There are two other important criteria of intentionality outlined by Chisholm. See, *Intentionality, Mind and Language*, pp. 33–34.

[15] We are aware that Heidegger lays much stress on "Angst as revealing Nothing." This kind of nothing is different from the "ontological nothing" which is the ground of all nullity.

[16] Details of this paper by Heidegger are found in Walter Biemel's *Martin Heidegger: An Illustrated Study*, trans. J. L. Mehta (London: Routledge and Kegan Paul, 1977), pp. 9–14, extracted from *On Time and Being*, pp. 74–80.

[17] Details of this can be found in my M. Phil. Dissertation *Authenticity in Heidegger's Philosophy: A Critique* (unpublished), University of Lagos, pp. 32–33.

[18] Biemel, *op. cit.*, p. 12.

[19] *Ibid.*, p. 13.

[20] *Ibid.*, p. 13.

[21] *Ibid.*, p. 14.

[22] *Ibid.*, p. 14.

[23] More will be said on this when we discuss "phenomenology as ontology."

[24] Heidegger, *Being and Time*, *op. cit.*, p. 51.

[25] *Ibid.*, p. 51.

[26] *Ibid.*, p. 51

[27] *Ibid.*, p. 51.

[28] *Ibid.*, p. 52.

[29] *Ibid.*, p. 53.

[30] *Ibid.*, p. 53.

[31] *Ibid.*, p. 54.

[32] Anthony Quinton, "The Concept of a *Phenomenon*," in Pivevic (ed.), *Phenomenology and Philosophical Understanding*, *op. cit.*, p. 2.

[33] Ibid., p. 2.

[34] Cf. Richard Schmitt, "Phenomenology," in *The Encyclopedia of Philosophy*, Vol. 6, 1967, p. 135.

[35] Anthony Quinton, *op. cit.*, p. 3 (quoted from *Critique of Pure Reason*, A249).

[36] *Ibid.*, p. 3.

[37] Edmund Husserl, *Ideas: General Introduction to Pure Phenomenology*, trans. W. R. Boyce Gibson (London: George Allen and Unwin Ltd., 1969) (5th impression), pp. 261–262.

[38] See Edmund Husserl, *The Paris Lectures*, trans. Peter Koestenbaum (The Hague: Martinus Nijhoff. 1970), p. xxvii.

[39] *Ibid.*, p. xxvii.

[40] *Ibid.*, p. 18.
[41] *Ibid.*, p. 23.
[42] *Ibid.*, p. lv.
[43] Cf. M. Heidegger, *Kant and the Problem of Metaphysics*, trans. J. S. Churchill (Bloomington and London: Indiana University Press, 1962), pp. 136–137.
[44] Cf. *ibid.*, p. 234.
[45] Richard Schmitt, "In Search of Phenomenology," *The Review of Metaphysics*, Vol. XV (1961–62), p. 468.
[46] Edmund Husserl, *The Idea of Phenomenology*, trans. W. P. Alston and C. Nakhnikan (The Hague, Martinus Nijhoff, 1970), pp. 11–12.
[47] M. Heidegger, "My Way to Phenomenology," quoted in Walter Biemel's *Martin Heidegger: An Illustrated Study, op. cit.*, p. 14.
[48] Cf. Richard Schact, "Husserlian and Heideggerian Phenomenology," *Philosophical Studies* Vol. 23m (1972), p. 293.
[49] Heidegger, *Being and Time, op. cit.*, p. 58.
[50] *Ibid.*, pp. 58–59.
[51] *Ibid.*, p. 59.
[52] *Ibid.*, p. 59.
[53] *Ibid.*, p. 60.
[54] *Ibid.*, p. 61.
[55] *Ibid.*, p. 61.
[56] *Ibid.*, p. 61.
[57] *Ibid.*, pp. 61–62.
[58] Richardson, *op. cit.*, p. 46.
[59] Martin Heidegger, *An Introduction to Metaphysics*, trans. Ralph Manheim (London: Yale University Press, 1959), p. 41.
[60] Heidegger, *Kant and the Problem of Metaphysics, op. cit.*, p. xv.
[61] Heidegger, *An Introduction to Metaphysics, op. cit.*, p. 41.
[62] Heidegger, *Being and Time, op. cit.*, p. 62.
[63] Vincent Vycinas, *Earths and Gods: An Introduction to the Philosophy of Martin Heidegger* (The Hague: Martinus Nijhoff, 1961), p. 9.
[64] *Ibid.*, pp. 9–10.
[65] *Ibid.*, p. 12.
[66] Herbert Spiegelberg, *The Phenomenological Movement, A Historical Introduction*, Vol. 1 (The Hague: Martinus Nijhoff, 1960), pp. 347–348.
[67] Christopher S. Nwodo, "Friedlaender Versus Heidegger: A-Letheia Controversy," *Journal of the British Society of Phenomenology*, Vol. 10: No. 2 (1979), p. 92.
[68] Cf. Heidegger, *Being and Time, op. cit.*, pp. 62–63.

ANTONIO ZIRIÓN QUIJANO

NOTES ABOUT THE PHENOMENOLOGY OF JOSÉ GAOS

For a certainty, José Gaos has been the Mexican philosopher who has contributed the most to the spreading of Phenomenology in Mexico.* Perhaps it is high time that his countless courses, essays and conferences devoted directly or indirectly to Phenomenology be remembered; part of his effort that it is certainly not necessary to call to mind are his translations of some of the main works of Husserl, including those two that Gaos calls "classics": *Logische Untersuchungen* and the First Book of *Ideen*. They are still present, as they have been during the last half century, to those who approach with more or less care the work of Husserl in all Spanish-speaking countries.[1]

It is much less known that Gaos maintained through all his career as a philosopher a very peculiar and certainly intense relation with Phenomenology. But for those who have read him, for those who have studied and commented on his works, it is rather obvious that, even though Gaos was not (in the very least) a follower of Husserl's philosophy (if there is such a thing), or of what he understood Phenomenology or Transcendental Idealism to be, he always kept and developed in his own philosophy several elements or features of a phenomenological sort or nature. It is a platitude among the experts to say that Gaos developed "phenomenological" studies or analyses, or studies which constitute "phenomenologies," or that he used the so-called "phenomenological method"[2] – to be true, abundantly and in central places of his investigations, we would add.[3]

We cannot find, however, in the literature on Gaos, a thorough reflection about the place and the sense of Phenomenology and the Phenomenological Method in his thought and in his work. In this paper I try to begin that reflection; but, as it will become apparent, it will not be possible to give it an end here.

Such reflection should consider, in the first place, two great questions: the first one asks for the concept that Gaos had of the Phenomenological Method and of Phenomenology, including both Gaos' idea or theory about them as well as his use of them. The second is the question about the role that Phenomenology or Phenomenological Method play within Gaos'

A-T. Tymieniecka (ed.), Analecta Husserliana, Vol. LIV, 331–344.

philosophy in a broad sense. But, in the second place, such reflection would also have to consider, together with those already mentioned, other more or less historical questions related to the filiation of Gaos' Phenomenology to the Phenomenology of Husserl or to other phenomenologies belonging to the wide phenomenological movement.

But before I start I would like to make clear an important point. This essay belongs to research work which is only in an early stage and which will embrace not only Gaos' Phenomenology, but the complete history of Phenomenology in Mexico. I offer here only some preliminary results of that research, and not in the order indicated in the program just hinted at. My most immediate interest has been to be able to make a clear demarcation between the Phenomenology of Gaos and that of Husserl. In particular, I have tried not to touch upon the philosophy of Gaos as a whole, or the history of his own relation with Phenomenology. Naturally, these limitations will have to be overcome in the complete investigation.

To get down to business, let me quote some incidental words of Gaos: "The names given to Philosophy seem to have in History a significance comparable to that the names given to the products of industry have in the commercial relations; perhaps Husserl's 'Phenomenology' has had a success which has been absent for the twin 'theory of objects' of Meinong due to the simple matter of onomastic appeal."[4] Well, perhaps. But, setting aside the theory of Meinong, what can be said, without "perhaps" and without a doubt, is that a very considerable part of the success of Phenomenology has not been the success of Phenomenology, but that of its name. In this sense, Gaos may be right. But we have to recognize that Phenomenology itself, until now, has failed.

It is evident that when we talk about Phenomenology today we do not talk about it "as the mathematician about mathematics," which is the way in which Husserl said that he talked about it.[5] It is obvious that, despite the diverse attempts to give it a presentation which would render it convincing, Husserl did not succeed in placing his Transcendental Phenomenology "on the sure road of a science." It may even be conjectured that he himself was convinced in the end that times were not ripe or well-disposed for the science on which philosophy should end up founded in a definitive way. "*Philosophy as a science*, as a serious, strict, even apodictically strict science – the dream is over."[6] These words of Husserl are frequently cited as if they expressed a courageous (although maybe mournful) awakening from a dream which is now abandoned to aim at more feasible goals; I think they should be read, instead, as the expression of a tragic verification of the historical defeat of Phenomenology. For we must not forget that to Husserl the defeat of Phenomenology would mean the defeat of philosophy itself.

As far as Gaos is concerned, he proclaims and adheres to the verdict of the times. "Husserl wanted to develop a science of the foundations of *all* sciences, and he developed one more philosophy about the principles of *all things*."[7] His "scientific philosophy" is "a monster of contradictions,"[8] and Phenomenology thus understood, characterized by the method of transcendental reduction, is finally labeled, somehow, and notwithstanding Husserl's manifest intentions, under the great name of metaphysics, and therefore sentenced and condemned to the same failure as that discipline, although at the same time, curiously enough, it is exalted and ennobled.[9] The fact that this sentence is still valid today, validates the historicism, the personism, the skepticism of Gaos. Still.

This should be enough, of course, to demarcate once and for all the Phenomenology or Phenomenologies of Gaos from the Transcendental Phenomenology of Husserl. But it is also important, in my view, to emphasize that that method of the Transcendental Reduction is the only one that can be called, from the point of view of Husserl, and in strict sense, the Phenomenological Method. And Gaos has stated, clearly enough, that this is not the method that he employs when he says he employs the phenomenological method.[10]

We would have to go through a good stretch of the history of philosophy in this century in order to understand why Gaos uses, in spite of his own words, this denomination.[11] That would be, among other things, the history of the success of the *name* of "Phenomenology." Such was its success, since its introduction by Husserl in his *Logische Untersuchungen* of 1900, that no later than 1950, Dietrich von Hildebrand, who had studied with Husserl in Freiburg in the first years of the century, could write: "The expression 'Phenomenology' has become today an equivocal expression."[12] Such was the rise of the so called "Phenomenology," that Max Scheler himself, an authority in "Phenomenology" if ever there were any, could say at the end of his life that "One really should no longer use the word 'Phenomenology.' After all, it does nothing but what philosophy has always done."[13]

I will try to synthesize the story in a few lines.

When Husserl explained for the first time in 1913 the method of Transcendental Reduction, it was too late to modify the idea of Phenomenology that his disciples and followers had formed themselves on the basis of *Logical Investigations*. In this work Husserl did not define clearly a science called Phenomenology nor a method called Phenomenological. The "Phenomenological" analyses carried out there were a sort of descriptive psychological analyses which benefited from several methodological proposals which for Husserl had (at least some of them) the character and the strength of principles: the principle of "freedom from presuppositions," the dictum of being guided by the things themselves, the rigorously descriptive practice, and above all, the employment of intellectual intuition (or categorial intuition, or, in particular, the intuition of essences), whose possibility and reality was revindicated in the *Investigations*. "Phenomenology" and the "phenomenological method" became recognized as the attitude and the method which was developed along the lines of those maxims and, above all, the method that employed that intellectual intuition, categorial or "direct," later also called eidetic, with the purpose of discovering essences and other ideal beings, objective necessities, synthetic *a priori* laws – not dependent, of course, upon the forms of subjectivity; in short, eidetic laws. In later works, all this approach is expounded under the title of "eidetic reduction" and constitutes, together with the "Transcendental Reduction," one of the two main methods of Transcendental Phenomenology.

With the departure of the disciples who were convinced of the sufficiency and independence of intellectual (or "eidetic") intuition, the separation of the senses of Phenomenology began. As time went on, many Phenomenologies and concepts of Phenomenology began to appear, all of them derived in more or less measure from the Phenomenology practised in *Logical Investigations*. "Realist Phenomenology," an important school even in our day, belongs to this plethora. Gaos, who uses for this "part" of Phenomenology the names of "eidetics" or "eidetic phenomenology," tells us that it became evident that "the substantial innovation and usable and lasting contribution of Phenomenology was in 'the Phenomenological method' – understood in the sense of the 'eidetic Phenomenology' detached from the 'rest' of Phenomenology. . . ."[14]

But how did Gaos himself understand Phenomenology or its method? Here there are a lot of questions to be formulated. In what measure is the phenomenological method of Gaos "eidetic"? In what measure does

Gaos accept essences, or, in general, ideal being, and in what measure are his analyses directed at essences or rather at ideas?

With that question, we have to ask this other: given that in several places Gaos attributes to Husserl the ultimate paternity of the phenomenological method, in what measure is Gaos still faithful to Husserl in matters of phenomenological method?

And finally, the question must be put that is the most important one from a philosophical point of view: is Gaos "a realist phenomenologist?", "an idealist phenomenologist?", "perhaps a neutral phenomenologist?"

The "rules" of the phenomenological method which Gaos applies and teaches seem to offer a natural point of departure. Gaos alludes to the rules in various places of his work, but where they are best explained is in his course *Del hombre* (On Man).

The first lesson of the course begins:

> The first major rule of the phenomenological method is to start with the *verbal expressions* that designate the phenomenon to which we intend to apply Phenomenology, in order to select an individual or specific *case* or *sample*, perceived or remembered or imagined, of the phenomenon designated by them, a case that should be so *without question*, and to proceed to the *descriptive*, conceptual and terminological *analysis* of the case or sample under the point of view of what is, with intellectual evidence, *essential* to it or to the phenomenon.[15]

The rule has two parts: the second, which refers to the selection of the individual or specific case, is intimately bound to the problem of the object in the phenomenological method, a problem which I will not tackle now. I only want to say at this point that this part of the rule may certainly be found (although in another formulation and not treated as a rule) among Husserl's prescriptions. On the other hand, the first part of the rule – "to start with the *verbal expressions* that designate the phenomenon . . ." – is not (and I believe it impossible that it could be) in Husserl. Gaos could have taken it from another phenomenologist, or it could have been a personal contribution to the method. It is also possible, though it is not easy to believe, that Gaos happened to believed that this prescription was implicit in Husserl, perhaps because he starts his phenomenological analyses, in *Logical Investigations*, by studying precisely the phenomena of expression. To start with the verbal expression that designates the object seems to have no other function in the phenomenological method, and above all in the phenomenological practice of Gaos, than that of achieving the identification or concrete determination, not so much of the object, but of the case or the sample of the object. I think that this point is not really very relevant, and has no great significance *in the lessons of Gaos*. But from another perspective it could be said that it is not easy to see why Phenomenology must limit itself to the analysis and description of objects or phenomena already named. If it could not by itself discover and name some objects, maybe not even its very name would have had the success that it had.

Perhaps a partial acceptance of this point may be found in Gaos himself when he claims that the method prescribes also giving technical terms to everything that is distinguished in the phenomenon.

> It is another rule [the third one] of the phenomenological method that of distinguishing also *terminologically* the phenomena which are really distinct, as an "expression" of having distinguished them *conceptually* due to their real distinctness.[16]

This third rule is also not found in Husserl, though in his idea of Phenomenology the function of conceptual clarification plays a role of first importance.

The second of the rules in *Del hombre* is stipulated as follows:

> . . . the second major rule of the phenomenological method: to dedicate oneself to the search for cases, *not* that *confirm* the results of the analysis of the first case; but that *invalidate* them, if cases such as these are found. For the search for confirming cases will never warrant the results against the possibility of invalidating cases; while the search for these cases, the invalidating ones, will guarantee the results as long as they are not found – a guarantee that cannot be absolute unless all *possible* cases have been found; but if they are found, they will force us to rectify, correct, perfect the results, and these will consist, in short, in an appropriate or inappropriate generalization of the concept of the phenomenon.[17]

It is not only that this second major rule is not in Husserl; under a certain interpretation, it even runs against the Husserlian doctrine of the eidetic reduction, and, moreover, it also runs against other phenomenological non-Husserlian theories of eidetic or intellectual intuition. To accept this second rule involves the risk of arriving at the confounding of this intuition, which for Husserl and many others allows a strict universality, with an inductive generalization. The point is delicate and asks for an analysis I cannot make here. But it also demands, of course, a better understanding of the ideas of Gaos about essences or ideal being and knowledge of it. The last of the rules has an intimate relation with this theme. Gaos expounds it thus:

> Phenomenology may be done upon the basis of cases or samples, which are *not* only *perceived*, but purely *imagined* – and even *imaginary*.

This rule (though not stipulated as a methodical rule) is the only one which not only is in Husserl, but was also developed in such a way as to give support in a certain sense to the very doctrine of eidetic reduction. We refer to the method of variation in fantasy. We cannot be sure that Gaos did not know it (though he gives credit to Husserl for the rule).[18] In any case, he did not formulate it at any place in his work, as far as we know. But what can be said, without doubt, is that he did not extract from it any important conclusion nor did it serve to mitigate the risk of inductivism in his own phenomenological method. If a sample plays a role as a mere variant in a potentially infinite chain of imaginary variation, in which the same essence remains always as the invariant, what is the need for searching afterwards for another sample which would come to invalidate the description that has been made? I think that a more detailed analysis would show that the Phenomenology of Gaos confuses, at least under a certain interpretation, the description of the sample or of the individual case with the description of the essence considered upon the basis of the intuition of the case.

Here we should have to take into account the long controversy Gaos waged against ideal being or against the Husserlian thesis of the existence of ideal being. However, we will not enter now into this polemic, which also has very intricate relations with themes that Gaos develops in his courses *De la filosofía* and *Del hombre*. But it is worthwhile to note here, for the time being, that, despite the persistent opposition of Gaos to accepting the ideal existence of essences beyond the concepts which express them (concepts which could be nothing more than a species of psychic products), some characterizations of his own (or another's) phenomenological investigations, in which he does not seem to be making historical concessions to Husserl, refer to the essences as objects worthy of knowledge, investigation, description, and analysis. Obviously, this does not bring to an end the problem of the "ontological status" of essences or of ideal being, so to speak. But it can contribute to orienting his treatment.

To take one example: "I once have tried to show how the essence of Philosophy and the essence of pride coincide phenomenologically feature by feature."[19] Another example: "There was a moment, *lato sensu*, in my life, when it seemed to me that the philosophical deception originated from a certain incompatibility between the essence of Philosophy and the 'essence' of contemporary life: Philosophy of Philosophy required the determination of this last 'essence,' starting from a 'Phenomenology' of contemporary life."[20]

Let us consider, finally, a third example which is also worthwhile to examine for other reasons with more care. Gaos refers to the series of descriptions and characterizations which constitute the so-called "Philosophy of Mexican Man" or "Philosophy of Mexicanness," and he says that these

> have employed the method which is considered the method par excellence to describe or characterize and define essences in the Philosophy of our day: the method of eidetic Phenomenology. Not the phenomenological method in the rigorous sense of Husserl's Phenomenology: the method of the eidetic-transcendental science of essences of pure phenomena of pure consciousness; but the phenomenological method in the broader and looser sense which spread from the strict Husserlian orthodoxy, stemming from within the school of Husserl itself and beyond, where the purpose is purely and simply to describe any object, directed by a vague intuition of what is essential in it, making descriptions not guaranteed by anything susceptible of objective or intersubjective verification, but only conducted by some "methodical" order.[21]

Now, according to Gaos, "if in spite of the employment of the phenomenological method, even in its rigorous sense, the greater part of the descriptions and characterizations of Mexicanness are more empirical than eidetic, it is because at this point the aporias of Phenomenology itself have worked even in the rigorous sense of Husserl."[22]

Which are these aporias? The first one is expounded as follows:

> The purpose is to define or to describe the essence of Mexican Man. To that end one must be looking at this essence. One cannot look at it but in the Mexicans themselves, at least in one of them. In order to see it in them one must know that they are Mexicans, insofar as they are different from the rest of human beings, not to say from the rest of beings, at all. And such a knowledge implies what is a Mexican – or that one is looking at the essence of Mexican Man. Vicious circle. Very early has eidetic Phenomenology in general come to find it.[23]

The second aporia is the one of the criterion of the essential. It is the question "of the essentiality of the whiteness of the plumage of the swan, invalidated by the swan of black plumage from Australia,"[24] but applied now to the case of the Mexican: is his clothing essential to the image of the "coiled" Mexican Man? "Is there – and this is a substantive question – a distinguishing criterion for those notes which make up essences, susceptible of universal objective-intersubjective validity or in any orders of facts and essences for any subjects?"[25]

I am not going to give a general response to these aporias, nor even to deny that they may involve in some sense authentic problems (perhaps not authentic aporias) with eidetic Phenomenology or with some particular eidetic Phenomenology. What I want to emphasize is that the reason why such descriptions and characterizations of Mexicanness or of Mexican Man may be or have been more *empirical* than *eidetic* does not in any way involve the possible aporias of Phenomenology. The explanation is, to my view, very simple: there is no essential relation (in the most strict sense) whatsoever between the determination or property of *being Mexican* and the other possible determinations or properties of the human individuals who possess or may possess that property, except perhaps the "property" of being able to be considered as persons. That property of the Mexican which I call *being Mexican* is not defined but by the rules which allow us to apply the name of Mexican to a person, and I suppose that in the context of the philosophy of the Mexican those rules may be found in the juridical stipulations that give to a person Mexican nationality. Trying to link this property with other characteristics (essential or not essential) of Mexican people (their clothing, their customs, their language, their psychological typology or pathology, etc.) is an enterprise as senseless as searching for the essence of the walls of a certain country in their color or their height or the material out of which they are made, or the way in which the sun shines on them at dusk. And this is true even if the executive authority of this country ordered that all the walls of the country be painted in the same color (and in the same shade of that color), or that all walls be made of the same material or be of the same height, or in the same disposition and situation in relation to the sun. Not even the fact that all swans in the world, supposing they were already known, were white, would make the whiteness of the swan be an essential property.

Here, of course, there are many distinctions and clarifications to be made. The complete answer has to also consider the problem of the "distinctive criterion of those notes which make up the essences," which of course not all phenomenologists, followers of Husserl or not, would solve in the same way. We believe that Husserl would appeal not only to his concept of founding (*Fundierung*), as Gaos thinks in this and other contexts,[26] but also, beyond it, to the theory of the stratification of the senses within the transcendental constitution of an object. In the field of realist Phenomenology, nearer to Gaos in some respects, Dietrich von Hildebrand, for instance, would appeal instead to his distinction between accidental essences, essences with a sense, and necessary essences.

But, in the end, essences or not, they are such of something, of an object. And we still do not possess clarity about which is, according to Gaos, the possible object (or objects) of Phenomenology.

> The method of the Anthropology of this course is in the first place a *phenomenological* method, for it deals with "phenomena" and not with "metaphysical objects," in the precise sense that the course will give to these expressions, as well as with phenomena which can be analyzed descriptively, or be analytically described, and this should be done before any other methodical operation possible and necessary for them is done.[27]

We now have to account for the concepts of "phenomenon" and "metaphysical object." Only if we gain clarity about them will we be able to solve the question of the object of the phenomenological method of Gaos. We need to know, for instance, if every object is given in the same degree and the same sense, and if the same kind of Phenomenology may be done with respect to all species of phenomena. And thus we would have to think over the question of the nature of the sample, the selection of which, according to Gaos, the first rule prescribes: whether it can truly be individual, that is, empirical, or specific; and what kind of object is the verbal expression "this room," which is the sample the identification of which begins the Phenomenology of verbal expression in *De la filosofía*, and what kind of object is the mimic expression "laughter," which begins the Phenomenology of mimic expression in *Del hombre*? What does "empirical" mean for Gaos? We have to think all this over also in connection with other cases. In what sort of experience are Mexicans given to us as Mexicans?[28] And of course we must not forget the relationship with phenomena as philosophy itself. And what of the phenomenon of negation?

Finally, what is the historic-philosophical meaning of the admission of some new phenomena which are not "phenomenic objects," but the phenomena of their "objectivity or objectivation," and that are conceived as the "phenomena par excellence?" What new concepts of Phenomenology are here announced? In *De la filosofía*, Gaos says,

> This Phenomenology [of presence] is the proper starting point of *Phenomenology par excellence*. The phenomena par excellence are *not* the *phenomenic objects* (beings), but the phenomena of the objectivity or objectivation of those objects, the phenomena of their *presence* (existence), their *appearing, disappearing, reappearing, disappearing forever*. Therefore, these phenomena make up what most properly may and should be called the "*phenomenicity*" of phenomena. The *Phenomenology of the phenomenicity*, or *of existence, and not of the "essences,"* is, then, Phenomenology par excellence.[29]

The study of all these concepts leads us to one of the central themes in

Gaos' conception of philosophy, as well as of the philosophy of philosophy. But at that threshold, before a swarm of questions, we must, for the time being, stop.

Universidad Nacional Autónoma de México

NOTES

* A Spanish version of this paper, with some changes and a different title, was read at the Simposio Internacional *Filosofía y Educación*, organized in homage to Fernando Salmerón on his 70th birthday, at Universidad Nacional Autónoma de México, November 13–17, 1995. This paper benefits from my first approach to Gaos' Phenomenology, in Zirión 1995.

[1] Gaos translated *Logische Untersuchungen* (see Husserl 1929), *Cartesianische Meditationen* (see Husserl 1942), and the First Book of *Ideen zu einer reinen Phänomenologie und phänomenologische Philosophie* (see Husserl 1949).

[2] See for instance Xirau 1969, p. 157; Xirau 1980, p. 305; Xirau 1994, p. 40; Yamuni 1989, pp. 10–11, 36, 45; Castro 1993, pp. 18, 84; Guy 1970, pp. 174, 185; Hernández Luna 1969, p. 74; Salmerón 1969, p. 105; Salmerón 1988, p. 255, and Salmerón 1994, p. 42.

[3] Gaos developed, according to his own testimony, the phenomenology of caress, of verbal expression, of the objects expressed by verbal expressions, of reason, of categories (existence, negation, the infinite), of presence or of phenomenicity, of human nature, of the history of philosophy, of traveling; according to the testimony of others, he also developed the phenomenology of "our life" (and, within it, the phenomenology of the phenomena of publicity and technocracy), of human time, of love; he also alluded to the metaphysical phenomenology of drunkenness in Heraclitus' Fragments, and to the phenomenology of the naive attitude and the natural view of the world in Heidegger, and, finally, he discussed the possibility of developing the phenomenology of affection and tenderness, of pride, of personal confession, of philosophy and of the philosopher.

[4] "Existencialismo y esencialismo," in Gaos 1987, p. 188.

[5] Husserl 1994.

[6] Husserl 1976, p. 508.

[7] Gaos 1961, p. 148.

[8] Gaos 1961, p. 145, note 4.

[9] Gaos 1961, p. 147, note 7; Gaos 1959, pp. 19–20; Gaos 1987, p. 299; Gaos 1982b, p. 382. On the nobility of metaphysics, see Salmerón 1994, pp. 46–47.

[10] See, for instance, Gaos 1992, p. 39: "The method is phenomenological in the sense of the Phenomenology founded by Husserl, but not in the sense of the idealist philosophy into which Husserl transformed it. . . ."

[11] Gaos is well aware of the situation. See his "Introducción a las 'Meditaciones cartesianas' de Husserl" in Gaos 1987, pp. 286ff.

[12] *Der Sinn philosophischen Fragens und Erkennens*, p. 90, cited by Wenisch 1987, p. 16.

[13] Cited by Spiegelberg 1978, p. 745.

[14] Gaos 1987, pp. 291–292.
[15] Gaos 1991, p. 43. It is also the beginning of *De la filosofía* (On Philosophy): "The phenomenological method prescribes starting with the verbal expressions which designate the object which will be subject to Phenomenology and form an example or sample of that object" (Gaos 1982b, p. 11).
[16] Gaos 1992, p. 73. And Gaos 1982b, p. 11.
[17] Gaos 1992, p. 47. In the same sense, Gaos 1982b: "But once the selected sample is analyzed and synthesized in this way, the phenomenological method prescribes us to '*generalize*' – looking carefully and loyally for samples, in order not to revalidate, but to invalidate the results of the analysis and synthesis of the selected sample, because only the results that are not invalidated by such samples will be able to be generalized to all samples of the phenomenon to which we are applying Phenomenology, in our case to all *verbal expressions*" (Gaos 1982b, p. 16).
[18] "All Phenomenology must be carried out on the basis of empirical cases or samples and it may be done on the basis of only one of them and even an imaginary or a fictitious one." (Gaos y Larroyo 1940, p. 75.) And he adds immediately: "all these are explicit teachings of Husserl."
[19] Gaos 1987, pp. 119–120.
[20] Gaos 1982a, p. 126.
[21] Gaos 1980, p. 84.
[22] *Ibid.*
[23] Gaos 1980, p. 85.
[24] Gaos 1980, p. 87.
[25] Gaos 1980, p. 88.
[26] Gaos 1980, p. 88, and Gaos 1992, p. 47.
[27] Gaos 1992, p. 39.
[28] I mention this example only to illustrate the nature of the problem. I am aware that Gaos concludes his discussion on the philosophy of the Mexican man with a negative answer. "Conclusion: it is problematic, at the very least, that a philosophy of the essence of Mexicanness be possible, and even more so for a philosophy about the Mexican man" (Gaos 1980, p. 90).
[29] Gaos 1982b, pp. 231–232.

REFERENCES

Castro, Octavio, *Estancias y visiones de un transterrado* (Xalapa: Universidad Veracruzana, 1993).

Gaos, José, *Discurso de filosofía y otros trabajos sobre la materia*. Xalapa: Universidad Veracruzana (Biblioteca de la Facultad de Filosofía y Letras, 1959).

Gaos, José, "Notas sobre Husserl," in *Anuario de Filosofía*, Año I (Mexico City), pp. 143–150, 1961.

Gaos, José, *En torno a la filosofía mexicana* (Mexico City: Alianza Editorial Mexicana, 1980).

Gaos, José, *Confesiones profesionales. Aforística. Obras Completas*, Vol. XVII. Prólogo y selección de la Aforística inédita por Vera Yamuni Tabush (Mexico City: UNAM, 1982a) (Nueva Biblioteca Mexicana, 85).

Gaos, José, *De la filosofía (Curso de 1960). Obras Completas*, Vol. XII. Prólogo de Luis Villoro (Mexico City: UNAM, 1982b) (Nueva Biblioteca Mexicana, 84).

Gaos, José, *Filosofía de la filosofía e historia de la filosofía. Obras Completas*, Vol. VII. Prólogo de Raúl Cardiel Reyes (Mexico City: UNAM, 1987) (Nueva Biblioteca Mexicana, 88).

Gaos, José, *Del hombre (Curso de 1965). Obras Completas*, Vol. XIII. Prólogo de Fernando Salmerón (Mexico City: UNAM, 1992) (Nueva Biblioteca Mexicana, 109).

Gaos, José, y Francisco Larroyo, *Dos ideas de la filosofía (Pro y contra la filosofía de la filosofía)*. (Mexico City: La Casa de España en México, Fondo de Cultura Económica, 1940.)

Guy, Alain, "El tiempo en la filosofía de José Gaos," *Diánoia* **XVI**(16) (1970), 172–186 (Mexico City).

Hernández Luna, Juan, "En torno a un curso sobre el historicismo del maestro José Gaos," *Cuadernos Americanos* **CLXVI**(5) (Sept-Oct,1969), pp. 74–80.

Husserl, Edmund, *Investigaciones lógicas*, trans. Manuel García Morente and José Gaos (Madrid: Revista de Occidente, 1929).

Husserl, Edmund, *Meditaciones cartesianas* (Mexico City: La Casa de España en México, 1942 ("Colección de textos clásicos de filosofía" del Centro de Estudios Filosóficos de la UNAM). (Segunda edición, aumentada y revisada, Fondo de Cultura Económica, 1986.)

Husserl, Edmund, *Ideas relativas a una fenomenología pura y una filosofía fenomenológica* (Mexico City: Fondo de Cultura Económica, 1949). (Second edition, 1962; third edition, 1986.)

Husserl, Edmund, *Die Krisis der europäischen Wissenschaften und die transzendentale Phänomenologie. Eine Einleitung in die phänomenologische Philosophie. Husserliana*, Band VI. Herausgegeben von Walter Biemel, 2. Auflage (The Hague: Martinus Nijhoff, 1976).

Husserl, Edmund, "Husserl an Metzger, 4. IX. 1919," in E. Husserl, *Briefwechsel*, Band IV: "Die Freiburger Schüler." *Husserliana Dokumente*, Band III. In Verbindung mit Elisabeth Schuhmann herausgegeben von Karl Schuhmann (Dordrecht/Boston/London: Kluwer Academic Publishers, 1994).

Salmerón, Fernando, "José Gaos: su idea de la filosofía," *Cuadernos Americanos* **CLXVI**(5) (September-October, 1969), pp. 102–129.

Salmerón, Fernando, "Jornadas filosóficas. La primera autobiografía de José Gaos," in Fernando Salmerón, *Ensayos filosóficos (Antología)*. Lecturas Mexicanas, Segunda Serie **109** (Mexico City), pp. 239–259 (Sept 1988).

Salmerón, Fernando, "Los estudios cervantinos de José Gaos," in Fernando Salmerón, *Los estudios cervantinos de José Gaos, seguido de un ensayo inédito de José Gaos y un comentario final de Carlos Montemayor* (Mexico City: El Colegio Nacional, 1994), pp. 3–50.

Spiegelberg, Herbert, *The Phenomenological Movement. A Historical Introduction*. Second edition, fifth impression (Phaenomenologica, 6) (The Hague/Boston/London: Martinus Nijhoff, 1978).

Wenisch, Fritz, *La filosofía y su método*, trans. Miguel García-Baró (Mexico City: Fondo de Cultura Económica, 1987).

Xirau, Ramón, "José Gaos o del valer la pena," *Cuadernos Americanos* **CLXVI**(5) (September-October, 1969), pp. 157–164.

Xirau, Ramón, "Los filósofos españoles 'transterrados,'" in *Estudios de historia de la filosofía en México* (Mexico City: UNAM, 1980), pp. 295–318.
Xirau, Ramón, "De Descartes a Marx: la historia de la filosofía en la obra de José Gaos," *Universidad de México* **XLIX**(521) (June, 1944), pp. 40–44.
Yamuni, Vera, *José Gaos, su filosofía*. Jornadas de la Facultad de Filosofía y Letras, Vol. 4 (Mexico City: UNAM. 1989).
Zirión, Q. Antonio, "The Marginal Notes of José Gaos in *Ideas I*," *Husserl Studies* **12** (1995), pp. 19–53.

ALBERTO ROSALES

LA FENOMENOLOGÍA EN LATINOAMÉRICA

Hablar de la Fenomenología en Latinoamérica implica naturalmente exponer su desarrollo a lo largo del tiempo y a través de la vasta geografía de nuestro continente. A diferencia de lo que puede ocurrir en otros lugares del planeta, esa tarea presenta entre nosotros grandes dificultades. De ello pudimos darnos cuenta hace ya un par de décadas, al colaborar con Herbert Spiegelberg en la nueva edición de su libro "*The Phenomenological Movement*" (1981).[1]

Carecemos en nuestros paises de visiones de conjunto acerca de la historia de la Fenomenología. Existen, ciertamente, trabajos que informan sobre su desarrollo en algunos paises y en algunos autores, pero ellos suelen ser del género encomiástico, que es bien conocido en nuestra vida cultural. O ellos no pasan de ser una mera mención de nombres, fechas y obras. Echamos de menos consideracioues criticas, que pongan de relieve lo bueno y lo malo en las producciones, o que interpreten el sentido de los desarrollos históricos. A todo ello se añade el relativo aislamiento en que se encuentran hasta ahora los movimientos fenomenológicos de los diversos países.

Nuestra exposición, limitada además en el tiempo, no puede pretender superar esas deficiencias. Al esbozar, al final, lo que podria ser el futuro del trabajo fenomenológico en Latinoamérica, hemos de proponer una posible perspectiva para interpretar la historia de ese movimiento.

Hasta donde es posible hacerse una visión de conjunto del desarrollo de la Fenomenología en Latinoamérica, en base de las investigaciones actuales, dispersas e incompletas, puede señalarse tal vez un período preparatorio, que abarca los años veinte y treinta de nuestro siglo, en México y Argentina. Luego viene una etapa de gran brillantez y vigor, sobre todo en México, desde fines de los años treinta hasta mediados de los años sesenta, cuando la Fenomenología pierde terreno frente a la filosofía analítica y al marxismo. En la Argentina la evolución ha tenido menos altibajos, pues si bien las ideas fenomenológicas no tienen tal predominio después de esa fecha, aún hoy en día se mantiene en ella un grupo importante de investigadores. En otros países, como

A-T. Tymieniecka (ed.), Analecta Husserliana, Vol. LIV, 345–355.

Colombia, Chile, Perú, Venezuela y Uruguay, el cultivo de la Fenomenología comienza hacia los años cuarenta y se prolonga durante unas tres décadas, hasta ser desplazado por el marxismo y la filosofía analítica, pero con grandes diferencias de un país a otro.

Ante todo hemos de dirigir una mirada al desarrollo de la Fenomenología en México, no sólo en señal de gratitud y homenaje hacia la tierra que nos acoje hoy tan generosamente en este congreso, sino también debido a la importancia de su movimiento fenomenológico. Como es bien sabido, la introducción de la Fenomenología en México fue preparada por la confrontación con el positivismo, iniciada por la llamada "Generación del Ateneo" (1908), cuyos pensadores más importantes fueron José Vasconcelos (1882–1959) y Antonio Caso (1883–1956), quienes se inspiraron inicialmente en Boutroux y Bergson. Bajo el signo de ese enfrentamiento con el positivismo se encuentra la recepción y el estudio de la Fenomenología por parte de Caso, quien fue el primero en divulgar en México, en sus obras y cursos universitarios, el pensamiento de Husserl, Scheler, Hartmann y Heidegger. Su libro *La Filosofía de Husserl* (1934) es la primera obra escrita en español sobre ese tema. La crítica de Husserl a la teoría económica del conocimiento es acogida por Caso en un capítulo de su libro más importante: *La existencia como economía, como desinterés y como caridad* (2 ed. 1941). Un fruto tardío de esa misma polémica contra el positivismo es su libro *Positivismo, Neopositivismo y Fenomenología* (1941). Por otra parte, Caso expone y discute la filosofía de los valores de Scheler y Hartmann en su obra *El concepto de la Historia y la Filosofía de los Valores*.

Los discípulos de Caso, Samuel Ramos (1897–1959), Adalberto García de Mendoza (1901–1956), José Romano Muñoz (1893–1967) y Eduardo García Maynez (1908–1993) prosiguieron, en algunos casos con contribuciones originales, la dirección de su maestro. Entre ellos debemos destacar las aportaciones de García Maynez y de Ramos al desarrollo de la Fenomenología en México. Si bien Ramos acoge en sus escritos y cursos universitarios las ideas de Husserl, Dilthey, Scheler, Hartmann y Heidegger, su pensamiento está influído sobre todo por el historicismo de Ortega y Gasset. Sobre esa base él dió un original giro antropológico hacia la circunstancia mexicana, con su libro *Perfil del hombre y la cultura en México*, que influyó decisivamente en el desarrollo de la Fenomenología en la siguiente generación, y de forma indirecta en la filosofía latinoamericana de tendencia nacionalista en general. Por su parte, García Maynez, quien fue discípulo directo de Hartmann a comienzos de los años treinta, revela en su vasta producción la influencia de su maestro, tanto por su adhesión a la teoría de los valores como entes objetivos, como por su esfuerzo en construir, en diálogo con la lógica matemática, una fundamentación del derecho a través de un sistema lógico-ontológico de principios. García Maynez expone la teoría de los valores de Hartmann en su *Ética* (1943) y en el libro *El problema de la Objetividad de los Valores* (1969).

El trabajo de Caso y de sus discípulos preparó el terreno para la segunda etapa del desarrollo de la Fenomenología en México, que se inicia a finales de los años treinta, con la llegada de un grupo de profesores españoles. De gran importancia para esa etapa han sido José Gaos (1900–1969), Joaquín Xirau (1895–1946), Eduardo Nicol (1907–1990), L. Recasens-Siches (1903–1977) y J. D. García Bacca (1901–1992), quien tras breve estancia en México y Ecuador se residenció en Venezuela.

Gaos ejerció desde entonces un papel decisivo en la marcha de la filosofía mexicana, por medio de sus publicaciones, de sus traducciones de obras capitales de la filosofía de tendencia fenomenológica y sobre todo a través de sus cursos universitarios. Su tesis doctoral en la Universidad de Madrid, titulada *Introducción a la Fenomenología, seguida de la Crítica del Psicologismo en Husserl* fue publicada en México en 1962. Sus traducciones han sido de decisiva importancia para el desarrollo del movimiento fenomenológico en el mundo de lengua castellana. Ya en España, a fines de los años veinte, Gaos había traducido junto con García Morente las *Investigaciones Lógicas* de Husserl. En México aparecieron luego las *Meditaciones Cartesianas* (I–IV, 1939) y el primer volúmen de las *Ideas* (1943), asi como *El Ser y el Tiempo* de Heidegger (1951) y los cuatro volúmenes de la *Ontología* de Hartmann (1954 ss.). Tanto a través de sus cursos universitarios como de su traducción de *El Ser y el Tiempo* Gaos contribuyó al gran influjo del pensamiento de Heidegger en la filosofía mexicana y latinoamericana durante las dos décadas siguientes. A pesar de ello, la tendencia personal de Gaos no estaba dirigida hacia la Fenomenología propiamente dicha, sino más bien hacia el historicismo de origen diltheyano y orteguiano, e incluso hacia un cierto escepticismo. Su obra sistemática titulada *De la Filosofía* (1962) es una "filosofía de la filosofía" que apunta en última instancia a aclarar los intentos fallidos de la razón humana al hacer metafísica. Esa tendencia tal vez contribuya a explicar la orientación antropológico-historicista de sus discípulos y su pronto abandono de la corriente fenomenológíca, a mediados de los años sesenta.

Entre los restantes profesores españoles exilados en México, J. Xirau fue autor de un libro sobre *La filosofía de Husserl* (1940) y E. Nicol, influído sobre todo por Heidegger y el existencialismo francés, publicó las obras *Historicismo y Existencialismo* (1950) *y Metafísica de la Expresión* (1957). Por otra parte, la obra del filósofo del derecho Recasens-Siches estuvo inspirada por Ortega y los pensadores fenomenológicos.

Bajo la guía de tales maestros, y sobre todo de Gaos, se formó un grupo de pensadores mexicanos, que asimiló principalmente la filosofía del primer Heidegger, el historicismo de Dilthey y de Ortega y el existencialismo francés. La tendencia antropológica de esas fuentes de inspiración sirvió de base a los pensadores del Grupo "Hyperión" para proseguir el camino indicado por Samuel Ramos hacia el esclarecimiento filosófico del hombre mexicano y de su cultura. Un punto culminante de esa tarea han sido las investigaciones de Leopoldo Zea sobre la historia de las ideas en México y en la América Latina en general. De su vastísima obra mencionamos como ejemplos sus libros *Ensayos sobre Filosofía de la Historia*, *Conciencia y posibilidad del mexicano*, así como *El pensamiento latinoamericano*. Emilio Uranga radicalizó la tendencia del Grupo Hyperión en su libro *Análisis del Ser del mexicano*, que pretende nada menos que una ontología del mexicano inspirada en Heidegger y Sartre.

A fines de los años cincuenta hasta mediados de los sesenta tuvo lugar una suerte de retorno provisional a Husserl, en los trabajos de los últimos discípulos de Gaos: L. Villoro, F. Salmerón y. A. Rossí. Inmediatamente después, todos ellos adoptan posiciones de crítica a Husserl, orientadas hacia la filosofía analítica.

En Argentina han sido sobre todo Francisco Romero (1891–1962) y Carlos Astrada (1894–1970) quienes, desde los años treinta, contribuyeron durante varias décadas, con su enseñanza universitaria y sus escritos, a divulgar las ideas fenomenológicas, a la vez que partieron de ellas para desarrollar un pensamiento propio. El libro capital de Romero es su *Teoría del Hombre* (1952). A partir de la idea husserliana de la intencionalidad de la conciencia, Romero concibe al hombre, con Scheler y Hartmann, como un ente que se trasciende en el conocimiento y en la acción, para objetivarse a sí mismo por esa vía e identificarse cada vez más con valores espirituales universales. Su *Lógica*, escrita en colaboración con Eugenio Puchiarelli, se adhiere expresamente a la orientación de Husserl y Pfänder. Por el contrario, Romero no simpatizó nunca con la derivación existencial de la Fenomenología.

Carlos Astrada, discípulo directo de Husserl, Scheler, Hartmann y Heidegger, evolucionó durante su vida a partir del pensamiento de sus maestros, sobre todo de Heidegger, a través de Nietzsche y Hegel, hasta recalar en su última época en el marxismo. A despecho de ese cambiante recorrido, su pensamiento tiene siempre como tema al hombre tomado en su radical finitud, una manera de ver que se inspira sobre todo en el primer Heidegger. De acuerdo con esa orientación, Astrada acoge y a la vez critica el pensamiento de Husserl en su libro *Idealismo fenomenológico y Metafísica existencial* (1936), así como en *Fenomenología y Praxis* (1967). Las ideas de Scheler y Hartmann sobre los valores son expuestas y criticadas en *La Ética formal y los valores* (1938) y en *El juego metafísico* (1942).

En buena parte gracias a la labor de esos maestros el pensamiento fenomenológico tuvo en la Argentina una vasta difusión. De ello dan testimonio, en primer lugar, libros y artículos que versan total o parcialmente sobre Husserl. A modo de ejemplo mencionamos entre sus autores a Eugenio Puchiarelli, Jacobo Kogan, Andrés Raggio y Antonio Aguirre. Este último es autor de una investigación en alemán reconocida internacionalmente sobre la reducción trascendental de Husserl: *Genetische Phänomenologie und Reduktion.* Puchiarelli, gran conocedor de la filosofía contemporánea, mantuvo viviente, con sus escritos y su enseñanza, durante las últimas décadas, el interés por la Fenomenología en la Argentina. Por otra parte, la obra de Carlos Cossío *La Teoría egológica del Derecho y el concepto juridico de la Libertad* atestigua la influencia de Husserl en el campo de la filosofía del derecho.

Como hemos dicho, el pensamiento de Heidegger, junto al de Scheler y Hartmannn, fue expuesto en Argentina desde los años treinta. Pero sólo después de la segunda guerra mundial fue ampliamente difundido, sea directamente por medio de la traducción de sus obras menores, sea indirectamente, como en toda Latinoamérica, a través de la Fenomenología francesa de Sartre y Merleau Ponty. Entre los numerosos autores argentinos que han hecho aportes al estudio de todos esos pensadores debemos nombrar a E. Puchiarelli, Adolfo Vásquez, Emilio Estiú, Vicente Fatone y Luis Juan Guerrero. A la generación siguiente pertenecen A. P. Carpio, Ricardo Maliandi, Mario A. Presas y Orlando

Pugliese. Maliandi es autor de una obra en alemán sobre Hartmann, titulada *Wertobjektivität und Realitätserfahrung*. Por su parte, Orlando Pugliese escribió Vermittlung und Kehre sobre el pensamiento de Heidegger.

Meintras que en otros países el cultivo del pensamiento fenomenológico se ha disminuído después de los años sesenta, él ha resurgido en la Argentina en las últimas décadas en torno a la Sociedad Argentina de Fenomenología y Hermeneútica y al Centro de Estudios Filosofícos de la Academia de Ciencias de Buenos Aires. Sus propulsores han sido Eugenio Pucciarelli y Roberto Walton. Un ejemplo de ese resurgimiento som los ensayos de Walton en torno a Husserl y Heidegger, reunidos recientemente en dos libros: *El fenómeno y sus cnfiguraciones y Husserl, mundo conciencia y temporalidad.*

La obra más representativa del pensamiento fenomenológico en Chile ha escrito numerosos ensayos sobre pensadores fenomenológicos, especialmente sobre Heidegger, Su libro *Filosofía sin Supuestos* recorre el camino que va de Husserl a Heidegger, al meditar sobre el lema "Zu den Sachen selbst," como un imperativo que exige la aclaración de todos los supuestos. Guillermo Hoyos Vásquez es autor de una obra sobre la teleología de la historia y de la intencionalidad en Husserl, que ha aparecido en la colección Phaenomenologica bajo el titulo de *Intentionalität und Verantwortung*.

En Perú, la divulgación del pensamiento fenomenológico fue iniciada en la década de los treinta, por obra de Carlos Cueto Fernandini y Felipe Alarco. En la generación siguiente, descuellan Alberto Wagner de Reyna y Francisco Miró Quesada. El primero de ellos, discípulo directo de Heidegger, publicó ya en 1938 una exposición titulada *La ontología fundamental de Heidegger*, que muchos leímos en aquella epoca, y a la cual siguieron otros trabajos sobre Husserl y Heidegger. Miró Quesada estudió en sus inicios la evolución de la Fenomenología en su libro *Sentido del movimiento fenomenológico* (1941). José Russo Delgado contribuyó al estudio y divulgación del pensamiento fenomenológico en sus cursos universitarios y con su obra sobre Heidegger, *El hombre y la pregunta por el Ser* (1963). Influencia fenomenológica puede encontrarse también en Walter Peñaloza, Augusto Salazar Bondy, César Góngora Perea y Victor Li Carrillo. Este último, discípulo directo de Heidegger y Fink, escribió sobre *Platón, Hermógenes y el lenguaje*, en la perspectiva de sus maestros.

La figura más destacada del pensamiento fenomenológico en el

Uruguay fue Juan Llambías de Acevedo. Su filosofía del derecho se inspiró en la teoría de los valores de Scheler, en la gnoseología de Hartmann y, a través de ellos, en la eidética de Husserl. Su libro más importante fue la *Eidética y Aporética del Derecho* (1940).

En Venezuela los estudios universitarios de filosofía renacieron, a fines de la década de los cuarenta, bajo la dirección de los profesores españoles Domingo Casanovas, Eugenio Imaz, J. D. García Bacca y Manuel Granell. Si bien todos ellos divulgaron, en una u otra dirección, la filosofía contemporánea, fue sobre todo García Bacca, a través de sus cursos universitarios y de libros como *Nueve grandes Filósofos contemporáneos y sus temas*, *Siete modelos del filosofar* y *Antropología filosófica contemporánea*, quien hizo conocer la filosofía fenomenológica. Sin embargo, el propio pensamiento de García Bacca, quien intentaba una extraña síntesis de la escolástica, la filosofía existencial, la lógica matemática y la ciencia, parece haber tenido escasa repercusión dentro de Venezuela o fuera de ella. Su *Metafísica*, aparecida en México, es un buen ejemplo de sus intenciones filosóficas. Por otra parte, Manuel Granell, quien desarrolló su pensamiento a partir de Ortega y en diálogo crítico con Heidegger, escribió un análisis del hombre histórico en su libro *La vecindad humana*. Entre sus discípulos, Ernesto Mayz Vallenilla, Federico Ríu y quien les habla, estudiamos además en universidades alemanas, y cultivamos, cada uno a su manera, las ideas fenomenológicas. Mayz Vallenilla, quien escribió inicialmente dos libros sobre el problema del conocimiento, en Husserl y Heidegger respectivamente, ha tratado luego de forjar una filosofía propia, que gira en torno al sujeto humano como creador de la técnica. En sus comienzos Ríu fue intérprete de los pensadores fenomenológicos, especialmente de Heidegger. Tanto él como Garcia Bacca giraron luego, a mediados de los años sesenta, hacia el marxismo. En lo que mi respecta, después de mi interpretación sobre la diferencia ontológica en Heidegger, que apareció en la colección Phanomenológica bajo el título de *Transzendenz und Differenz* (1970), he tratado de apropiarme criticamente de las teorías fenomenólogicas y de aproximarme por esa via a soluciones propias.

Hemos esbozado en líneas generales las etapas de la Fenomenología en Latinoamérica, su pasado y lo que podría llamarse su situación presente. ¿Qué futuro podemos esperar de ella entre nosotros? ¿Cuáles son sus posibilidades? Ciertamente, el contenido concreto del futuro en la historia humana nunca puede anticiparse en su mayor parte. Lo nuevo puede dar un giro imprevisible a la marcha de las cosas. Sin embargo, como apunta el mismo Husserl, la experiencia del pasado y del presente nos permite anticipar posibilidades que constituyen un marco general determinado, dentro del cual puede presentarse lo relativamente nuevo y desconocido, hasta ahora indeterminado. Para concluir, quiero señalar cuatro grupos de posibilidades, arraigadas en nuestro pasado, que podrían ser el futuro de la Fenomenología en Latinoamérica.

1. Ciertamente, el cultivo y estudio de la Fenomenología podría continuar existiendo, como suelen existir la mayor parte de las corrientes filosóficas que han echado raíces entre nosotros. Es perfectamente posible seguir conservando la tradición fenomenológica a través del estudio de los textos originales y de su exegesis en tesis universitarias, en libros y artículos de revistas. Esa labor conservadora podría alcanzar incluso altos niveles de erudición y sería deseable que así fuera, pues todo ello tiene una función preparatoria indispensable para empuñar y realizar posibilidades más altas.

2. En efecto, sería insensato creer que el número y la frecuencia de las publicaciones filosóficas que aparecen en un país, el número de sus cátedras de filosofía, la ronda de sus congresos, simposios y discusiones, constituyen ya una prueba de la existencia y vigor de una vida filosófica. Buena parte de las informaciones que hemos presentado antes sobre el movimiento fenomenológico en Latinoamérica son de ese tipo. Bien sabemos que todo ese ajetreo cultural puede constituir una grandiosa apariencia, que encubre más bien una radical probreza de ideas y decisiones filosóficas. Auténtica vida filosófica existe sólo allí, donde algunos pensadores, quizas algunos pocos y tal vez sólo uno de ellos, es capaz de apropiarse de su tradición y avizorar nuevos proyectos, a partir del diálogo entre los problemas que esa tradición trae consigo y los frenómenos a los cuales se enfrenta el pensar. La Fenomenología podría tener esa posibilidad en Latinoamérica, si aquellos que la estudian, en vez de limitarse a interpretar y comentar los grandes autores, buscaran apropiársela criticamente y, siguiendo el lema husserliano de ir "a las cosas mismas," tratasen de repensar sus respuestas, aprender de sus errores y avanzar más allá de sus flaquezas.

3. La ausencia de una apropiación crítica de las grandes concepciones filosóficas europeas está enraizada entre nosotros en la falta casi general de la crítica. Se suele entender a la evaluación crítica como una negación malintencionada de las obras de los colegas y se pasa por alto que ella, según lo sugiere la propia palabra "crítica," requiere no sólo señalar lo malo, sino también reconocer lo bueno en una obra.

Por ello es hasta ahora sumamente arduo, adquirir una visión fundada de la historia de la Fenomenología en nuestros países. En la filosofía latinoamericana falta, en general, una concepción propiamente filosófica de su propia historia. Cuando hablabamos al comienzo sobre las dificultades que se topa todo aquel que quiera tratar esa historia, hemos anunciado el esbozo de una posible perspectiva para esa interpretación. Antes de concluir quisieramos detenernos un momento para considerar esa posibilidad.

Creo que se nos aceptará sin mayores objeciones que aparte de la historia cultural de la filosofía, debe haber también una interpretación propiamente filosófica del acaecer mismo de nuestra disciplina. La primera es una subespecie de la ciencia histórica y procede de acuerdo con la metodología de las ciencias humanas empíricas. La filosofía, por su parte, tiene entre otras tareas una aclaración de si misma y por ello también de su propia historia. Tal interpretación filosófica de esa historia sólo es posible en cada caso dentro de un proyecto filosófico determinado y está expuesta, por ello, a todos los riesgos de la filosofía misma. Una interpretación semejante de la historia de la filosofía puede encontrarse, por ejemplo, en autores tan diversos como Aristóteles y Husserl mismo.

Una interpretación filosófica de los hechos de la historia de la Fenomenología en Latinoamérica presupondría una perspectiva en la cual esos hechos puedan ser vistos e interpretados. Para sugerir al menos de qué tipo de perspectiva se trata, podemos partir de un acontecimiento característico de la recepción de la Fenomenología en Latinoamérica: la lucha entre positivismo y antipositivismo. ¿Cómo podría explicarse ese conflicto? ¿Y cómo puede explicarse la recepción de la Fenomenología en nuestros países? Con frecuencia se cree que tales hechos pueden explicarse recurriendo a las influencias que un pensador ha recibido de otros. Pero recibir una influencia o acoger una nueva idea, como la Fenomenología, es algo que presupone ya una manera de pensar. El positivismo pudo penetrar y echar raíces en Latinoamérica, en el siglo 19, y retornar luego en décadas recientes, porque en ella se había formado, desde el siglo 18 una tradición moderna, en especial, del pensamiento ilustrado. Y es porque nuestros países habían crecido, desde la conquista y colonización, en una tradición metafísica aristotélico-escolástica, que había y hay en ellos ojos y oídos para la filosofía en general, para el positivismo y luego para la Fenomenología, para la metafísica y para la anti-metafísica.

Tales hechos son bien conocidos y se suele mencionarlos como datos de los cuales no se puede sacar nada más. Y ello es cierto, mientras los historiadores de la cultura, que escriben también sobre la historia de la filosofía, no se pregunten qué implica tal tradición metafísica. ¿Pero qué es y qué buscaba el primer comienzo de la metafísica occidental? ¿Por qué va con ella, como su misma sombra, un pensamiento anti-metafísico? ¿Qué giro experimenta esa metafísica en la edad moderna, y por qué la metafísica de lo suprasensible culmina en cierto sentido, y a la vez parece liquidarse, a partir del idealismo alemán? ¿Por qué irrumpe entonces el positivismo? ¿Qué sentido tienen desde ese giro epocal los intentos metafísicos posteriores, entre ellos los de tendencia fenomenológica?

Todas esas preguntas sugieren, sin duda, una comprensión de la filosofía en total. Esas preguntas parecen ciertamente superfluas para aquellos que creen que lo preguntado está todo muy claro y que basta echar mano de cualquier manual para saber la respuesta. Pero ellas son lo más importante para el verdadero filósofo, el cual no se contenta con prejuicios en boga o con medias verdades. La perspectiva dentro de la cual habría que introducir y contemplar los hechos de nuestra vida filosófica latinoamericana, sería, según esto, la perspectiva de la historia de la metafísica. Yo mismo he intentado recientemente interpretar en ese contexto a la filosofía relativista nacionalista latinoamericana en mi ensayo "Vías y extravíos del pensamiento latino-americano."

¿Qué frutos tendría esa manera de mirar? Ante todo ella nos permitiría saber cuál ha sido el verdadero sentido de la presencia de la Fenomenología en Latinoamérica. Entonces sería posible evaluar, qué pensadores han tenido realmente idea de lo que estaban haciendo y cuáles no han hecho otra cosa que vestirse a la moda de su época. Entonces podrá saberse qué autores y cuáles obras repiten tan sólo doctrinas recibidas y cuáles divisan nuevos problemas y respuestas. A la luz de esa interpretación posible se revela, además, que una visión sinóptica de la Fenomenología en Latinoamérica como la que hemos tenido que desplegar en lo anterior, sólo puede ser una información bibliográfica para una investigación futura.

Finalmente, la inserción y enjuiciamiento de nuestra filosofía latinoamericana dentro de la historia de la metafísica está lejos de negar y empequeñecer la peculiaridad y novedad de nuestras producciones. La novedad y particularidad sólo son visibles sobre el trasfondo de una tradición y de un pasado y en contraste con ellos.

4. No sólo ha faltado hasta ahora una perspectiva en la que los fenomenólogos podrían abordar esa tarea posible, sino también una previa evaluación crítica de las aportaciones de uno u otro pensador. Esa carencia de crítica respecto de obras determinadas, así como la opacidad que reina en torno a nuestro pasado filosófico, son señales de un fenómeno más amplio: la ausencia de una "intersubjetividad" filosófica. A diferencia de lo que ocurre en otros países con vida filosófica, lo que piensa o escribe el filósofo individual en Latinoamérica es, por lo general acogido amablemente en un limbo, donde no hay buenos ni malos, ni reprobación ni reconocimiento. Las aportaciones hechas en el pasado sobre un tema determinado no sirven de peldaños necesarios para las investigaciones que se hacen hoy sobre ese mismo tema. Los productos del trabajo intelectual parecen no ser compartidos por todos los investigadores. Ni estos ni sus producciones parecen formar parte de un mismo mundo filosófico común. Ciertamente, esa situación no puede ser eliminada por decretos y resoluciones. Pero ella podría cambiar un tanto si un grupo de pensadores, unidos por la Fenomenología, en el sentido más amplio de ésta, contribuyera a constituir una verdadera comunidad filosófica en Latinoamérica.

Ojala que la visión de lo que la Fenomenología ha sido hasta ahora pueda inspirar nuevos bríos para el trabajo del porvenir!

Simón Bolóvar
Caracas

NOTE

[1] Cf. el Prefacio del Zotor. Además: Outline of the History of Phenomenology in lbero America, en Phenomenology Information Bulletin, Vol. 7, 1983.

PART SIX

ORTEGA Y GASSET'S PHILOSOPHY OF HUMAN SPIRIT AND CULTURE

Guadalajara's State House.

MARÍA DEL CARMEN PAREDES-MARTÍN

CULTURE AND LIFE ORIENTATION IN ORTEGA Y GASSET

1. THE NEED FOR CULTURE

In Ortega y Gasset's *Misión de la Universidad*, one reads "there is always a system of living ideas that represents the highest level of time, a fully updated system." (J. Ortega y Gasset, *Obras Completas* [Madrid: Alianza/Revista de Occidente, 1983], Vol. IV 342; henceforth *MU*. All citations herein are to the *Complete Works*.) This system is culture. This is how Ortega expresses "an unavoidable need" (IV 341) that characterizes human life as such. "As humans, we cannot live without ideas" (IV 321) and living cannot be understood as anything more than an unceasing activity that progressively constructs our existence. Right away, attention to this activity gives us an image of urgency and immediacy, reflected in the solving of the problems most directly affecting us. However, if we pursue this further we soon see that, as Ortega stated early on (*Meditaciones de Quijote*, I, 1914; henceforth *MQ*), everything has a double dimensionality of surface and depth that makes us go beyond the first impression. Indeed, after the pressures of concrete tasks, a view of depth reveals to us the existence of a plan in life, according to which one chooses one or another activity and a set of convictions that serve as a basis for that plan and lend sense to our actions. The project, the conviction and the sense are only manifested through concrete tasks that give content to our lives.

The fact that our life projects or plans might not persist for long, might be modified or might be completely changed, also depends on the system of ideas supporting them, although it does not affect the principle that all life, at each moment, develops according to a plan that justifies it. "*That is, all life – whether it wishes it or not – needs to justify itself*" (IV 342). Thus, the objective aspect of culture is directly related to the constant justification that life makes of itself. However, this would not be possible if previously there were not a subjective aspect, i.e., that humankind should have formed a series of ideas about what things are, about the world and about itself. Both aspects are found together in culture in its truest sense; that is, leaving on one side what culture

A-T. Tymieniecka (ed.), Analecta Husserliana, Vol. LIV, 359–375.

represents as an "ornament" (IV 321), we turn to the role its plays in all human life.

To briefly illustrate the need for culture, let us establish the possible limits between two extreme poles of the relationship between culture and life: on one hand, the pole that represents the understanding of culture as a counterpart to nature and, on the other, the notion of culture as a dimension inherent to the plenitude of life. In principle, "culture is always the negation of nature" (*Renán*, I 460), both at the individual and at the collective level. In this polarity, "nature" represents the basest level of human life, with its proclivity to abandon itself to instincts. The corrective element to that primitive spontaneity comes from culture, involving the introduction of a certain order and normativity in the originating impulses. In this sense, culture appears as an autonomous power whose content to a large extent is inherited in the form of norms and rules of conduct.

This simplified way of introducing the fact of culture does not necessarily imply that culture is something mechanically added to life – a superstructure imposed on forms of animal behaviour, as it were. On the contrary, it makes us consider to what extent human life always moves within the scope permitted by the polarity mentioned above. Humankind forms part of nature because it is included within nature and yet, at the same time, it is always opposite nature in the sense that to some extent it is extraneous to it (cf., for example, *El mito del hombre allende la técnica*, IX 620). In sum, the influence of culture is subject to the limits marked by the unavoidable dimension of nature within humankind. In consequence, "culture is not all of life but only the moment of sureness, firmness, the moment of clarity" (*MQ*, I 355) that must be reached to clearly demonstrate the best meaning of things.

But if culture is a dimension of life, then its meaning can be addressed from a perspective other than the merely negative one derived from the polarity we have mentioned. The moment of clarity is demanded by life itself as a need to unfold all the virtualities locked up in it. In this respect, we should recall that for Ortega "the lives of women and men . . . have a transcendental dimension in which, so to speak, a person comes out of itself and participates in something which is not itself, something which is beyond itself" (*El tema de nuestro tiempo*, III, 166; henceforth *TNT*). This transcendental dimension is formed of thought, will and the feelings and emotions related to art, religion, law, science and all the other aspects of culture. In general terms, Ortega always held this basic idea that even in its immanence human life is not limited to the circle of so-called immediate needs but rather "in it, what is immanent is a transcendence beyond itself" (ibid.) towards a dimension beyond nature.

Now, although we find hints of this fundamental idea throughout Ortega's work – regardless of the criticism of culture, which we shall address later on – what does display important variations is the interpretation of the content of culture as a dimension of life. In the work we have just cited, culture is the product of certain life functions that point to a trans-life sense. Thus, the human acts that create culture are in themselves vital phenomena, although their results become independent from life and acquire value according to objective laws. The influence of Nietzsche is patent in this view of culture as a vital phenomenon. On other occasions, Ortega offers a different view of the same phenomenon, one less bound to biologistic tenets, although the differences in interpretation do not necessarily respond to different periods of his writing. Indeed, one also sees different interpretations of culture appearing in the same work, depending on the different contexts being explored.

Apart from the explicit emergentism of some of Ortega's texts, it is interesting to note the importance of the *relationship* between culture and life, as an indispensable element of any interpretation of both terms. Within this relationship, the polarity between culture and nature perhaps offers the most schematic view, one which for its more correct evaluation demands the complementary view of the integration between both elements. This is because in the relationship between culture and life, as in all dynamic units, the force of dispersion is as important as that of incorporation (see for example, *España invertebrada*, III 53 ff; henceforth *EI*).

This leads us to understanding the need for culture within the scope of the reciprocal action of integration and disintegration within which human life develops, not linearly and immediately but rather through a series of mediations and frequent contradictions. Indeed, we should note that in Ortega's work the need for culture is posed quite early on in connection with the theme of mediation, which specifically impels the movement of integration and disintegration in our life project. "All cultures consist of the solving of contradictions. Barbarism, by contrast, is blindness towards contradiction that allows us to access only one of the terms" (*Espectador II*, II 144). Culture therefore contributes not only to enabling the solving of conflicts but, above all, to the ability to put into practice the imperative of the understanding of the conflict itself, of its terms and of its internal contradiction. When this understanding is lacking, it is replaced by an attitude unable to take into account the different elements of the contradiction, a unilateral attitude that for this very reason self-affirms itself from a lack of culture or from an "anaemic culture" (see II, 87).

The foregoing does not imply that culture provides an all-embracing view of things and gives us a bird's eye view of problems. Rather, culture allows us the choice of a point of view (see *TNT*, III 179) from which we will be able to obtain a more accurate perspective of things and, through it, the meaning or *logos* of what in all appearance is merely a contradictory event. The need to choose a point of view forms part of the formation of the subject and the choice should be in consonance with the reality of things; that is, with the perspective or the definitive being of the world (see I 321) since only life is not a *particular* point of view; rather it encompasses all points of view (*Ideas y Creencias*, V 385; henceforth *IC*).

In the interrelationship between culture and life, we see, on one hand, that culture can be understood not as just another ingredient of life but rather as the dimension that gives plentitude to life and, on the other, that life also manifests itself in its creativity as a principle of culture and of its criticism. Hence the need for culture to answer a fundamental demand of life itself and its development. Ortega holds that "all needs, if empowered, are finally converted into a new area of culture" (*MQ* I 320) in which the concept of the human being attains its most genuine meaning as "an individual in humanity" (*La pedagogía social como programa político*, I 512).

As we shall see, culture is called upon to provide the normativity required so that the reality of existence will fit into objective values, rising above the impulse of spontaneous living. In this sense, culture is not only a human need, as we have been saying, but also an imperative in the strict sense. From his very earliest writings, Ortega often calls upon the notion of "salvation" to indicate through his expression "let us save ourselves in things" (I 131, 517, etc.) an idea that has both an intellectual and a moral interpretation. Nevertheless, in the general characterization of culture, Ortega's "let us save ourselves in things" acquires a very precise specification in the sense that "culture is what saves us from the shipwreck of life, allowing people to live without their lives being a meaningless tragedy or radical debasement" (*MU* IV 321). The order that culture introduces into daily life fulfills – or should fulfill – this salvific role and this becomes even more necessary since culture is also conceived as the production of things – science, morals, art – that is, of universal realities.

2. THE PRINCIPLE OF LIFE ORIENTATION

Culture as human production and culture as a system of ideas are therefore the two interpretations that might summarize what has been said until now. From both points of view, the reference to humankind as the subject on which all cultural works are based is clear.

Firstly, "culture is born of the living background of the subject and is life *sensu stricto*, spontaneity, subjectivity" (*TNT*, III 172) from which it will be consolidated in norms, preferences and values; that is, it will become objectivized in the different dimensions integrating human life. This process of objectivization implies that cultural works acquire independent value as they depart from the subject and the particular circumstance that enabled them to be brought into existence. However, this fate of culture, which is necessary for it to attain consistency in itself, involves the risk that on becoming separated from the individual it may enter into conflict with the life from which it arose. This is why Ortega warns that "culture only lasts while it is receiving a constant vital flow from the subjects. When this transfusion is interrupted, and culture departs, it soon dries up and becomes hieratic. Culture thus has its moment of birth – its lyrical moment – and its moment of ankylosis – its hieratic moment" (III 173). Precisely one of the causes of a cultural crisis is ankylosis of the predominant culture at a given moment.

Secondly, culture as a system of ideas about the world and humanity responds, on the one hand, to the task of effectively directing human existence according to the perspective of the best values, since only in the perspective of what is most valued "are things justly valued" (*MQ*, I 338). On the other hand, this concept is quite opposed to the assumption – very often unexamined – that culture is something superfluous with respect to the other fundamental dimensions of life. One of the reasons that allows us to understand the intrinsic value of culture is the fact that in such a system of ideas the presence of convictions forming the basic underlying structure of all human life is unquestionable. Thus, the ideas of culture are not merely ideas or "idea-occurrences," according to a well-known Ortegan expression (see *IC*, V 384), but rather ideas which we have for dealing with the world; "Living is struggling with something – with the world and with oneself." However, that world and that "oneself" that the individual meets already appear under the guise of an interpretation, of "ideas about oneself and about the world" (V 384). The strength of these ideas lies in their vitality, in the fact that they remain connected to the flux that originated them. Therefore, we are actually dealing with "living ideas," which no individual should lack.

With respect to this latter aspect, culture is expected to serve as a "path" or guide with which to orient oneself in one's existence. This function of the system of ideas of culture also represents a principle of social organization and has considerable ethical content into which the Ortegan point of view that culture should be in the service of life is condensed. The double imperative according to which "life should be cultural, but culture should be vital" (*TNT*, III 169) would express the content of this function very succinctly. This is because not only should culture not depart excessively from the subjectivity that brought it into being but also because a cultural effort is not reduced to being an interpretation of life – even though this is very important – but rather such an effort is converted into effective knowledge, into knowledge about orientation. I shall return to this point below.

Besides the individual convictions that form the world of each individual at a given stage of his or her life, in a world "that is valid in itself in spite of and against our acceptance of it" (*En torno a Galileo*, V 35), this property is what clearly distinguishes collective from individual convictions. Accordingly, if individuals' convictions represent the subjective moment of the human being-world relationship, collective convictions have intersubjective and epochal value: that is why they structure the "ideas of the times," the "spirit of the moment." "A conviction of mine, however firm, is only valid for me. But the ideas of the time, ambient convictions, are held by an anonymous subject, who is no one in particular, just society" (*ibid.*). The validity of convictions in themselves is what makes them function as a source of objectivity, which informs social life against individual attitudes.

One thus sees with respect to the specificity of collective convictions and with respect to the convictions of subjects, that the former should not be considered as a mere numerical extension of the latter. In the former the objectivity of the valid world in society is shown; the latter encompasses the particular subjectivity of each subject. Ortega highlights this difference between the objectivity of the former and the subjective nature of the latter, inasmuch as collective beliefs, convictions and uses not only have the reality lent to them by our possible adhesion but also the reality seen in the fact that they are imposed upon us: "*the fundamental sociological phenomenon that is validity* which appears, not only in opinion, but in all uses . . . does not consist in more or less individual adhesion . . . but it is precisely (social) use because it is imposed on them. Because of this, everything social is a reality other than the individual" (*El hombre y la gente*, VII 265).

However, this difference becomes attenuated from the moment at which validity is not maintained outside individuals' attitudes, thus conserving a reference for the subjects that accept or reject such social uses and convictions. This, therefore, implies that the objectivity of the world of collective life is also mediated by the subjectivity of individuals, who feel positively or negatively interpellated by current convictions. In any case, the fact that most individual convictions are nourished by the social world and coincide with those of the collective, and that Ortega applies the same notion of interpretation to both, suggests that we can never escape from subjectivity, although an individual is seen to be formed intersubjectively from the very start. In so far as they are to be identified as such, cultural crises imply an awareness of crisis – firstly collective, and hence intersubjective – the validity of which as a crisis also requires recognition of the individuals forming the social group wherein that culture prevails.

According to Ortega, to understand how a cultural crisis is produced, and in general any historical crisis, the following should be taken into account: "culture is no more than the interpretation that people give to their lives, the series of more or less satisfactory solutions that they invent to solve their problems and needs in life." Both those of material nature and of so-called spiritual nature should be included in these terms. Once such solutions for true needs have been created, they are also true solutions; they are ideas, evaluations, enthusiasms, ways of thinking, styles of art, of law, that emerge sincerely from humankind's radical background depending on whether its authentic nature was true at the beginnings of a culture. However, the creation of a repertoire of cultural principles and norms brings with it a constituent and, in fact irremediable, drawback. Precisely because an effective solution has been created, precisely because it "is there," following generations do *not* have to create it but rather receive it and develop it. However, reception that saves the effort of creation has the disadvantage of inviting vital inertia (*En torno a Galileo*, V 77). The delay in culture with respect to the needs of the time is generally one of the causes of the vital inertia found at the core of all cultural crises. This misfit or phase-out is especially exacerbated when no model or norm in the past is recognized for the future, as Ortega understands the situation in Europe during the first few decades of the twentieth century. Moreover, since the term "culture" does not denote an atemporal concept, since all human activity is historical activity and as such subject to change, attention to cultural variations is indispensable for making a diagnosis of a crisis situation.

In sum, as stated by Ortega in 1933, each crisis is a historical change that does not affect just certain aspects of the human world but inverts them completely, affecting their very foundations. The individual is left "without a world" (V 70) and from here we see that to live in crisis is equivalent to living "completely lost" (V 59). These expressions, however, do not indicate a material absence of culture in the sense that it is normal for culture to become complicated and become multiplied and that the educated individual in a crisis should have an overloaded culture in times of crisis. What is lost then is not the culture learnt nor the ideas about the world that such culture transmits, but rather the orientation of life itself. In sum, one is dealing with the loss of the meaning of culture as orientating knowledge. Thus, for Ortega "historical crisis" and "cultural crisis" are fundamental forms that can be adopted by human life. Accordingly, during the crisis the individual does not *know* what to do because s/he does not *know* what to think about him/herself and the world. From both the individual and collective viewpoints, times of crisis negatively show that culture as an orientating knowledge is the most suitable counterconcept to that of culture as refinement.

3. THE GENERATIONS

The sense of culture is not usually lost suddenly and neither is it lost in absolute terms. What is typical of the changes that give rise to a crisis is that certain anthropological-temporal trends or rhythms are developed, among which the passage of one generation to the next and the level of a given period with respect to other periods are striking. The sense of these changes imposes a rhythm of depersonalization and collectivization that Ortega interprets negatively in the sense that one is dealing with a progressive loss of true existence and an increasing lack of interest in the main problems of the times.

For Ortega, "what each generation calls 'our time' always has a certain level" (*La rebelión de las masas*, IV 156; henceforth *RM*) that characterizes "vital time," that is, it has the level of a period, which is also historical time. Thus, each period was a vital impression of itself based on comparison with other periods. It is in this way, looking to the past, that each period is aware of its peculiarity as a mature or young period, a period featuring plenitude or decadence. The same happens in individual lives: each individual is aware of the relationships between his/her own life and the level of the period through which that life travels. Accordingly, the requirement that the vital system of ideas should have validity or should represent "the highest level of the period" – as mentioned at the beginning of this work – is essential for the desired orientation to occur, since not just any idea about humankind and the world is valid; rather, ideas must correspond to the demands of these in order to become oriented in the present.

Those who are not "up with the times" – those below this level – "are condemned to a lower life, a more difficult life, painful and harsh" (*MU*, IV 342, 322ff). This is what Ortega denounces, from different points of view, as the generalized situation in the Europe of his time: the anachronism of the dominant culture with respect to the level of the problems being posed. Despite this, such a situation is not new since it has recurred many times in history, even though the crises have followed different courses and have had different origins in each period. With respect to his own time, Ortega states the following: "It is not fullness of time, and yet it rests on all the times there have been and on all known fullness . . . it believes itself to be greater than the others and at the same time believes itself to be a beginning, without being sure that it is not an agony" (*RM*, IV 162). This vital feeling of belonging to a period, both ascendent and agonic – as Nietzsche also felt about his own time – is a characteristic of the awareness of crisis manifested in the thought of our philosopher.

Let us now explore the role played by the category of generation in the self-feeling of a period and in cultural variations. We shall do so because it is well known that heroic and brilliant individuals, "the great individuals," have merited considerable attention in Ortegan analyses. Despite this, however, we do understand that, when a people is attempting to deal with historical crisis systematically, significant changes are not advanced by isolated individuals, however relevant they may be, but rather by large social groups. The changes that Ortega is interested in are not those affecting this or that more or less concrete manifestation of culture; rather, he is interested in those that transform the deep substrate of "the radical sensation about life" (*TNT*, III 146). In other words, historical changes are those that transform the world of a collectivity and, eventually, the structure of the vital drama of individuals. Such transformations confer on the different generations a distinctive seal.

Accordingly, a generation represents "the variations in vital sensitivity that are decisive in history" (*TNT*, III 147) since each configures "an integral mode of existence" (*En torno a Galileo*, V 39) that has its own and typical characteristics that are distinct with respect to any previous generation. According to this, what makes up the most genuine historical reality of each period is the vital sensitivity that joins the different members of a generation, above and beyond their individual differences and discrepancies.

The reality of historical change lies in the fact that each generation alters the world or vital horizon it finds was made by previous generations; it integrates into itself both what is its own and what it has received, producing a result that is always different as regards novelty and tradition. Of course, it is very important to bear in mind that what determines the *need* for change lies at the most fundamental plane of the temporal nature and finiteness of human life, although the visible rhythm of changes is signalled by the greater or lesser degree of homogeneity or heterogeneity between what is new and what is old. At the same time this fosters the possibility or impossibility of finding a common principle in the thoughts and feelings of one generation and others. Since the main mission of each generation is to construct its existence for itself, since thought acquires consistency and concreteness in the actions of individuals and, indeed, since the relationship of the individual with his/her period is *mediated* by things, the vital sensitivity of each period is configured in a historical task that provides real content to the physiognomy of the period.

However, there are generations that do not fulfill their task because most individuals prefer to "slot into" what has been inherited rather than construct their own works. Such self-falsifying generations are clear harbingers of periods of crisis, and culture is therefore not realized in them. Concerning his own generation, Ortega feels that "it has not been often that human beings have lived so out of contact with themselves, and perhaps Humankind has never supported so docilely forms that are not proper to it, hand-downs from other generations that do not respond to its inner beatings" (*TNT*, III 151). The dissociation between effective life and culture that comes from the past is no more than a consequence of the fact that the mission called for by the current time has not been fulfilled. Thus, rather than speaking of "a failure in culture," we should speak of a crisis, of "the most radical crisis in modern history" (III 172, 186) when the established culture no longer offers an answer to the current world.

Apart from this overall consideration of the role of generations, from the point of view of the individual, all periods of crisis are periods of transition, evidenced by the difficulties inherent in breaking away from a given system of ideas to search for another system of ideas that has not yet been constructed. To carry out this transition implies the creation of a new horizon, an opening to another vital perspective, which earlier generations did not know. Therefore, in a situation of crisis one should at least take into account – along with the specific components of each moment – the traditional culture, which provides an "already given" interpretation of the world, the new vital sensitivity that struggles to emerge and the way in which individuals assume the task of living.

The failure in culture does not first stem from that fact that its objective value proves to be insufficient at a given time but rather from the fact that the members of a generation fail to review it critically. Thus, in all cultural analyses there is a factor of responsibility that is left unattended, that becomes marginal. Here, I am referring to a task that has been overlooked or poorly executed, a task that although corresponding to the collective members of a generation, should, and this is characteristic of Ortega's thought, be advanced by the most relevant individuals.

4. THE FUNCTION OF MINORITIES

The fact is that a generation, like all human social groups, has the structure of a minority and a mass, of select and common individuals, who conserve their identity thanks to a permanent distance between each other. For Ortega, this is a natural inequality or, more accurately, a difference that crosses the plane of the juridical forms of the state and of the principles of positive law, since it is an anthropological difference. From here it follows that "the primary social fact is not the mere gathering of a few individuals, but rather the articulation . . . , organization in the tandem of directors and directed. . . . This implies in some the capacity to direct; . . . in others a certain intimate facility for being directed" (*IC*, III 94ff).

Indeed, this natural and social difference – which is of course "of historical degree" (*TNT*, III 148) – first demands of the minority the responsibility of playing a directing role within the social group and, correlatively, for most it involves a receptive response to the example of the best. This means that the principle of responsibility presiding over the ethics of Ortega – and most of Ortega's analyses involve the ethical problem – is split into two dimensions. This is so because although Ortega demands of each individual a constant effort to attain a true existence and to promote the growth of a responsible ego against an impersonal and empty ego, such a requirement does not have the same content for all; what is required depends on whether one is one of the few who can command or whether one is one of the many of common ability.

Let us remember that the distinction between minorities and masses does not refer to social classes, whatever criteria is used for dividing these up – economic status, education, power, etc., – but rather to the ways adopted by an individual "to be." This human typology corresponds to two forms of life, the noble and the common, respectively characterized by effort and inertia and manifested in all forms of individual existence. Obviously, being a select individual and living a noble existence forged by demands on self is qualitatively different from being a common individual and living a vulgar life, dominated by the law of least effort. This difference, both anthropological and ethical, has important social consequences that are expressed in ways of doing things and in different functions. Again, the presence of Nietzsche pervades all this.

Moreover, because this distinction is prior to all social stratification, it does not coincide with such groupings either since all social classes or groups have this structure of minority and mass according to which both elements need each other reciprocally. It would be naive to assume that the minority would be able to perform its functions and have influence alone, outside the attitude of the mass. Ortega says that "it is completely mistaken to assume that the enthusiasm of the masses depends on the merits of the directors. Strictly, the truth is quite the opposite:

the social value of the directors depends on the capacity for enthusiasm of the mass" (*EI*, III 92).

Despite this, the qualitative difference between these two classes of individuals bears with it the fact that the role of the minority is determinant in the development of either a crisis situation or an hour of historic plenitude. Furthermore, for Ortega what is understood as society cannot endure without an excellent minority (see III 97), since he considers societies to be vehicles for perfection. The spiritual force that propels this desire for perfection is to be found in eminent individuals and its practice results in exemplarity and the orientation of others towards what is better; this is essential if the majority is not to be dispersed.

From this point of view, all cultural crises have seen desertions of the minorities, who represent the rhythm of ascendent life and the level of the times. This desertion occurs in different orders since Ortega sometimes uses the category of select minority according to what individuals are for themselves – and hence with anthropological meaning – and at other times according to the functions to be played by the best endowed minority, that is, attending to its social projection. With respect to what this desertion of the minorities represents in general terms, both individually and socially, in *El tema de nuestro tiempo* he writes, "It is inadmissible that the people obliged to assume the responsibility of our time by the relevant intellectual conditions should live, like the vulgus, drifting, caught up in the superficial vicissitudes of each moment, without searching for a rigorous and broad orientation in the course of history" (III 152).

The first responsibility of select individuals is to serve as models in life for the rest. This is decisive but at the same time something that occurs naturally when outstanding individuals act as such, by virtue of the greater demands they make on themselves and on their actions. "The select individual selects his/herself by demanding more than others. Select therefore means a privilege of pain and effort. Select individuals are all those who on a level of perfection and demands aspire to a higher level of demands and perfections" (*Goethe desde dentro*, IV 487). The responsibility of achieving a higher goal in such demands, which are never abstract but in consonance with the requirements of the period, is a kind of destiny that falls upon the select individual by virtue of his/her own idiosyncrasy.

However, all individuals are faced with the permanent possibility of rejecting their vital task. And one of the causes of social decadence and cultural crisis can be found in the abandonment of the destiny reserved for the best. Among the motives that give rise to this situation, it is worth mentioning the confusion of culture and a progressive sophistication that gradually distances individuals from themselves and from their vocations. Once again, the weight of individual attitudes becomes patent in the problems of society since the consequences of such attitudes have intersubjective value and an objective meaning. Ortega points out that "in times of false crisis, pretended positions are frequent" (*En torno a Galileo*, V 70) and this lack of authenticity brings with it the fact that whole generations falsify themselves. The disorientation in life brought about by the multiplying effect of such falsifications is described with the metaphors of "drifting" and "shipwreck." In sum, culture as an adulterated institution as compared with culture as creation is the result and antithetical expression of human life as crisis.

5. THE HUMAN PROTOTYPE OF CRISIS IN THE TWENTIETH CENTURY

If it is difficult to exaggerate the importance of the notion of the select minority, neither is it possible to forget that each historical period obtains its unmistakable profile from the state of the relations between the minority and the mass. This issue is explicit from the very first writings of Ortega, for example, in *Vieja y nueva política, Prospecto de la "Liga de educación política española"* or *Misión política de las minorías intelectuales*, and we can indeed say that this thesis orchestrates highly relevant aspects of his thought. All interindividual action, in the widest acceptance of the term, is developed in a minority-mass interrelationship. And since this is so, the mass alone also plays an important role in the characteristics of historical situations and, as is well known, a very singular role with respect to the crisis of the twentieth century.

The mass individual is the average individual who predominates in public life from the First World War: "a type put together very fast, built out of just a few and rather poor abstractions, and who, for the same reason, is identical right across Europe. It is to this type that we owe the sad aspect of asphyxiating monotony pervading the whole continent. This mass individual is the individual previously emptied of all his/her own history, with no feelings for the past and, accordingly, docile to all the so-called 'international' disciplines" (*RM*, IV 121).

The revolt of such individuals consists of a spiritual hermetism which, on the one hand, converts them into beings delighted with themselves and, on the other, shuts them off from any call to a higher level of existence.

From this characteristic of "being shut off," which Ortega considers as "an absurd way of being an individual" (IV 185), there derives a generic action that points to the goal of deposing excellent individuals and imposing a generalization of the mediocre; in sum "the right to vulgarity" (IV 148, 188). Leaving on one side the impression of aestheticism produced by some of Ortega's formulations, and focusing ourselves above all on the seriousness of his point of view, we may deduce that it is not the overwhelming presence of the masses that irritates him, but rather the fact that the deposing and abolition of the excellent suppresses all desire for social perfection and eliminates from the individual's heart the will to achieve higher goals. One is therefore dealing with an ethical impoverishment, a collective demoralization.

In *España Invertebrada*, Ortega writes, "when in a nation the mass refuses to be a mass – that is, to follow the directive minority – the nation breaks down, society breaks up and social chaos imposes itself; that is, a historical disarticulation is initiated" (III 93). But the case of the rebel mass individual is even more complex since here one is not merely dealing with the lack of obeisance to a minority. This individual believes him/herself to be in the possession of "ideas" that are not really ideas at all. And neither do they afford culture. The system of ideas that condenses the state of a culture is situated at a higher level than that of any majority opinion.

In agreement with this requirement, the rebelliousness of the mass individual consists – beyond his/her refusal to be directed by better individuals – of a blindness towards truth and a refusal to accept cohabitation under norms that are not the fruit of his/her subjective inclinations. Therefore, what is being denounced is not simply a hermetism towards others but a hermetism towards the truth that others can show me. At this point, Ortega corrects the subjectivism of other proposals and states his case in favour of a certain Platonism of ideas. To have an idea is to admit "an orb of intelligible truths" (IV 190); it is to give oneself up to its objective content and accept the existence of a model. According to this, rebelliousness also consists in being prepared to recognize the need for life models and therefore to feel comfortable within the limits of satisfaction itself.

Ortega finds the newest factor of his period in this absence of a call to normativity in the different dimensions of life. He is very clear on this point: "There is no culture when there are no principles of civil law to appeal to. There is no culture where attention is not paid to certain of the latest intellectual positions that can be referred to in disputes. There is no culture when trading laws under which one can seek respite do not preside over economic relations. There is no culture when aesthetic issues do not recognize the need to justify works of art" (IV 188ff). This text mentions aspects that are essential in the Ortegan conception of culture, which can be present in moments of crisis but which can only be formally considered as culture if they are stripped of an order of norms that one can adjust oneself to.

Ortega offers us a detailed description of the "rebellion of the masses" but, above all, he is interested in analyzing its underlying motives. Among the causes of the rebelliousness of the mass individual the most outstanding characteristic factor is the attitude of the individual to technology and history. Indeed the main elements of the world Ortega is thinking about are democracy and technology (see IV, 177 and 197ff). The increase in welfare for the average individual of any social class derives from this, but does not lead – in Ortega's view – to a recognition of the subjective and objective potentialities that it offers but rather to a mystification of the true value of civilization, because the fruit of technical thought is considered to be something already given and "natural."

This attitude is the clearest symptom of the indifference of the individual towards the efforts of intelligence, towards the experimentation and industry that render technology possible. An inversion thus occurs between the originating world, traditionally called "nature," and the human world, which is the world of culture, in terms of the following paradox: the individual lives in a civilized world but s/he is a primitive being, a *Naturmensch* (IV 196), who lives far from the tenets of his/her civilization.

Ortega's criticism is directed at an aspect of the psychology of the average individual living in an advanced society who bases his/her rebelliousness on completely ignoring the human labour that sustains the reality of the world. At this point Ortegan thought is very directly related to that of other philosophers of our time, such as Marcuse or Sartre, and of course, Heidegger and the Frankfurt School.

The "primitivism" of this attitude towards technology appears clear from Ortega's point of view, since he connects technology, science and culture as manifestations of the same creative impulse. The importance he gives to this behaviour is justified by the fact that for Ortega, technology and experimental science are the typical products of the culture of our time. The function of technology consists in increasing the complexity of the world in a convergence of theory and practice that, in general, humankind uses for its own benefit. Accordingly, the mass individual that Ortega describes is not "*homo faber*" – the creator of technology – nor "*homo sapiens*" – the creator of forms – but the new alienated individual who is unable to understand the principles of this new environment that alters both man and nature.

This primitivism of the rebellious mass individual of the twentieth century – but not of the masses of all periods – has as a consequence a reification of the technical world and hence of part of our cultural world, which also leads to a reification of the individual him/herself who benefits from it. This is so because the true meaning of technology lies not in the immediate welfare that it produces nor in the artefacts populating our world that derive from it, but rather in the fact that it offers the construction of a new world made by and for humankind and in the fact that it presupposes human intentionality and a teleology governed by certain norms.

It is true that Ortega also links the development of technology with the phenomenon of the alienation of humankind from the world (see *El mito del hombre allende la técnica*, IX 618ff) but in turn, and inseparably, it is technology itself that is the first source of facilities allowing humankind to become released from the limitations of the environment. Thus, it is not possible to merely establish an antithesis between culture and technology; rather it is necessary to search for the roots of the crisis in the attitude of the individual towards the technical world in which s/he lives.

University of Salamanca

HAROLD RALEY

THE EVOLUTION OF PHENOMENOLOGY: ORTEGA, HEIDEGGER, AND THE ANALYTICAL VERSUS EMPIRICAL DISTINCTION IN MARÍAS

In his book *The Tragic Sense of Life* (*Del sentimiento trágico de la vida*) (1913), Miguel de Unamuno asked whether life is to be lived or to be understood, for he saw in the question an antithetical choice between heroic irrationalism and suffocating reason. To which José Ortega y Gasset soon responded in his *Meditations on Quixote* (*Meditaciones del Quijote*) (1914), that it is to be understood by living, for life in the Ortegan sense was nothing less than an all-embracing and superior reason in process, what he would later refer to variously as vital, historical, or narrative reason. Within its wide boundaries, circumscribed like an island in a vast ocean, abstract rationalism appears as a special subset of reason.[1]

Naturally, this did not settle the matter at hand. This crisis of reason was to continue for many years to come. Indeed it could be argued, perhaps not without some slight exaggeration, that the postures taken in these two classic works mark the beginning of Spanish philosophy in the twentieth century.

As for Unamuno, *The Tragic Sense of Life* would prove to be his definitive philosophic statement and he would have little of substance to add to his arguments. On the other hand, in 1914 Ortega stood poised at the beginning of his system. As Julián Marías puts it, "In 1914–16 Ortega was in possession of the nucleus of his philosophy. Its expansion, extension, and articulation would occuy the rest of his life, some forty years of creative effort."[2]

But Unamuno and his fellow exponents of irrationalism and *Lebensphilosophie* were not the only irritants to which Ortega responded. Also at issue was what he considered to be the unsatisfactory phenomenology in Husserl's *Ideen zu einer Reinen Phänomenologie und Phänomenologischen Philosophie*, also published in 1913.

For a brief historical moment in 1911–12, the instrumentality of Husserlian phenomenology alone seemed to hold for Ortega the promise of transcending irrationalism and of surpassing the neo-Kantianism of Germany and the positivism that predominated in the rest of Europe. Ortega says of himself and his classmates at Marburg: ". . . phenome-

A-T. Tymieniecka (ed.), Analecta Husserliana, Vol. LIV, 377–388.

nology for us was not a philosophy: it was . . . a stroke of good fortune."[3] Yet not long afterward, he goes on to say that he abandoned phenomenology "at the very moment of receiving it." However, this comment must be weighed against Ortega's subsequent procedure. I have argued elsewhere that if Ortega ceased to proclaim phenomenology as a method, it was only because he had assimilated it permanently into his own.[4]

Husserl's procedure is too well known to need more than a summary here. He begins with the *epoche*, that is, the phenomenological reduction, or suspension, of postulates or beliefs offered by common sense or science with respect to the value or transcendence of the intuited object. Next he proceeds to the *eidetic reduction*, which is the perceptive generalization whereby the conscious *Erlebnis* or intuition acquires universal validity. Thus purified, perception can then point to a description of objects as they are constituted in cognition. To put it another way, in their reduced condition phenomena allow us to perceive them as cognoscible manifestations or acts. They differ from "natural" seeing by becoming a seeing that is also a *cogitatio*.[5]

Husserl then discovers to his own satisfaction what had been implicit since the beginning of the phenomenological procedures: *the transcendental ego*. In his *Cartesianische Meditationen* (1929) he explains that the objective world, the only one that exists for him phenomenologically, derives its existence and significance from the ego insofar as it is a *transcendental ego*.[6]

Up to this point Ortega and his companions from Marburg appear to adhere to the Husserlian doctrine. But certain reservations and doubts appear at the juncture with pure immanent consciousness. Professor Nelson Orringer has argued compellingly that Ortega and his companions were unwilling to return with Husserl to subjectivity but instead were bent on applying his method to things, to the phenomena about them.[7]

In 1913 Ortega, now back in Spain, offers the objection that if consciousness amounts to the consciousness of reality, then it can hardly be also a reality itself, or as he puts it, it cannot at once be both the arena and the content of reality.[8] He goes on to state, though not to insist at length, that consciousness itself was merely a hypothesis derived from Descartes and that to embrace it is to restrict phenomenology to a subset of Cartesianism. For Ortega, so-called "pure consciousness" is never the primary reality. In order for it to exist at all, I must cease to live and act spontaneously and turn deliberately, "intentionally," as it were, to the memory of what has happened to me. Thus the *epoche*, or suspension, eidetic or otherwise, is impossible in the strictest sense because it always involves a further act of execution on my part.

It must be pointed out, however, that even as Ortega veers away from Husserl, his indebtedness to Husserlian phenomenology remains evident. He states, for instance, in a very Husserlian tone: "Before they can be thought of as real or unreal, objects are objects, that is, presence immediately apparent to consciousness."[9]

But this was not Ortega's final position on the matter. He goes on to argue that not only is consciousness not the absolute reality ("relative to nothing," as Husserl might say), but that it is not reality at all. The statement "consciousness of" does not indicate a primary phenomenon, for what I find before me is not consciousness but things. Or to be more phenomenologically accurate, I find myself with things. The primary datum is, therefore, I with things, doing something with them, something which in the broadest sense we call living. Hence the primary phenomenon is not a thing nor can it be reduced to the being and likeness of things. Ortega calls this primary phenomenon "radical reality."

But we have no warrant to go to the other extreme and think that this radicality is the only or absolute reality. It arises from the fact that, as Husserl states with other aims in mind, ". . . anything worldly necessarily acquires all the sense determining it, along with its existential status, exclusively from my experiencing, my objectivating, thinking, valuing, or doing, at particular times – notably the status of an evidently valid being is one it can acquire only from my own evidences, my grounding acts."[10] In succinct language, Ortega expresses a similar view of this radicality: "It simply means that in the happening called *life* every other reality, even those that claim to transcend life, is given to each person in the form of presence, indication, or symptom. It is, therefore, the root of all other realities, and only for this reason is it radical."[11]

The primordial methodological and even epistemological problem with this assertion, which becomes at once also a tension between style and hermeneutics, may be stated as a question: how can one think through this existential process that Heidegger would call *Dasein*, and Ortega, "radical reality," if it consists primarily not in my reflective thought but in my protean participation?

Neither Heidegger nor his contemporaries such Karl Jaspers or French philosopher Gabriel Marcel appear to have resolved this tension between concept and expressive form. I have stated elsewhere my opinion that the celebrated "difficulty" of the Heideggerian style lies in the fact that his language is unable to capture the movement of *Dasein* toward its self-revelation as *aletheia.*[12] Arguing along this line, we could say that the very inadequacy of language constitutes one of the prime conditions of the *Umwahrheit* which in common experiences tends to obscure the *Dasein.* This would mean, in turn, that in a curious way the common enigmas of being-in-the-world have as their literacy correlate the stylistic difficulties of talking about them at all.

In Heidegger this initial and fundamental linguistic inadequacy is further exacerbated by the equally troubling fact that despite his insistence on the analytical modulations of being, or *Daseinanalytik* (*Seinsinn, Seinweise, Zuhandensein,* etc.), as far as I know, he never gets around to telling us what Being itself really is. Indeed he goes so far as to remind us that "Being and the structure of being lie beyond every entity and every possible character an entity may possess."[13] This means that both Heidegger and his readers find themselves at an unwieldly distance from the object of their attention. Indeed, the distancing is so great that a deceptive immobility appears to envelop the *Dasein.* The reader is unable to perceive its internal movement and thus cannot anticipate the intellectual process of the great thinker. Unable to discover the intimate and personal dialectical pathway on which Heidegger has embarked, we seem to have no choice but to assume the role of admiring but passive spectators of the impossibly remote spectacle of reticent Being.

Compare this static remoteness in Heidegger to Ortega's dramatic narrative of radical reality in *Meditations on Quixote.* Yet we must guard against mistaking these differences of perspective for views of different realities. A convincing claim can be made that both thinkers wrestle with the same protean reality of human life and in very discrete ways they do so from the same level, what Ortega often referred to as the "height of the times."

Just what are the similarities and differences between Heidegger and Ortega? Perhaps they can be partially summarized as follows: first, both theories rely on the phenomenological descriptive method (without resorting to the *epoche*). Second, while the *Dasein* itself is not human life but the mode of being (*Seinsweise*) of the entity that we are, it nevertheless operates in the arena of life and presupposes it. Third, whereas Heidegger intended the *Daseinsanalytik* to be propaedeutic to a metaphysical treatment of being in general, and as the theory of radical reality is already a metaphysics, both are analyses of those structures and properties common to all human life.[14] These structures include or imply, among others, corporality, circumstantiality, historicity, phenomenological perspectiveness, the real "embeddedness" or radicality of realities that appear within my life, the implication of sociality, language, relationships, and, in the Ortegan line of thought at least, the elucidating instrumentality that coaxes all this to clarity and shows it in its connectedness: vital or historical reason.

Here we may summarize certain fundamental propositions of the Ortegan theory so as to set the stage for a leap to the empirical level of life. To begin with, the true connection or reality is the system of life itself; therefore philosophy need not take upon itself the burden of inventing any other, textual or otherwise. Reality in its most concrete form is my life, for I can begin from no other vantage point. If I say "I live," then the concretion and connectedness of reality are subsumed in the statement and in the acts and actions of "my" living. Thus, as Marías will repeat many times in his writings, to live is to apprehend reality in its true systematic connectedness.

But this is also the same definition he gives of reason, and therefore he can argue that life is the concrete form of reason.[15] Looking at this proposition from another angle, in order to go on living I need to orient myself in the world in which I find myself, in the world into which I have been thrown, as Heidegger might say. I need to know what to cling to and what to avoid; in short, I must make sense of reality by grasping its true connections. Thus Marías goes on to say that life is possible only through reason. This is what is meant by *razón vital*, "vital reason." (Perhaps a better English translation would be "living reason.") To put it another way, a way that constitutes a definitive response to Unamuno and the proponents of *Lebensphilosophie*, life is not possible without reason, but on the other hand, neither is reason possible without life, for its actual functions are performed only by living. Lest this appear to be a retreat into idealism, it bears repeating that this living reason is the apprehension of reality in *its* connectedness, not in mine, but in the arbitrary connections I might impose on it.

Therefore we find ourselves in a peculiar situation from which there is no easy escape: the only real life is individual, it is mine, the "mine" of each person. It is singular, temporal, circumstantial, and, as Ortega always maintained, it yields to understanding only by telling it. Yet, as

Marías says, "... I cannot understand, narrate, or tell *my* life except from the structure of life per se, which does not have reality."[16] Both Ortega and Heidegger had analyzed the structure by pointing out its necessary requisites. Therefore the universality derives from this necessity, for such conditions must be given in each life and are thus common to all. From this reasoning we can go on to state that "human life" can now be called "human life in general."

Naturally, the task of reaching this general conclusion does not rest incumbent on each individual person. I find that the analysis of life and its significance, its planes and hierarchies, its permissible zones and forbidden areas, as well as the world realities embedded within it, has already been done by society. Therefore, reality as I encounter it presupposes a prior interpretation; it is intrinsically historical. Marías is then led dialectically to argue that Adamism in all its forms, from claims of naive objectivism to outright primitivism, amounts to an abstract error.

But what about the human person, what we once called simply man, in this intrinsic theory of human life? Here he makes his appearance – or reappearance – not as an anecdotal afterthought of philosophy, nor simply as a theme of unexamined human experience, but as *the empirical structure of human life*. The analytical structures, those necessary for human life to begin with, may be subsumed in Ortega's formula, I am I and my circumstance, but for actual meaning they must be fleshed out with circumstantial content. Otherwise, they are mere *leere Stellen*, or "empty spaces." Like an algebraic formula, they do not in themselves offer real knowledge but only the means of manipulating unreal quantities. Yet they derive not from abstract thought but from an analysis of actual life in its concreteness. This is why Marías calls it an "intrinsic" and "necessary" theory.

The question is whether one can pass directly from this abstract theoretical plane to the concrete life of "flesh and blood," as Unamuno might say. For Marías the answer is no. As early as 1947, in his *Reason and Life*, he had begun to glimpse an intermediate zone hitherto ignored by philosophy. The conviction grew as he produced such works as *Generations: A Historical Method*, "Human Life and its Empirical Structures," and "Psychiatry Seen from Philosophy."

In order to explain this intermediate zone, Marías gives three examples of very different realities: a pentagon, an owl, and the man Cervantes. In order to understand a pentagon, that is, an ideal object, a definition and perhaps an illustration will suffice. In the case of an owl, a description of its physical characteristics, its habitat, and other zoological data would be necessary. But in the case of Cervantes, a minimal biography or "story," must be told if we are to get at his reality as a person. In the narration of his life we are told that he was born at a certain date and place, that he held various posts, that he fought at Lepanto and lost the use of an arm, that he married a certain woman, that he wrote certain novels and plays, and that he died.

Whatever may be said of Cervantes presupposes the analytical structures common to all human life. But this knowledge alone is not enough. Marías points out that other assumptions complement the bare analytical data. We read that Cervantes lost the use of an arm at Lepanto and we realize that he had a certain corporeal structure that included arms. We are told that he married Catalina de Palacios, which means that he was a male and could have a precise amorous and sexual relationship with a woman – and she with him. We learn that in his old age he wrote *Persiles*, and we come to realize that he experienced aging and death.

None of these features receive more than marginal treatment in the analytical theories put forth by Ortega. His general metaphysical theory of human life virtually ignores sexuality, age, anatomy, and mortality. And though he insists on individual life, he seldom mentions the term "person." For his part, Heidegger speaks at length on "Being unto death" and describes the field of human existence in terms of mood, understanding, and speech; yet it is somewhat remarkable that he describes human existence without even mentioning the term "man" at all. He presupposes that we know already what man is. In any case, we cannot ascribe human existence to the same mode of being as things.

Yet for Marías this general phenomenological description of human life or existence rests on a range of unmentioned assumptions without which neither the story of Cervantes nor any other biography would make any sense whatsoever to us. They constitute predominant elements of what Marías calls the "communicability of circumstances," which permits our narrative transmigration to, and understanding of, other lives.

Though customarily overlooked, these silent assumptions have not gone completely unnoticed in philosophy. Aristotle spoke of the "essential" (*kath auto*) and the "accidental" (*kath symbebekos*) qualities of being. He noted that whereas *logos* was essential to man, being Athenian or blond or musically inclined was not. Yet even though they are not essential, such properties are found only in mankind. This idea was to reappear in the medieval Schoolmen.

But Marías insists that this idea of such accidental "properties" is remote from the area of reality he describes as the "empirical structure." First of all, the *idion*, or *proprium*, refers to being in general, including, of course, the zoological being we call man. But for Marías it is not a question of things but of human life, which is not a thing at all but a dramatic reality of a very different order and which must be understood in a very different way. Furthermore, instead of appearing as "accidental" features, the elements of the empirical structure are characterized by being structural. This apparent tautology directs our attention to a capital fact: this systematic character of life is what permits our understanding of it. In theory, human life could occur in other forms (a common theme of science fiction) without violating the general, analytical requisites that Ortega and Heidegger describe; in fact it occurs in a specific bodily and mortal way that undergirds what we call "experience of life." Marías describes the empirical theory in these terms: "To it belong all those determinations which, without being ingredients of analytical theory, are not chance, coincidental, or factual events or contents of Cervantes' life, but empirical though structural elements, and therefore previous to each concrete biography and with which we can count, elements which function as an *assumption* of that biography."[17]

The constants of the analytical theory exhibit the convenience of unwavering stability. To take an example, if we lived in other worlds, Ortega's formula "I am I and my circumstance" would probably still hold true, but the empirical and corporeal contents would be different from what we know here, just as they differ significantly among the countless multitudes of cultures and unnumbered individual lives.

From this condition two assertions present themselves: first, the empirical structure appears as the field of all possible human variation in history. For instance, we apprehend the world sensorially – sight, hearing, touch, etc. – in an empirical manner that we all know and take for granted. Thus, while this sensorial quality could arguably be thought of as intrinsic to the analytical theory, its empirical forms are subject to modification. Think, for instance, of how technology has expanded our senses far beyond the old natural range. In a similar way, consider how the empirical mortality of human life has fluctuated enormously in different eras and civilizations.

The second point to be made is that among the necessary requisites of the analytical theory we must acknowledge the requirement of empirical manifestation. In other words, as we move beyond an abstract consideration of life, we find that it necessarily appears in one actual, empirical form or another. Or, to use Marías' words, ". . . empirical structure, taken overall, is *one* ingredient or requisite of the analytical structure of human life."[18]

Here Marías turns to a convenient feature of the Spanish language, its dual verbs of being, *ser* and *estar*, in order to illustrate his empirical theory. Whereas the English verb "to be" and its counterparts in nearly all other Western languages (*ser*, *être*, *sein*, etc.) may refer to both real and unreal states, qualities, or conditions, the second Spanish verb *estar* focuses on being somewhere in an actual situation. Marías says that it refers necessarily to reality. Cognate to but with a much broader range than the English "stand" and "stance," or the Latin *stare* from which it derives etymologically, *estar* is the -stance of circum*stance*.

I find myself in life; it is "where" I am (*estoy*), and this is the broadest meaning of the Ortegan "circumstance." I find myself circumstantially with things, yet not as another thing, not simply as existence or being but as a series of projects forming an all-encompassing task we call *living*. It is not enough, therefore, to say simply, "I live" (*yo vivo*); one must say "I am living" (*estoy viviendo*). And this on-going quality, which at once makes us aware that we are living from the past and toward the future, has structural implications, a structure that Marías describes as biographical and summarizes with this statement: "When we become interested in the *biographical structure* of *estar* – that is, when we consider *estar* in a simultaneously biographical and structural way, we arrive at an indispensable concept for a theory of man as empirical structure of life, for an anthropology in the strict sense of the word: the concept of *installation*."[19]

The scope of these comments does not permit more than a gesture toward the forms of worldly installation in the theory of Marías. They include, among others, the *sexuate* condition (the neologism is his); age in the sense of temporal and generational stages of life; race in a historical rather than ethnic meaning; and language with its embedded interpretations of reality and thus its empirical function as a primary metaphysics.

At the same time it must not be assumed that these installations are static. The frontal and futuristic (or "facial") movement of life is implied in the expression "I am living" (*estoy viviendo*). Thus I project my life from the installations in which I am living, and the tension between installation and projection can best be expressed in terms of vectors of varying intensity and direction. Marías refers to them as the vectorial structure of life.

In summary, which naturally is far from a conclusion of the matter, we can say that a full examination of the empirical structure as the concrete form of circumstantiality leads to an anthropology, and from there to the vast possibilities of personhood which Marías has been exploring in his most recent writings.[20] On the other hand, as we have seen, the analysis of human life as the primary or radical reality is either a metaphysics per se, as in the case of Ortega, or else serves as its propaedeutic in the thought of Heidegger. In any case, both theoretical levels can be seen in the title of what may prove to be Marías' major contribution to philosophy: *Metaphysical Anthropology: The Empirical Structure of Human Life.*

Houston Baptist University
Texas

NOTES

[1] Ortega never made overt mention of Unamuno's work, but it seems obvious that beginning as early as 1907–08 he was increasingly dismayed and later irritated by what he saw as Unamuno's irrational *boutades*. He wrote in September of 1908: "El espíritu de Unamuno es demasiado turbulento y arrastra en su corriente vertiginosa, junto a algunas sustancias de oro, *muchas cosas inútiles y malsanas*" (*Obras completas*, I, p. 118 [emphasis mine]). Ortega later confided to Julián Marías regarding the latter's work on Unamuno: "For you he is a theme. You do not realize that I have spent my life struggling with him" (*Memorias*, I, 351 [my translation, as are subsequent ones, unless otherwise indicated]). The most comprehensive and ambitious description of vital reason appears in *El tema de nuestro tiempo* (*Obras completas*, III, pp. 141–203).

[2] "En 1994–16 posee Ortega el núcleo de su filosofía. Su expansión, dilatación, articulación, serán la obra del resto de su vida, de cuarenta años de esfuerzo creador" (*Ortega: Las trayectorias*, p. 122).

[3] ". . . [L]a fenomenlogía no fue para nosotros una filosofía: fue . . . una buena suerte" (*Obras completas*, VIII, p. 42). Ortega associates himself with his German classmates Nicolai Hartmann and Heinz Heimsoeth and makes clear their dissatisfaction with the neo-Kantianism imposed on them at Marburg by Paul Natorp and Hermann Cohen.

[4] In "Husserlian 'Reduction' seen from the Perspective of Phenomenological 'Life' in the Ortegan School" (*Analecta Husserliana*, **XXXVI**, pp. 371–385).

[5] Of course, as Ortega points out in *Meditations on Quixote*, even so-called "natural" seeing is also a form of *cogitatio*: "If there were only a passive seeing, then the world

would be reduced to a chaos of luminous dots" (*Obras completas*, I, p. 336). To illustrate Ortega's point in another way, if we see before us what we call an "elderly man," it becomes apparent upon further consideration of our perception that the "elderly" qualities presuppose an assumption of prior ages in the life of the person. We know he is elderly because of the implicit comparison of more youthful phases of his life. Yet these earlier phases are not included in our visual sense. Ortega would like to say that we add them at the instant of visual reception and that seeing includes both a passive dimension, i.e., the sensory reception, and an active seeing that interprets as it sees.

[6] E. Husserl, *Cartesian Meditations*, p. 26.

[7] See Nelson Orringer, *Ortega y sus fuentes germanicas*, p. 39.

[8] *Obras*, I, p. 256.

[9] Ortega goes on to note: "The feature that distinguishes phenomenology as something new consists of elevating to the level of a scientific method the attention paid to such experience (*lo vivido*) within this plane of the immediate and patent as such" (*Obras completas*, I, p. 256).

[10] *Cartesian Meditations*, p. 26.

[11] *Obras completas*, VIII, p. 274.

[12] "A la altura de los tiempos: el nivel de la filosofía en Heidegger, Marcel y Ortega," *Dos centenarios filosoficos: Martin Heidegger – Gabriel Marcel*, pp. 77–92.

[13] *Being and Time*, p. 62.

[14] Ortega reproaches Heidegger for the latter's failure to explain specifically what he understands by *Sein*. It was but one instance, according to Ortega, of Heidegger's failure to go to the "root" of philosophical problems: "Not to have done this is what separates me profoundly from Heidegger, notwithstanding the fact that I admire his undeniable genius. . . ." But he goes on to admit their common starting point: "But I can accept almost none of his positions apart from those common to all of us whose starting point is *living human reality*" (*Obras completas*, VIII, p. 275 [emphasis mine]).

[15] Since I intend to follow Marías' reasoning in developing the "empirical" phase of this study, I prefer to rely on statements made in his seminal work, *Metaphysical Anthropology*. See especially pp. 58–60.

[16] *Metaphysical Anthropology*, p. 60.

[17] *Metaphysical Anthropology*, p. 74.

[18] *Metaphysical Anthropology*, p. 76.

[19] *Metaphysical Anthropology*, p. 82.

[20] In *La mujer en el siglo XX* (1980), *La mujer y su sombra* (1986), *La felicidad humana* (1987), *La educación sentimental* (1992), *Mapa del mundo personal* (1993), *Razón de la filosofía* (1993), and *Tratado de lo mejor* (1995).

REFERENCES

Heidegger, Martin, *Being and Time*. Translated by John Macquarrie and Edward Robinson (New York: Harper, 1962).

Husserl, Edmund, *Cartesian Meditations*. Translated by Dorion Cairns (New York: Collier Books, 1962).

Husserl, Edmund, *Ideen zu einer Reinen Phänomenologie und Phänomenologischen Philosophie* (The Hague: Martinus Nijhoff, 1952).

Marías, Julián, *Memorias*. I (Madrid: Alianza Editorial, 1988).
Marías, Julián, *Metaphysical Anthropology*. Translated by Frances M. López-Morillas (College Park: Pennsylvania State University Press, 1971).
Ortega y Gasset, José, *Obras completas*. 11 Vols. Madrid: Revista de Occidente, 1961–69.
Unamuno, Miguel de. *Del sentimiento trágico de la vida* (Madrid: Aquilar, II, 1958).

FRANCISCO JAVIER MASSA RINCÓN

DOING, THE COMMON ROOT OF "ETHICS" AND "METAPHYSICS" IN THE PHILOSOPHY OF JOSÉ ORTEGA Y GASSET

The object of this paper is to establish that ethics provides the foundation for metaphysics, both being understood not in response to the *quaestio ontica* (relative to the specific, factually existent individual) but rather in response to the *quaestio ontologica* (human life as radical reality).

Establishing so extraordinary a statement requires a series of preliminary clarifications that define the meaning of the terms "ethics" and "metaphysics." There is a key term (that of existence), which shows the way to proposed founding. To put it briefly: ethics lays the foundation for metaphysics to the extent that it is the prerequisite for existence. The meaning of this cryptic thesis becomes clearer when we say that: metaphysics, understood according to the definition ventured by Aristotle (*zetoumene episteme*), but freed from the scholastic-Aristotelian perspective that has surrounded it practically until the present century, fits perfectly with the projective mission that "metaphysical doing" consists of for Ortega. Metaphysics, from this perspective, should not be understood as an ambit which more or less transcends the human being, supposedly inhabited by a fundamental being which is the giver of ultimate meaning to reality: rather, although without excluding this possibility, metaphysics is a projective doing that life itself, inasmuch as it is a connective reason for reality, demands in order to fulfill this rational requirement. Metaphysics, then, is not an ambit, but an itinerary – an itinerary with no assumptions, given that the doing which shapes it does not admit them.

Thus understood, metaphysics requires a principle to endow it with this final necessary radicality. This principle is ethics, from which possibility acquires effectivity. Ethics is not, like metaphysics understood in the traditional manner, the *terminus* of our intellectual operations. Neither is it, like the metaphysics understood in the new sense postulated, an itinerary, but rather its possibility. The key to everything must be sought in the doing which life consists of and with which, inevitably, I find myself. Doing seems to point to the existent. To consider this in

A-T. Tymieniecka (ed.), Analecta Husserliana, Vol. LIV, 389–394.

and of itself is to fall into realism, and that is a supposition. Rather, doing executes the effectiveness of the possible as such: it invents (*in-veniat* = to make come) its being. Idealism? Not at all. There is no room here for any dealing with reality that comes about by executive means in the idealist consciousness.

Is Ortega the first to have uncovered the ethical condition of metaphysics? Formally yes, intuitively no. In a manner analogous to Aristotle discovering the "thought that one thinks to one's self" and attributing it to God, Leibniz effects something similar regarding the ontological priority that doing (as the ethical structure of human life) has over its metaphysical concretion.

"*Nihil existit sine majore existendi quam non existende ratione.*" This formula is in the important work published by G. Grua under the title *De contingentia.*[1] It distinguishes between necessary and contingent truths, but all share the connection between the subject and the predicate, that is, the property expressed in the principle of inclusion. This makes it possible to prove necessary truths, and establishes the verification of contingent truths as the ideal attainable to man only after an infinite hypothetical analysis.

The formulation of the question is to be found in this context. Although it does not contain the verbal predicate "esse," due to the fact that one of the fundamental meanings of the "sum" is precisely that of existing, this version reiterates the specific value of the principle of sufficient reason in the ambit of existents. The affirmation that "nothing exists without more reason to exist than not to exist" presupposes the generating principle that everything possible tends towards existence.[2] On a first level the only requisite for existence is that of logical possibility. Everything that is not internally contradictory has a reason for its existence, which is its natural tendency towards the same. Due to the mere fact of not being internally contradictory, it has more reason to exist than not to exist. However, this reason does not suffice to make its existence effective. Not all possibilities exist.

On a second level there is a limiting principle of reconcilability. Not all possibles are reconcilable with each other. This generates diverse series of reconcilables factually exclusive of each other. Only one of them can enter into existence. The principle of sufficient reason establishes a primary logical requirement directed at existents. Which of the series of reconcilables finally achieves existence? What is the selection criteria at work among all that, which has "more reason to exist than not to exist?"

"*Semper illud fieri, quod plus rationis habet.*" This other principle, that of sufficient reason, appears in the same context as the previous formulation. According to Leibniz himself, it is the opposing, or positive, version of "*nihil fieri sin ratione*": "And nevertheless we accept at once as universal and general that which reason and experience confirm (to the extent that it is given to us to penetrate into the meaning of things), that principle implanted in our spirit by God that nothing happens (*fieri*) without reason, and, contrariwise, that which has most reason always happens (*fieri*)."[3]

Here the selection criteria previously sought is expressed. From among the diverse series of reconcilables that which "has the most reason" comes into being. The choice is not only between logical reconcilables, but also between reconcilables that are factually exclusive ("*ex oppositis*"). Thus having a reason to exist does not suffice. For this reason to be enough, one must have a greater reason than other possibles.

The criteria for the comparison of reasons is that of essence, that is to say, the wealth of phenomena brought to the final result of the existent, which is this world. In the Leibnizian exposition this choice is identified with God's morally necessary decision to choose what is best for existence. The moral necessity stems from the fact that the decision of the will is informed by an exhaustive knowledge. Consequently, the will, directed towards the good, cannot but choose what is best.

We can conclude that in Leibniz, existence is forged in the will. Ortega's formula "life gives much to do"[4] seems to point to something similar. Nevertheless, the existence that I *might* have is not the same as that which, for instance, a table has, willy-nilly. With this consideration we enter into a hermeneutic of existence which, applied to life, radically problematizes its essence because it must be performed in circumstances foreign to it:[5] "I am unique, my essence is mine alone, and it must be acted out in the other. Thus existence does not coincide with essence, with being." Man, having to live outside of himself, among things, among other men, brings no pre-formed essence to his self-execution ("nothing substantive has been given to man. He has to do everything for himself"[6]). It is precisely this "being linked to that which we are not yet" that makes man different from all other beings and poses the root of man's problem of being.

What Ortega means to affirm with this is our life's strict character of uniqueness. It is in no way intended to affirm that *my life* is the only thing that really exists – among other reasons because *my life*, in the strict executive sense, is the only thing that, in its existential fulfillment, makes the coming to pass of concrete-existents-homogeneous-among-themselves possible. Uniqueness as a primary category simply indicates the inexorably unchangeable (which has already been pointed out, there being no possible homogeneity between my essence and the essence of things, since I myself lack it), the untransferable character of our life.

Speaking of metaphysics as the doing which is "the search for a radical certitude in radical reality" only makes sense in this context.

If we agree to define this peculiar quest as the science that investigates "the first," we will not hesitate to agree with Ortega on the constituents of its primordial notes, to wit, that . . . "*the philosopher is ignorant of his object and of himself knows only: 1) that he is no other object; 2) that he is an integral object, that he is the authentic all, that which omits nothing and, consequently, that alone which is enough for himself.*"[7]

It is this vision of the object of philosophy that leads Ortega to consider that this science has always left behind it a repertoire of unquestioned beliefs – among them the belief in being,[8] the belief that things have a being. The problem does not arise because man "lives in the belief in being" or, to put it in a less anthropologizing manner, that "man lives in being" (this not being, strictly speaking, a problem since it belongs to what Ortega calls *prephilosophy*); the problem arises, rather, from the moment at which one begins to philosophize, since philosophy cannot be founded on belief in being, though it may have to account for it, if it does not want to be a penultimate philosophy without radicality, that is to say, something other than philosophy.

"Problematicity will figure among the primary metaphysical categories," warns Rodríguez Huéscar in his *Perspectiva y Verdad*,[9] analyzing a text of Ortega's from 1910: *Adán en el Paraíso*. Regarding this we may ask ourselves the meaning of that radical problematicity of life on this initial level of Ortegan thought. Ortega shows it to us through the concept of awareness. "Becoming aware of something," he writes, "is not knowing it, but merely noticing that something is before us."[10] *Awareness* then, is the act of "noticing" or, as Ortega would say later, of grasping something. "But 'to become aware of' is not yet to know – and I underline the 'yet,'" elaborates Rodriguez Huéscar, "because it is essential. A dark spot in the distance, on the horizon, what could it be? Could it be a man, a tree, a church tower? We do not know: the dark spot awaits, it *aspires* to be made out: before us we have not a thing, but a problem. What we notice, through awareness, is not then anything, it is not even a thing at all, but a problem. Consequently, becoming aware is not just a not knowing, or being ignorant of, but also a concomitant *knowing that we do not know*, and, finally, a petition to know that which we do not know. The 'petition' seems in this instance to come from that which is not known, which *aspires* to be made out. *Awareness* is, then, in this instance, consciousness of a problem."[11]

It is in this sense that we must understand the statement made by Ortega in *Qué es Filosofía?* regarding "the essential determination of the I": "*I am not but that which I think I am*" which, far from outlining any kind of relativism is simply proof that life is, radically, essentially consciousness (understood not in the idealist sense but rather in the sense of "to give account and reason of oneself"), since life is also radically and essentially a problem; consequently the meaning of this affirmation is that life is *consciousness of a problem* and, as Rodriguez Huéscar points out, because of this it is also a problem of consciousness "inasmuch as in this, its radical problematicity, its essential structure and ethical consistency will be established."[12]

University of Barcelona

NOTES

[1] G. Grua, *De contingentia*, 1686, 302–306.

[2] The best example of how far Ortega is from this understanding of existence as a "requirement of the possible" is in his *Meditaciones de Quijote, Obras Completas*, I (Madrid: Alianza Ed., 1983). In Leibniz X is possible if the sum of notes that make up the concept of X are not logically self-contradicting. In Ortega's meditation on the forest (op. cit.) the fundamental idea is that the *vital* reality forest, that is to say the forest *as forest* – not as "set of trees," or land, or aggregate of wood – consists of *possibility*. Ortega introduces an ontological subject of the utmost importance with this appraoch, to wit, that of *realities* which consist of *possibilities*. The entire paragraph is devoted to the clarification of this idea: "It is a path where we could detain ourselves; it is a place of many springs from whence we hear a feeble murmur in the arms of silence, and which we could discover in a few steps; they are verses of songs sung in the distance by birds perched on some branches under which we could venture. The forest is a sum of possible acts of ours which, upon fulfillment, would lose their true value" (op. cit., p. 331). The

last sentence is decisive: in it appear *substantivized* (not abstracted) possibilities, and the consistence of reality as such is based upon them.

[3] "*Et tamen illud simul tenere in universum vel generatim, quod et ratione et experimentia ipsa (quantum nos in res penetrare datum est) firmatur, insitum divinitus menti nostrae principium, nihil fieri sine ratione, et ex oppositis semper illud fieri, quod plus rationis habet*" (Grua, op. cit., 304–305).

[4] José Ortega y Gasset, *Unas lecciones de metafisica, Obras Completas*, XII, *Leccion* III, p. 39.

[5] *Ibid.*, p. 57.

[6] José Ortega y Gasset, *Ensimismamiento y alteracion, Obras Completas*, V, p. 301.

[7] José Ortega y Gasset, *Qué es Filosofía?*, *Obras Completas* VII, p. 308.

[8] Ortega understands "belief" as the foundation of my life, that which enables me to live; man *is* in "belief"; belief is not something man *has* – having would apply rather to ideas, that is, to more or less complex psychological processes, in short, to the *quaestio ontica* – and that is not the point.

[9] See Antonio Rodriguez Huéscar, *Perspectiva y Verdad*, Madrid: Alianza Editorial, 1985.

[10] José Ortega y Gasset, *Adán en el Paraíso, Obras Completas* I, p. 475.

[11] Rodriguez Huéscar, op. cit., p. 52.

[12] *Ibid.*, p. 53.

ANTONIO ZIRIÓN QUIJANO

THE HUSSERL DICTIONARY AND THE GUIDE FOR TRANSLATING HUSSERL INTO SPANISH

My purpose here is to inform the scholarly community about two lexicographical projects that I have undertaken on Edmund Husserl's work. The first of these is one that I have called, provisionally, the Husserl Dictionary. The second one is the compilation of a glossary which has the purpose of serving as a guide for translating Husserl into Spanish.

To be able to appreciate the motivation behind both projects, which is, in my opinion, not only a personal one, but also one philosophically and pedagogically motivated, I think it is worthwhile to get started with a review of the difficult situation I found when I began to study Husserl in Mexico more than fourteen years ago.

At that time there were no professors who would formally expound Husserl's thought or who would teach Phenomenology, neither in the Facultad de Filosofía y Letras of the Universidad Nacional Autónoma de México – where I studied – nor in any other university in Mexico City or the whole country. José Gaos, who was a translator of three of Husserl's major books and a professor of Phenomenology among many other things, and in some sense a phenomenologist himself, had passed away twelve years before. Fernando Salmerón and Luis Villoro, who once had had considerable closeness to Phenomenology, were already far from Husserl and his Phenomenology and did not teach about it anymore. Teaching on Husserl throughout the country came down to a few isolated lessons or lectures, to insufficient mentions and readings in Introduction to philosophy courses, and to much too brief allusions in courses on the history of philosophy in the twentieth century.

Furthermore, there were no accessible editions of the very few works in Spanish of an introductory or general character about Husserl and Phenomenology. By the way, there were no accessible editions either of well-known translations of some of his most important works: Gaos' translations of *Ideen I* and *Cartesianische Meditationen*, Villoro's translation of *Formale und Transzendentale Logik*, Elsa Tabernig's translation of *Philosophie als strenge Wissenschaft* and Otto Langfelder's translation of *Phänomenologie des inneren Zeitbewusstseins*.

With the exception of a 1967 edition of *Investigaciones lógicas* that

A-T. Tymieniecka (ed.), *Analecta Husserliana, Vol. LIV*, 395–405.

I got by miracle, I studied all these works in library copies and in bound photocopies. And of course, none of the various introductory and brief works written by Husserl were yet translated.

Obviously, this situation was not at all suitable to the alleviation of the already great difficulty intrinsic to the study of Husserl's thought. But there were yet other factors: one of them was the absolute absence of incentives coming from a medium where Husserl was, although nominally revered by some, really ignored by the majority and practically unattended to by all; and the other was the additional and unnecessary difficulty created by the lack of consistency and uniformity in the existing translations of Husserl. For that reason, the study of Husserl was, for the average student without a solid knowledge of the German language, an undertaking almost impossible to accomplish, a prospect that alone was capable of defeating the few Husserlian vocations that emerged, even before these could become established or even be recognized as such.

This was true, at least, for Mexico in 1980. Since then, a few of the conditions pointed out have changed, but the general situation is almost the same. It would be interesting to find out to what degree the situation is comparable in the rest of Latin America. There are common problems, to be sure, as for instance in what I have said about translations. Maybe to the optimists it would not sound too shocking to verify that to date only about the 24% of the *Husserliana* collection has been translated into Spanish. If we add the translation of *Erfahrung und Urteil*, not included in *Husserliana*, that percentage grows to 27%; so there remains to be translated only 73% of the corpus, if we think that it is worthwhile to translate the whole. But it has to be said that I am not including the *Textkritische Anhänge*, which no doubt would make the percentage of the not yet translated grow to more than 90%. There is another fact that might be revealing: of all existing translations, almost 78% were made before 1962, in the 33 years from 1929 to 1962; in the last 31 years, in contrast, what has been translated is only somewhat more than the 22% of everything translated. And I will not talk, in order not to lament it, about the almost absolute lack of translations of what is called secondary literature (here would be included series as valuable as *Phaenomenologica* or *Analecta Husserliana*, so we need not go too far), the translation of which has to be a substantial part of an authentic phenomenological culture. But let us put this and its somber implications aside.

I mentioned a moment ago the intrinsic difficulty of the study of

Husserl's thought. I think, in effect, that independently of any unfavorable contingent conditions, the process of familiarizing oneself with Husserl's Phenomenology is a painful process *per se*. I suppose that this will be evident for those who have been in front of a group of students who lack any previous knowledge of Husserl or of his works. In what concerns myself, when I embarked on my job as a teacher and became aware that not only had it been difficult to learn something about Husserl, but that to teach about him was also a tough business, I became convinced that the teaching effort, as important as it was, was not enough.

The scant labor of translation that I have done on some introductory works by Husserl[1] had as its first intent to make that process less painful to the student. But very soon (precisely in 1984) there occurred to me the idea of making what I now call a Husserl Dictionary, and much later, during work on that and almost naturally, there arose the idea of a Glossary for translation.

Even though I first had the idea of the Husserl Dictionary in the last months of 1984, many diverse activities prevented me in the following years from carrying out that project much further than to outline the general design of the dictionary and to write an essay or prototype entitled "Husserl Dictionary in Miniature." This miniature of a dictionary was made without the help of computers, and was based on only one, and a brief one, of the works of Husserl: *Die Idee der Phänomenologie* (*Husserliana* II), and its counterpart in Spanish *La idea de la fenomenología* (Madrid: FCE, 1982). In fact, I could not begin work on the dictionary until 1991.

But what is this project about? The Husserl Dictionary is, or will be, what I call a "contextual dictionary," that is, a dictionary that, under the terms or concepts selected as entries, gives the passages or excerpts of the works that form the corpus upon which the dictionary is made, passages in which those terms or concepts are defined or clarified, or at least used in such a way that their sense may be fixed, be it relatively so. To the mere selection and grouping of the passages several kinds of annotations may be added, which normally have the function of linking groups of passages or of terms, explaining exceptions, etc., as well as giving specifications about the various senses or acceptations of a term or group of terms, etc.

This kind of "contextual" dictionary has been made upon the works of many philosophers. All, or at least all Kant students or scholars, know for instance the *Kant Lexikon* of Rudolf Eisler (Berlin: 1930), which in some respects has served me as a model. And it should be clear, already because of the comparison with this illustrious antecedent, that, in spite of my calling it a "contextual" dictionary, the Husserl Dictionary has in fact very little to do with a concordance.

A concordance is a list – that can be automatically and easily obtained with a computer – of some or all of the occurrences of some or all of the terms within a work. Normally, a concordance gives, beside the reference for each occurrence (work, page, line), the occurrence itself, that is, the term within a context of variable size (almost always a line, sometimes two or three, before and/or after the term, etc.); but this size is fixed, that is, it is the same for all the terms of the concordance.

I want to clear up that I have nothing at all against concordances – which, for sure, have also been made (and continue to be made) for the classical works of the main philosophers in the history of philosophy. On the contrary, I think they are very respectable works and very useful in some fields for some kinds of research. I simply want to distinguish the idea of the Husserl Dictionary from the idea of a concordance. And the main difference is that the dictionary requires a selection, hierarchization and arrangement, first of the terms, but also of the passages (or contexts), and almost as a consequence of this, it does not imply the same size of the passage or context for every term, since this size will be determined by the nature of the passage itself in relation to the term or concept that it serves to "define" or clarify. In this sense, the Husserl Dictionary has more resemblance to a subject index than to a concordance; it is just a subject index that not only lists the references, but presents the passages themselves, the texts themselves.

Of course, this does not mean that the computer may be put aside. The computer is used practically in almost all phases of the project, from the capture and collation of the texts themselves to the extraction of the passages pertaining to a particular concept, the compilation of the vocabulary of entries, et cetera, et cetera. And the computer is used in the project in such a way that perhaps it would be of interest to treat this theme by itself in a specialists' session. I will refer, though in a superficial way, to this point and to the distinct steps of the job. But I have to explain first another of the features that the dictionary will have: to wit, that of being a bilingual dictionary.

With much generality, the main purpose of the dictionary is to facilitate the study of Husserl for Spanish-speaking students, and I seek to do this by putting within reach of the student of any one of the (already translated) works of Husserl what is nuclear or substantial in the remainder of his works: that is, exactly, passages from other works, and also the same work, where Husserl treats the themes that the student is interested in, in any particular case.

But, because of a series of reasons which it would be tedious to list here – one of which will be brought to light immediately – I am interested in taking or bringing this student (that is, the student who is reading his Husserl text *translated into Spanish*) to the original texts of Husserl, to the texts in the German language of Husserl. That is why the Husserl Dictionary will be bilingual, and this is why it should consider as its basic corpus at the very least all the works of Husserl already translated into Spanish (and in principle, also all those to be translated in the future, with no limit in time), as well as their corresponding originals in German. In this way, he who would consult it will have in every instance the possibility of referring to the very words of Husserl. Independently of any other considerations, I think this feature has an important pedagogic value, and this is for me its chief merit.

I may go now into the promised allusion to the more technical questions. Even though there still are some technical parts of the project which are not completely solved, the general plan of the job is decided. The capture of the texts is the first step. Naturally, a scanner is used in this and an optical character recognition (OCR) program; but the capture accomplished with the scanner is not the only one that is made: because of motives concerning division of time and human resources, another capture is made by hand, and this with the object of collating both versions and getting a clean text. This collation is effected semi-automatically with the *Nota Bene* program, a potent word processor expressly designed for academic work. The most recent version of *Nota Bene* still works under the DOS operative system (but I understand that a Windows version is in preparation).

Once we have the clean text, or the clean texts, the German and the Spanish, the most important part of the whole work may begin, that is, the selection and marking of concepts and passages. Both operations are effectuated simultaneously upon both texts and in this *Nota Bene* is also used. I do not need to say that this selection, and the careful reading that it requires, are purely and exclusively human labors (and philosophical labors, I should add).

The following steps of indexing, compilation, grouping, et cetera, which are rather more complex and which I am not going to expound on in detail, are carried out with the help of *Nota Bene Orbis* and *Nota Bene Tabula*, which are two other auxiliary modules of *Nota Bene* that permit it to create what the program calls "text-bases" (data bases with a text content, simply told) and concordances upon them, respectively.

The last result will be an alphabetical list of terms grouped conceptually and under which there will be found, arranged and grouped according to their sense and signification, the passages or the texts themselves that define them, or clear them up, etc. Or, better, two lists, the German and the Spanish, one beside the other. But, what I have said notwithstanding, I want to emphasize that the final format of the dictionary (the way to present it, so to speak) is not yet decided. Precisely, to have in an electronic medium the lists of which the finished dictionary will consist allows us to think over the possibility of very different forms of presentation or publication, including publication on disc or in other electromagnetic media, or the possibility of editing the dictionary as a sort of analytic apparatus along with the publication on disc of the works of Husserl that compose the corpus of the dictionary. But I am not going to enter here in more details about that. It is time to expose how the idea of a glossary for translation arose out of the project of the Husserl Dictionary.

Being a professional translator, so to speak, I have nevertheless a great distrust of translators. It is not that I am convinced that a translator is always a traitor; the translator may well be essentially a traitor, as is sometimes said; but the traitor to one cause may be a hero to another cause (at least in cases such as this). The problem with translators is not only the very common, unfortunately, lack of professionalism, incompetence, irresponsibility, superficiality, but also the fact that there are different and differing so to speak "schools" of translation, guided by opposite principles or by diverse sensibilities concerning language (here I mean mainly the Spanish language). What has happened, then, is that Husserl translations into Spanish are to a great extent lacking in consistency, and, as a group, lacking in uniformity.

Somewhat arbitrarily, I call "consistent" the translation that renders always the same expression (with the same sense) of the original language (the *Ursprache*) with the same expression in the language into which it is translated (the *Zielsprache*), and different expressions with different expressions; and I call "uniformity" this "consistency" when extended to various works or texts. (These terms could no doubt be replaced by those of intra-textual uniformity and inter-textual uniformity, respectively.)

The goal of facilitating the study of Husserl in Spanish, that is the main goal of the Husserl Dictionary, could not ignore this situation, and should thus propose for itself the prior goal of contributing to the consistency and uniformity of the translations of Husserl into Spanish. Obviously, if this point were not attended to, the dictionary would simply reflect the lack of uniformity of the existing translations, which would be catastrophic, because it would only serve to prolong the reigning terminological confusion. There was no other solution that presented itself but this one: to modify the passages on the Spanish side of the dictionary to make them consistent and uniform. That would be a very long road, certainly; but as long as I could not find a shorter one, I decided to begin walking the long one.

This is a road that is really longer still. There have been, and there are still, no criteria nor serious discussions among Husserl translators, or even among specialists in Husserlian terminology, and therefore there have been no decisions as to the terms or expressions that more aptly translate the majority of the most important concepts in his works, nor concerning the ways to solve a series of very peculiar problems of translation in view of the peculiar senses in which Husserl uses some words.

There arose the idea of elaborating a Glossary-Guide for translating Husserl into Spanish based on the discussion and interchange of ideas of a task force. At this point, mention of the antecedent of the well-known and rightly renowned *Guide for Translating Husserl* by Dorion Cairns is unavoidable. Cairns' *Guide* has a repertory of about 3,215 main terms (the entries in boldface), plus 1,774 second acceptations or compound phrases or idioms, with a total of 4,989 expressions for which a translation or several translations are given. As you know, Cairns not only proposes his own translations into English, but includes other translations already done into English and into French, and that is why he calls his guide, in the first line of his Preface, a "multilingual glossary." Moreover, he includes occasionally, as references, terms which are found in works translated and published into English by other philosophers: Hume, Kant, Stuart Mill, etc.

On my part, I initiated the work on the guide for translating Husserl into Spanish in 1992. I will not tell here all the history of the last three years in which I have tried to gather a team of translators and specialists, and to organize a meeting during which we could discuss the kind of criteria and problems which I mentioned above. But for several reasons, some personal or circumstantial, extrinsic, others intrinsic, this working system failed. But there is something still there, and I want to refer briefly to it, and how I am planning to renew the working project based on it.

What we have is a computerized data base which includes among its fields:

FIRST: The complete Cairns' *Guide for Translating Husserl*, with all his contributions. In effect, the inventory of terms in the Cairns' guide was intended to serve, from the beginning, as a basic list upon which we would work. In the data base, a record has been assigned to every acceptation, every sense, every compound word or phrase, and their respective translations.

SECOND: To the Cairns' corpus, some 134 entries have been added, either by me or by some of the members of the informal team. There remains open, of course, the possibility of eliminating terms of the Cairns list.

THIRD: Then we have proposals for translations into Spanish for many of the terms or expressions of the data base, provided by members of the team which is at present dissolved. Initially, each member of the team was invited to propose his translations for the terms on Cairns' list. I received, and included in the data base, only the lists of Mario A. Presas (Argentinian, translator of *Cartesianische Meditationen*), Julia Valentia Iribarne (also Argentinian, who has published a book on "intersubjectivity in Husserl" which includes some passages translated from *Husserliana XV*), and Agustín Serrano de Haro, working in close association with Miguel García-Baró (both from Spain; García-Baró has translated *Husserliana II* and the Fifth Cartesian Meditation – to complete José Gaos' translation of the first four). Of course, here I include also my own proposals for many terms.

FOURTH: Then we have the translations to many of the terms in the data base which are found in the works of Husserl already published in Spanish. See: the translation of Manuel García Morente and José Gaos of *Logische Untersuchungen*, the already mentioned translations of Gaos, Luis Villoro, Elsa Tabernig, and Otto Langfelder, then the Jas Reuter translation of *Erfahrung und Urteil*, the Jacobo Muñoz and Salvador Mas translation of *Die Krisis*, the very recent translation by César Moreno and Javier San Martín of the lessons on *Grundprobleme der Phänomenologie*, and also the just mentioned translations by García-Baró and Presas and my own translations of *Die Pariser Vorträge* and *Der Encyclopaedia Britannica Artikel*. I am not counting, though they will be included in due time, the translations actually in process: the one of *Ideen II* that I am about to complete, and that of *Ideen III*, which is being done by Luis Eduardo González. It has to be recognized that up to now the incorporation of (the terms of) all these translations into the data base has not been done in a systematic way, and it is far from being complete.

FIFTH: We have also, lastly, anticipated fields, and to a very limited degree filled them up, for the translations found in standard bilingual dictionaries (German-Spanish), or the definitions of the terms in German dictionaries.

SIXTH: We have also foreseen, though it be perfectly empty, the field for the final selection, for the term selected to translate each expression in the base. This will be the last step in the whole process.

This is then what we have. The data base is made with some of the auxiliary modules of the *Nota Bene* word processing program, called *Nota Bene Ibid. Plus*. I think it worthwhile to mention an advantage that *Ibid. Plus* shares with all the modules of *Nota Bene*: all of them may be used as a part of the normal functioning of the main word processor, without having to close it first. In the case of *Ibid. Plus*, this feature allows one to have access to the data base and consult it at any time, while one is working in word processing.

Now, this data base, the main tool of the work, may grow practically in indefinite measure and in various respects: it can grow indefinitely as to the information added to each field; it can grow indefinitely as to the number of records; and it can grow also in a great measure as to the number of fields that could be added. Because of this capacity to shelter new fields, this very data base may serve as a fundamental structure for several distinct projects, even a project, for instance, of assembling a glossary for the translation of Husserl into all Romance languages, which would be a fine way of crowning the idea of Cairns of making a multilingual glossary.

But at the moment I am not aspiring to so much. I would be content with the development of the more humble project of completing a guide for translating Husserl into Spanish, with contributions of the many translations already done and of translators and specialists, and have in this way a canon similar to the Cairns' *Guide*, partially adopting its structure but also modifying it in several respects.

Having discarded the idea of forming a fixed task force, I am planning now to work out a kind of newsgroup system on the Internet network, sending my proposals for translations and my argumentation to those interested in knowing them, and accepting comments and proposals from them. I am right now working on the design for such a system, which in principle does not need to be too complex, and on the mechanics of the work that it will involve. Although there are still many indefinite or obscure points, I hope that the phenomenological community of the world (but primordially of the Spanish-speaking countries) will receive this same year my invitation to take part in the work to be carried out using this new system.

Once the glossary-guide is finished, and I think I am not too optimistic in thinking that a first version of it could be ready in 1998, the next step will be to make uniform, based in it, the translation of the passages of the Husserl Dictionary already selected. This making uniform will not be in any way thoroughgoing, but it will indeed comprise at least all terms selected as entries of the dictionary, though in all passages. Only in this way will the Husserl Dictionary be able to fulfill its task of facilitating the study of Husserl for the Spanish-speaking students, and also fulfill the additional task of serving as a basic reference work for research into the arduous and vast work of Husserl. I think I am not too pessimistic if I say that the Husserl Dictionary, or a first version of it, will not be ready before 2000.

But science, these are Husserl's words, is an infinite task. . . . Well, it is science that is at stake here, or at least I so believe. I believe in effect that Husserl carried out the most serious project in this century to transform philosophy into a rigorous science, and I believe moreover, in all seriousness, that philosophers (or, with more caution, phenomenologists) should not simply pass on to other things before they have reviewed the virtues and scope of this project or before having extracted from it all its potentialities. And I believe also, to make matters worse, that this cannot be done if we do not at last bring to its crescendo (and we are far from doing so) our assimilation of the work of Husserl, the published work and the work still to be published. And I believe that this means, in good measure, to undertake the task of clarifying concepts, of hierarchizing the texts themselves and their passages, and of ordering and gathering the interpretations. And I believe, lastly, that, for Spanish-speaking people, all this implies the task, in a certain sense lateral and in a certain sense preliminary, of agreeing on criteria for translation and bringing uniformity to the translations – the task, in the end, of creating a canon of translation.

As for myself, I only hope that, with the technical works that I have undertaken, the day will sooner arrive in which for the study of Husserl in Spanish there remain only the difficulties proper to the "things themselves," that is to say, those of the texts themselves (already correctly translated, etc.) and the ideas they embody.

Universidad Nacional Autónoma de México

NOTE

[1] *Las conferencias de París. Introducción a la fenomenología transcendental*, and *El artículo de la* Encyclopaedia Britannica. Both published by Instituto de Investigaciones Filosóficas of UNAM (Mexico City) in its series Cuadernos (Nos. 48 and 52, in 1988 and 1990, respectively).

This book is a sequel to:

Volume LII: Phenomenology of Life and of the Human Condition, Book I: *Laying down the Cornerstones of Its Field*
Volume LIII: Phenomenology of Life and of the Human Condition, Book II: *The Ontopoietic Outburst in Creative Virtualities*

and to:

Volume XV: Foundations of Morality, Human Rights and the Human Sciences
Volume XX: The Moral Sense in the Communal Significance of Life
Volume XXII: Morality within the Life- and Social World. Interdisciplinary Phenomenology of the Authentic Life in the 'Moral Sense'
Volume XXXI: The Moral Sense and Its Foundational Significance: Self, Person, Historicity, Community. Phenomenological Praxeology and Psychiatry
Volume XXXV: Husserlian Phenomenology in a New Key. Intersubjectivity, Ethos, the Societal Sphere, Human Encounter, Pathos
Volume XLV: Z. Zhai: The Radical Choice and Moral Theory. Through Communicative Argumentation to Phenomenological Subjectivity

INDEX OF NAMES

A-T. Tymieniecka (ed.), Analecta Husserliana, Vol. LIV, 407–411.

Analecta Husserliana

The Yearbook of Phenomenological Research

Editor-in-Chief

Anna-Teresa Tymieniecka

The World Institute for Advanced Phenomenological Research and Learning, Belmont, Massachusetts, U.S.A.

1. Tymieniecka, A-T. (ed.), *Volume 1 of Analecta Husserliana.* 1971 ISBN 90-277-0171-7
2. Tymieniecka, A-T. (ed.), *The Later Husserl and the Idea of Phenomenology.* Idealism – Realism, Historicity and Nature. 1972 ISBN 90-277-0223-3
3. Tymieniecka, A-T. (ed.), *The Phenomenological Realism of the Possible Worlds.* The 'A Priori', Activity and Passivity of Consciousness, Phenomenology and Nature. 1974 ISBN 90-277-0426-0
4. Tymieniecka, A-T. (ed.), *Ingardeniana.* A Spectrum of Specialised Studies Establishing the Field of Research. 1976 ISBN 90-277-0628-X
5. Tymieniecka, A-T. (ed.), *The Crisis of Culture.* Steps to Reopen the Phenomenological Investigation of Man. 1976 ISBN 90-277-0632-8
6. Tymieniecka, A-T. (ed.), *The Self and the Other.* The Irreducible Element in Man, Part I. 1977 ISBN 90-277-0759-6
7. Tymieniecka, A-T. (ed.), *The Human Being in Action.* The Irreducible Element in Man, Part II. 1978 ISBN 90-277-0884-3
8. Nitta, Y. and Hirotaka Tatematsu (eds.), *Japanese Phenomenology.* Phenomenology as the Trans-cultural Philosophical Approach. 1979 ISBN 90-277-0924-6
9. Tymieniecka, A-T. (ed.), *The Teleologies in Husserlian Phenomenology.* The Irreducible Element in Man, Part III. 1979 ISBN 90-277-0981-5
10. Wojtyła, K., *The Acting Person.* Translated from Polish by A. Potocki. 1979 ISBN Hb 90-277-0969-6; Pb 90-277-0985-8
11. Ales Bello, A. (ed.), *The Great Chain of Being* and *Italian Phenomenology.* 1981 ISBN 90-277-1071-6
12. Tymieniecka, A-T. (ed.), *The Philosophical Reflection of Man in Literature.* Selected Papers from Several Conferences held by the International Society for Phenomenology and Literature in Cambridge, Massachusetts. Includes the essay by A-T. Tymieniecka, *Poetica Nova.* 1982 ISBN 90-277-1312-X
13. Kaelin, E. F., *The Unhappy Consciousness.* The Poetic Plight of Samuel Beckett. An Inquiry at the Intersection of Phenomenology and literature. 1981 ISBN 90-277-1313-8
14. Tymieniecka, A-T. (ed.), *The Phenomenology of Man and of the Human Condition.* Individualisation of Nature and the Human Being. (Part I:) Plotting

the Territory for Interdisciplinary Communication. 1983
Part II see below under Volume 21. ISBN 90-277-1447-9

15. Tymieniecka, A-T. and Calvin O. Schrag (eds.), *Foundations of Morality, Human Rights, and the Human Sciences.* Phenomenology in a Foundational Dialogue with Human Sciences. 1983 ISBN 90-277-1453-3
16. Tymieniecka, A-T. (ed.), *Soul and Body in Husserlian Phenomenology.* Man and Nature. 1983 ISBN 90-277-1518-1
17. Tymieniecka, A-T. (ed.), *Phenomenology of Life in a Dialogue Between Chinese and Occidental Philosophy.* 1984 ISBN 90-277-1620-X
18. Tymieniecka, A-T. (ed.), *The Existential Coordinates of the Human Condition: Poetic – Epic – Tragic.* The Literary Genre. 1984 ISBN 90-277-1702-8
19. Tymieniecka, A-T. (ed.), *Poetics of the Elements in the Human Condition.* (Part 1:) The Sea. From Elemental Stirrings to Symbolic Inspiration, Language, and Life-Significance in Literary Interpretation and Theory. 1985
For Part 2 and 3 *see below* under Volumes 23 and 28. ISBN 90-277-1906-3
20. Tymieniecka, A-T. (ed.), *The Moral Sense in the Communal Significance of Life.* Investigations in Phenomenological Praxeology: Psychiatric Therapeutics, Medical Ethics and Social Praxis within the Life- and Communal World. 1986
ISBN 90-277-2085-1
21. Tymieniecka, A-T. (ed.), *The Phenomenology of Man and of the Human Condition.* Part II: The Meeting Point Between Occidental and Oriental Philosophies. 1986 ISBN 90-277-2185-8
22. Tymieniecka, A-T. (ed.), *Morality within the Life- and Social World.* Interdisciplinary Phenomenology of the Authentic Life in the 'Moral Sense'. 1987
Sequel to Volumes 15 and 20. ISBN 90-277-2411-3
23. Tymieniecka, A-T. (ed.), *Poetics of the Elements in the Human Condition.* Part 2: The Airy Elements in Poetic Imagination. Breath, Breeze, Wind, Tempest, Thunder, Snow, Flame, Fire, Volcano... 1988 ISBN 90-277-2569-1
24. Tymieniecka, A-T., *Logos and Life.* Book I: Creative Experience and the Critique of Reason. 1988 ISBN Hb 90-277-2539-X; Pb 90-277-2540-3
25. Tymieniecka, A-T., *Logos and Life.* Book II: The Three Movements of the Soul. 1988 ISBN Hb 90-277-2556-X; Pb 90-277-2557-8
26. Kaelin, E. F. and Calvin O. Schrag (eds.), *American Phenomenology.* Origins and Developments. 1989 ISBN 90-277-2690-6
27. Tymieniecka, A-T. (ed.), *Man within his Life-World.* Contributions to Phenomenology by Scholars from East-Central Europe. 1989
ISBN 90-277-2767-8
28. Tymieniecka, A-T. (ed.), *The Elemental Passions of the Soul.* Poetics of the Elements in the Human Condition, Part 3. 1990 ISBN 0-7923-0180-3

Analecta Husserliana

29. Tymieniecka, A-T. (ed.), *Man's Self-Interpretation-in-Existence.* Phenomenology and Philosophy of Life. – Introducing the Spanish Perspective. 1990
ISBN 0-7923-0324-5
30. Rudnick, H. H. (ed.), *Ingardeniana II.* New Studies in the Philosophy of Roman Ingarden. With a New International Ingarden Bibliography. 1990
ISBN 0-7923-0627-9
31. Tymieniecka, A-T. (ed.), *The Moral Sense and Its Foundational Significance: Self, Person, Historicity, Community.* Phenomenological Praxeology and Psychiatry. 1990 ISBN 0-7923-0678-3
32. Kronegger, M. (ed.), *Phenomenology and Aesthetics.* Approaches to Comparative Literature and Other Arts. Homages to A-T. Tymieniecka. 1991
ISBN 0-7923-0738-0
33. Tymieniecka, A-T. (ed.), *Ingardeniana III.* Roman Ingarden's Aesthetics in a New Key and the Independent Approaches of Others: The Performing Arts, the Fine Arts, and Literature. 1991
Sequel to Volumes 4 and 30 ISBN 0-7923-1014-4
34. Tymieniecka, A-T. (ed.), *The Turning Points of the New Phenomenological Era.* Husserl Research – Drawing upon the Full Extent of His Development. 1991 ISBN 0-7923-1134-5
35. Tymieniecka, A-T. (ed.), *Husserlian Phenomenology in a New Key.* Intersubjectivity, Ethos, the Societal Sphere, Human Encounter, Pathos. 1991
ISBN 0-7923-1146-9
36. Tymieniecka, A-T. (ed.), *Husserl's Legacy in Phenomenological Philosophies.* New Approaches to Reason, Language, Hermeneutics, the Human Condition. 1991 ISBN 0-7923-1178-7
37. Tymieniecka, A-T. (ed.), *New Queries in Aesthetics and Metaphysics.* Time, Historicity, Art, Culture, Metaphysics, the Transnatural. 1991
ISBN 0-7923-1195-7
38. Tymieniecka, A-T. (ed.), *The Elemental Dialectic of Light and Darkness.* The Passions of the Soul in the Onto-Poiesis of Life. 1992 ISBN 0-7923-1601-0
39. Tymieniecka, A-T. (ed.), *Reason, Life, Culture, Part I.* Phenomenology in the Baltics. 1993 ISBN 0-7923-1902-8
40. Tymieniecka, A-T. (ed.), *Manifestations of Reason: Life, Historicity, Culture.* Reason, Life, Culture, Part II. Phenomenology in the Adriatic Countries. 1993
ISBN 0-7923-2215-0
41. Tymieniecka, A-T. (ed.), *Allegory Revisited.* Ideals of Mankind. 1994
ISBN 0-7923-2312-2

Analecta Husserliana

42. Kronegger, M. and Tymieniecka, A-T. (eds.), *Allegory Old and New.* In Literature, the Fine Arts, Music and Theatre, and Its Continuity in Culture. 1994 ISBN 0-7923-2348-3
43. Tymieniecka, A-T. (ed.): *From the Sacred to the Divine.* A New Phenomenological Approach. 1994 ISBN 0-7923-2690-3
44. Tymieniecka, A-T. (ed.): *The Elemental Passion for Place in the Ontopoiesis of Life.* Passions of the Soul in the *Imaginatio Creatrix.* 1995 ISBN 0-7923-2749-7
45. Zhai, Z.: *The Radical Choice and Moral Theory.* Through Communicative Argumentation to Phenomenological Subjectivity. 1994 ISBN 0-7923-2891-4
46. Tymieniecka, A-T. (ed.): *The Logic of the Living Present.* Experience, Ordering, Onto-Poiesis of Culture. 1995 ISBN 0-7923-2930-9
47. Tymieniecka, A-T. (ed.): *Heaven, Earth, and In-Between in the Harmony of Life.* Phenomenology in the Continuing Oriental/Occidental Dialogue. 1995 ISBN 0-7923-3373-X
48. Tymieniecka, A-T. (ed.): *Life. In the Glory of its Radiating Manifestations.* 25th Anniversary Publication. Book I. 1996 ISBN 0-7923-3825-1
49. Kronegger, M. and Tymieniecka, A-T. (eds.): *Life. The Human Quest for an Ideal.* 25th Anniversary Publication. Book II. 1996 ISBN 0-7923-3826-X
50. Tymieniecka, A-T. (ed.): *Life. Phenomenology of Life as the Starting Point of Philosophy.* 25th Anniversary Publication. Book III. 1997 ISBN 0-7923-4126-0
51. Tymieniecka, A-T. (ed.): *Passion for Place. Part II.* Between the Vital Spacing and the Creative Horizons of Fulfilment. 1997 ISBN 0-7923-4146-5
52. Tymieniecka, A-T. (ed.): *Phenomenology of Life and the Human Creative Condition.* Laying Down the Cornerstones of the Field. Book I. 1997 ISBN 0-7923-4445-6
53. Tymieniecka, A-T. (ed.): *The Reincarnating Mind, or the Ontopoietic Outburst in Creative Virtualities.* Harmonisations and Attunement in Cognition, the Fine Arts, Literature. Phenomenology of Life and the Human Creative Condition. Book II. 1997 ISBN 0-7923-4461-8
54. Tymieniecka, A-T. (ed.): *Ontopoietic Expansion in Human Self-Interpretation-in-Existence.* The I and the Other in their Creative Spacing of the Societal Circuits of Life. Phenomenology of Life and the Creative Condition. Book III. 1997 ISBN 0-7923-4462-6
55. Tymieniecka, A-T. (ed.): *Creative Virtualities in Human Self-Interpretation-in-Culture.* Phenomenology of Life and the Human Creative Condition. Book IV. 1997 ISBN 0-7923-4545-2

Analecta Husserliana

56. Tymieniecka, A-T. (ed.): *Enjoyment*. From Laughter to Delight in Philosophy Literature, the Fine Arts and Aesthetics. 1997 ISBN 0-7923-4677-7

Kluwer Academic Publishers – Dordrecht / Boston / London

MIX
Papier aus verantwortungsvollen Quellen
Paper from responsible sources
FSC® C105338

If you have any concerns about our products,
you can contact us on
ProductSafety@springernature.com

In case Publisher is established outside the EU,
the EU authorized representative is:
Springer Nature Customer Service Center GmbH
Europaplatz 3, 69115 Heidelberg, Germany

Printed by Libri Plureos GmbH
in Hamburg, Germany